# The Future of Post-Human Chess

# The Future of Post-Human Chess

## A Preface to a New Theory of Tactics and Strategy

Peter Baofu

Viva Books

New Delhi | Mumbai | Chennai | Kolkata | Bengaluru | Hyderabad | Kochi | Guwahati

**First Indian Edition 2012**

**VIVA BOOKS PRIVATE LIMITED**

- 4737/23 Ansari Road, Daryaganj, New Delhi 110 002
  E-mail: vivadelhi@vivagroupindia.net, Tel. 42242200
- 76, Service Industries, Shirvane, Sector 1, Nerul, Navi Mumbai 400 706
  E-mail: vivamumbai@vivagroupindia.net, Tel. 27721273, 27721274
- Jamals Fazal Chambers, 26 Greams Road, Chennai 600 006
  E-mail: vivachennai@vivagroupindia.net, Tel. 28290304, 28294241
- B-103, Jindal Towers, 21/1A/3 Darga Road, Kolkata 700 017
  E-mail: vivakolkata@vivagroupindia.net, Tel. 22836381, 22816713
- 7, Sovereign Park Apartments, 56-58, K. R. Road, Basavanagudi, Bangalore 560 004
  E-mail: vivabangalore@vivagroupindia.net, Tel. 26607409, 26607410
- 101-102 Mughal Marc Apartments, 3-4-637 to 641, Narayanguda, Hyderabad 500 029
  E-mail: vivahyderabad@vivagroupindia.net, Tel. 27564481, 27564482
- Beevi Towers, First Floor, SRM Road, Kaloor, Kochi 682 018, Kerala
  E-mail: vivakochi@vivagroupindia.net, Tel: 0484-2403055, 2403056
- 232, GNB Road, Beside UCO Bank, Silpukhuri, Guwahati 781 003
  E-mail: vivaguwahati@vivagroupindia.net, Tel: 0361-2666386

***www.vivagroupindia.com***

*Published by arrangement with*
Cambridge International Science Publishing
7, Meadow Walk, Great Abington
Cambridge CB1 6AZ, UK



ISBN : 978-81-309-2001-6

Published by Vinod Vasishtha for Viva Books Private Limited, 4737/23, Ansari Road, Daryaganj, New Delhi-110 002. Printed and bound by Syndicate Binders, Noida.

To Those in the Future World
Beyond Both the Science of Chess and the Art of Chess

# Books also by Peter Baofu

**Category I: Social Sciences and Related Fields**

- 34. *The Future of Post-Human War and Peace* (2010)
- 33. *The Future of Post-Human Law* (2010)
- 32. *The Future of Post-Human Organization* (2009)
- 31. *The Future of Post-Human Mass Media* (2009)
- 30. *The Rise of Authoritarian Liberal Democracy* (2007)
- 29. *Beyond the World of Titans, and the Remaking of World Order* (2007)
- 28. *Beyond Capitalism to Post-Capitalism* (2005)
- 27. Volume 1: *Beyond Democracy to Post-Democracy* (2004)
- 26. Volume 2: *Beyond Democracy to Post-Democracy* (2004)
- 25. *The Future of Capitalism and Democracy* (2002)

**Category II: Natural Sciences and Related Fields**

- 24. *The Future of Post-Human Geology* (2010)
- 23. *Beyond Cosmology to Post-Cosmology* (2010)
- 22. *The Future of Post-Human Personality* (2010)
- 21. *The Future of Post-Human Sexuality* (2010)
- 20. *The Future of Post-Human Urban Planning* (2009)
- 19. *The Future of Post-Human Engineering* (2009)
- 18. *The Future of Post-Human Unconsciousness* (2008)
- 17. *The Future of Complexity* (2007)
- 16. *Beyond Nature and Nurture* (2006)
- 15. *The Future of Post-Human Space-Time* (2006)
- 14. *The Future of Post-Human Consciousness* (2004)

**Category III: Formal Sciences and Related Fields**

- 13. *The Future of Post-Human Formal Science* (2010)
- 12. *The Future of Post-Human Language* (2009)
- 11. *The Future of Post-Human Geometry* (2009)
- 10. *The Future of Post-Human Mathematical Logic* (2008)
- 9. *The Future of Information Architecture* (2008)

**Category IV: The Humanities and Related Fields**

- 8. *Beyond Ethics to Post-Ethics* (2010)
- 7. *The Future of Post-Human Martial Arts* (2009)
- 6. *The Future of Post-Human Creative Thinking* (2009)
- 5. *The Future of Post-Human Knowledge* (2008)
- 4. *The Future of Aesthetic Experience* (2007)
- 3. *Beyond Civilization to Post-Civilization* (2006)
- 2. Volume 1: *The Future of Human Civilization* (2000)
- 1. Volume 2: *The Future of Human Civilization* (2000)

# CONTENTS

## Part Three: Tactics



## Part Four: Conclusion

# TABLES

*Category III: Visions on the Mind*



*Category IV: Visions on History*

*Category IX: Visions on Culture*

# FOREWORD

For insiders in the world of chess, it may come as a surprise when a consummate researcher like Dr. Peter Baofu undertakes an investigation into the nature of chess. And, for the more casual chess players as well as the non-chess players, this exploration of chess contains some significant insights.

The revelation (in Chapter One) that nearly a billion people either play chess or know how to play the game in one of its variations commands our respect and attention and raises the question, What is the power and attraction of a game that has fascinated humankind for centuries?

Dr. Baofu's analysis begins with an overview of significant developments in the history of the game and posits the next steps for consideration by serious readers. Accolades to Dr. Baofu for taking on this unusual challenge and presenting us with a futuristic vision for additional musings.

*Sylvan Von Burg*
*School of Business*
*George Washington University*

ACKNOWLEDGMENTS

Like all other previous books of mine, this one is written to challenge conventional wisdom with a novel way of thinking, or more specifically in the current context, a new theory to understand the future fate of chess.

In light of this political incorrectness, this book receives no external funding nor help from any formal organization or institution—as this is something that I often stressed in all my previous books.

My only reward is that joyful discovery of something new not conceived before in history.

Furthermore, I greatly appreciate the foreword by Sylvan von Burg at George Washington University School of Business.

In any event, and as always, I bear the sole responsibility for the ideas presented in this book.

# ABBREVIATIONS

ALD = Peter Baofu. 2007. *The Rise of Authoritarian Liberal Democracy: A Preface to a New Theory of Comparative Political Systems*. Cambridge, England: Cambridge Scholars Publishing, Ltd.

BCIV = Peter Baofu. 2006. *Beyond Civilization to Post-Civilization: Conceiving a Better Model of Life Settlement to Supersede Civilization*. NY: Peter Lang Publishing, Inc.

BCPC = Peter Baofu. 2005. *Beyond Capitalism to Post-Capitalism: Conceiving a Better Model of Wealth Acquisition to Supersede Capitalism*. NY: The Edwin Mellen Press.

BCOS = Peter Baofu. 2010. *Beyond Cosmology to Post-Cosmology: A Preface to a New Theory of Different Worlds*. Cambridge, England: Cambridge International Science Publishing, Ltd.

BDPD1 = Peter Baofu. 2004. Volume 1. *Beyond Democracy to Post-Democracy: Conceiving a Better Model of Governance to Supersede Democracy*. NY: The Edwin Mellen Press.

BDPD2 = Peter Baofu. 2004. Volume 2. *Beyond Democracy to Post-Democracy: Conceiving a Better Model of Governance to Supersede Democracy*. NY: The Edwin Mellen Press.

BEPE = Peter Baofu. 2010. *Beyond Ethics to Post-Ethics: A Preface to a New Theory of Morality and Immorality*. Charlotte, NC: Infomration Age Publishing.

BNN = Peter Baofu. 2006. *Beyond Nature and Nurture: Conceivng a Better Way to Understand Genes and Memes*. Cambridge, England: Cambridge Scholars Publishing, Ltd.

BWT = Peter Baofu. 2007. *Beyond the World of Titans, and the Renaking of World Order: A Preface to a New Logic of Empire-Building*. Cambridge, England: Cambridge Scholars Publishing, Ltd.

FAE = Peter Baofu. 2007. *The Future of Aesthetic Experience: Conceiving a Better Way to Understand Beauty, Ugliness and the Rest*. Cambridge, England: Cambridge Scholars Publishing, Ltd.

FC = Peter Baofu. 2007. *The Future of Complexity: Conceiving a Better Way to Understand Order and Chaos.* London, United Kingdom: World Scientific Publishing Co.

FCD = Peter Baofu. 2002. *The Future of Capitalism and Democracy.* MD: The University Press of America.

FHC1 = Peter Baofu. 2000. Volume 1. *The Future of Human Civilization.* NY: The Edwin Mellen Press.

FHC2 = Peter Baofu. 2000. Volume 2. *The Future of Human Civilization.* NY: The Edwin Mellen Press.

FIA = Peter Baofu. 2008. *The Future of Information Architecture: Conceiving a Better Way to Understand Taxonomy, Network, and Intelligence.* Oxford, England: Chandos Publishing (Oxford) Limited.

FPHC = Peter Baofu. 2004. *The Future of Post-Human Consciousness.* NY: The Edwin Mellen Press.

FPHCHESS = Peter Baofu. 2010. *The Future of Post-Human Chess: A Preface to a New Theory of Tactics and Strategy.* Cambridge, England: Cambridge International Science Publishing, Ltd.

FPHCT = Peter Baofu. 2009. *The Future of Post-Human Creative Thinking: A Preface to a New Theory of Invention and Innovation.* Cambridge, England: Cambridge Scholars Publishing, Ltd.

FPHE = Peter Baofu. 2009. *The Future of Post-Human Engineering: A Preface to a New Theory of Technology.* Cambridge, England: Cambridge Scholars Publishing, Ltd.

FPHFS = Peter Baofu. 2010. *The Future of Post-Human Formal Science: A Preface to a New Theory of Abstraction and Application.* Cambridge, England: Cambridge Scholars Publishing, Ltd.

FPHG = Peter Baofu. 2009. *The Future of Post-Human Geometry: A Preface to a New Theory of Infinity, Symmetry, and Dimensionality.* Cambridge, England: Cambridge Scholars Publishing, Ltd.

FPHGEOL = Peter Baofu. 2010. *The Future of Post-Human Geology: A Preface to a New Theory of Statics and Dynamics.* Cambridge, England: Cambridge International Science Publishing, Ltd.

FPHK = Peter Baofu. 2008. *The Future of Post-Human Knowledge: A Preface to a New Theory of Methodology and Ontology.* Oxford, England: Chandos Publishing (Oxford) Limited.

FPHL = Peter Baofu. 2009. *The Future of Post-Human Language: A Preface to a New Theory of Structure, Context, and Learning.* Cambridge, England: Cambridge Scholars Publishing, Ltd.

FPHLAW = Peter Baofu. 2010. *The Future of Post-Human Law: A Preface to a New Theory of Necessity, Contingency, and Justice.* Cambridge, England: Cambridge Scholars Publishing, Ltd.

FPHMA = Peter Baofu. 2009. *The Future of Post-Human Martial Arts: A Preface to a New Theory of the Body and Spirit of Warriors.* Cambridge, England: Cambridge Scholars Publishing, Ltd.

FPHML = Peter Baofu. 2008. *The Future of Post-Human Mathematical Logic: A Preface to a New Theory of Rationality.* Cambridge, England: Cambridge Scholars Publishing, Ltd.

FPHMM = Peter Baofu. 2009. *The Future of Post-Human Mass Media: A Preface to a New Theory of Technology.* Cambridge, England: Cambridge Scholars Publishing, Ltd.

FPHO = Peter Baofu. 2009. *The Future of Post-Human Organization: A Preface to a New Theory of Communication, Decision-Making, and Leadership.*.Cambridge, England: Cambridge Scholars Publishing, Ltd.

FPHP = Peter Baofu. 2010. *The Future of Post-Human Personality: A Preface to a New Theory of Normality and Abnormality.* Cambridge, England: Cambridge International Science Publishing, Ltd.

FPHS = Peter Baofu. 2010. *The Future of Post-Human Sexuality: A Preface to a New Theory of the Body and Spirit of Love-Makers.* Cambridge, England: Cambridge Scholars Publishing, Ltd.

FPHST = Peter Baofu. 2006. *The Future of Post-Human Space-Time: Conceivng a Better Way to Understand Space and Time.* New York: Peter Lang Publishing, Inc.

FPHU = Peter Baofu. 2008. *The Future of Post-Human Unconsciousness: A Preface to a New Theory of Anomalous Experience.* Cambridge, England: Cambridge Scholars Publishing, Ltd.

FPHUP = Peter Baofu. 2009. *The Future of Post-Human Urban Planning: A Preface to a New Theory of Density, Void, and Sustainability.* Cambridge, England: Cambridge Scholars Publishing, Ltd.

FPHWP = Peter Baofu. 2010. *The Future of Post-Human War and Peace: A Preface to a New Theory of Aggression and Pacificity.* Cambridge, England: Cambridge Scholars Publishing, Ltd.

# • PART ONE •

## *Introduction*

# CHAPTER 1
## *INTRODUCTION—THE FUN OF CHESS*

Chess, essentially, is not just a game of humans but a set of mathematical rules that obey basic laws of physics.
—Arne Moll (2009)

## A Didactic Tale of Two Chess Sensibilities

Is chess really so mechanical that it is, for some chess scholars nowadays, "a set of mathematical rules that obey basic laws of physics"—especially when the chess engine famously known as "IBM Deep Blue" beat the world chess champion Gary Kasparov in a series of chess matches in 1997?

This popular chess sensibility can be contrasted by an opposing one in the older days, when many computer-skeptics "found it very, very hard to believe chess engines could ever replace or imitate the best of human chess thinking" or "believed the strongest human grandmasters would always be able to outplay computers when all tactics had been drained out of the position." (A. Moll 2009)

Contrary to these two opposing chess sensibilities on the nature of chess and its future (together with other views as will be discussed in this book), the strategy and tactics used in chess are neither so physically mechanical nor artistically subjective to the extent that their respective defenders would like us to believe.

But this unorthodox challenge to the two opposing chess sensibilities (and other views) does not logically imply that chess is therefore an uninteresting board game, or that chess studies are useless. Of course, neither of these two extreme views is reasonable either.

Instead, this book provides an alternative (better) way to understand the future of chess, especially in the context of strategy and tactics—while learning from different approaches in the literature but without favoring any one of them (nor integrating them, since they are not necessarily compatible with each other). Thus, this book offers a new theory to go beyond the existing approaches in the literature on chess in a new way not conceived before.

If successful, this seminal project is to fundamentally change the way that we think about chess, from the combined perspectives of the mind, nature, society, and culture, with enormous implications for the human future and what I originally called its "post-human" fate.

## Chess and Its Variants

A good place to start is to define the word "chess" and to understand its variants in history.

The word "chess" is from "Middle English *ches*, from Anglo-French *escheks*, *eschés*, plural of *eschec*" to mean "check," which is to expose an an opponent's king to an attack. (MWD 2010)

In Western chess, for instance, chess is "a board game played between two players. It is played on a chessboard, which is a square-checkered board with 64 squares arranged in an eight-by-eight grid. At the start, each player controls sixteen pieces: one king, one queen, two rooks, two knights, two bishops, and eight pawns." (WK 2010)

### The Value of Each Chess Piece

Each piece has a value relative to those of other pieces, as shown below: (WK 2010l)

- Pawn = 1
- Knight = 3
- Bishop = 3
- Rook = 5
- Queen = 9
- King = (undefined)

The reason for the king to be "undefined," because "the king…cannot be captured, let alone traded, during the course of the game. Some early computer chess programs gave the king an arbitrary large value (such as 200 points or 1,000,000,000 points) to indicate that the inevitable loss of the king due to checkmate trumps all other considerations." (WK 2010l; D. Levy 1991)

Other chess scholars, however, gave a different value to the king; for instance, "[i]n the endgame, when there is little danger of checkmate, the fighting value of the king is about four points." (WK 2010; E. Lasker 1934)

## The Movement of Each Chess Piece

The rules of the game are that each piece can move only in a certain way, as defined below: (WK 2010)

- "The King"—"moves one square in any direction, including diagonally. A player may not make any move which allows the king to be captured. If such a move is not possible the game ends with a checkmate. Along with the rook, the king is also involved in castling move."
- "The Rook"—"can move any number of squares along any rank or file, but may not leap over other pieces. Along with the king, the rooks is also involved during the castling move."
- "The Bishop"—"can move any number of squares diagonally, but may not leap over other pieces."
- "The Queen"—"combines the power of the rook and bishop and can move any number of squares along rank, file, or diagonal, but it may not leap over other pieces."
- "The Knight"—"moves to any of the closest squares which are not on the same rank, file or diagonal, thus the move forms an 'L'-shape two squares long and one square wide. The knight is the only piece which can leap over other pieces."
- "The Pawn"—"usually moves one step straight forward, but it may move two steps straight forward on its first move unless its path is obstructed. However, the only way a pawn may capture is by moving one square diagonally forward. The pawn has two special moves, the en passant capture, and pawn promotion."

In other words, if one were to change the rules of the game, it would subsequently be played differently (as revealed in other games with different rules as will be discussed in this book).

## The Algebraic Chess Notation for Recordng Moves

There are many different notations for recording moves of these different chess pieces.

For illustration, one popular notation is known as "the algebraic chess notation," and some good examples are shown below. (WK 2010 & 2010f; WCF 2010)

- K = "king"
- Q = "queen"
- R = "rook"
- B = "bishop"
- K = "knight"
- 1 to 8 = "rank" (along the vertical axis)
- a to h = "file" (along the horizontal axis)
- x = "capture"
- 0-0 = "kingside castling"
- 0-0-0 = "queenside castling"
- + (or #) = "checkmate"
- P (or =) = "stalemate"
- h = "helpmate"
- s = "selfmate"
- r = "reflexmate"
- 1-0 = "White won"
- 0-1 = "Black won"
- ½-½ = "draw"
- ! = "a good move"
- !! = "an excellent move"
- ? = "a mistake"
- ?? = "a blunder"
- !? = "an interesting move that may not be best"
- ?! = "a dubious move, but not easily refuted"
- ser- = "series"

For instance, with this illustrative system of algebraic chess notation in mind, "Qg5" means "queen moves to the g-file and 5th rank (that is, to the square g5)," "Ngf3" refers to "knight from the g-file moves to the square f3," "e4" means "pawn moves to the square e4," "Bxf3" refers to "bishop captures on f3," etc. (WK 2010)

In any event, "[t]he object of the game is to checkmate the opponent's king, whereby the king is under immediate attack (in 'check') and there is no way to remove or defend it from attack on the next move." (WK 2010)

## Different Variants of Chess in the World

Western chess "emerged in Europe during the second half of the 15th century after evolving from an older game (Shatranj) of Indian origin. Aspects of art are found in chess composition....The tradition of organized competitive chess started in the 16th century. The first official World Chess Champion, Wilhelm Steinitz, claimed his title in 1886....Chess is a recognized sport of the International Olympic Committee, and is led by the FIDE. Today, chess is one of the world's most popular games, played by millions of people worldwide at home, in clubs, online, by correspondence, and in tournaments." (WK 2010)

Besides Western chess, there are other variants. For instance, there are chess games which are "played with a different board, special fairy pieces or different rules. There are more than two thousand published chess variants, the most popular being xiangqi in China and shogi in Japan." WK 2010; D. Pritchard 2000 & 1994)

However, many chess variants can be classified in accordance to the following noticeable categories. (WK 2010)

- "Direct predecessors of chess"—in the tradition of chaturanga [developed in GuptaIndia] and shatranj [later adopted in Sassanid Persia, which in turn was brought to late-medieval Europe]" (WK 2010 & 2010a) This is so, although chess historians are still not very sure about "[t]he place of its origin," because "some say it was in Afghanistan and some say it was in India." (WCC 2009)
- "Traditional national or regional chess variants"—like xiangqi [in China], shogi [in Japan], janggi [in Korea] and makruk [in Thailand], which share common predecessors with Western chess" (WK 2010)
- "Modern variants of chess"—like "Chess960, where the initial position is one selected randomly from a possible 960 starting positions," so "[t]his random positioning makes it more difficult to prepare the opening play in advance" (WK 2010; E. van Reem 2001)
- "Variants with returning figures"—such as "bughouse chess or crazyhouse." (WK 2010)

Of course, there can be other variants, but these four categories here suffice for the illustration of the varieties of chess.

# Chess, and the Relationship between Strategy and Tactics

Yet, all chess variants have one fundamental feature in common, in that they are all played with a combination of strategy and tactics, and it is no wonder that "[t]heoreticians have developed extensive chess strategies and tactics since the game's inception." (WK 2010)

Let's examine what the strategy and tactics in chess refer to, as shown below.

This understanding of strategy and tactics is not only useful in chess (as summarized in *Table 1.1*) but also outside chess (as summarized in *Table 1.2* and elaborated further in Chapter 4, the concluding chapter).

## Strategy

The term "strategy" in chess refers to the "evaluation of chess positions and with setting up goals and long-term plans for the...play." (WK 2010)

For instance, "players must take into account numerous factors such as the value of pieces on board, control of the center and centralization, the pawn structure, the king safety, the control of key squares or groups of squares (for example, diagonals, open-files, and dark or light squares), etc." (WK 2010)

In general, chess strategy can be divided into three main parts (or stages), namely, (a) "opening," (b) "middlegame," and (c) "endgame," as described below.

### *Opening*

The word "opening" in chess refers to "the group of initial moves of a game (the 'opening moves'). Recognized sequences of opening moves are referred to as openings and...are catalogued in reference works such as the *Encyclopaedia of Chess Openings*." (WK 2010)

Chess openings can be quite diverse, as they "vary...widely in character from quiet positional play (e.g. the Réti Opening) to very aggressive (e.g. the Latvian Gambit). In some opening lines, the exact sequence considered best for both sides has been worked out to 30–35 moves or more. Professional players spend years studying openings, and continue doing so throughout their careers, as opening theory continues to evolve." (WK 2010; S. Collins 2005)

Yet, all chess openings have "fundamental strategic aims," and good examples are outlined below. (WK 2010; S. Tarrasch 1987)

- "Development"—"To place (develop) the pieces (particularly bishops and knights) on useful squares where they will have an optimal impact on the game."
- "Control of the center"—"Control of the central squares allows pieces to be moved to any part of the board relatively easily, and can also have a cramping effect on the opponent."
- "King safety"—"Keeping the King safe from dangerous possibilities. A correct timing for castling can often enhance this."
- "Pawn structure"—"Players strive to avoid the creation of pawn weaknesses such as isolated, doubled or backward pawns, and pawn islands...and to force such weaknesses in the opponent's position."

It is often said, however, that "[m]ost players and theoreticians consider that White, by virtue of the first move, begins the game with a small advantage," but even here, there is no guarantee, since "Black usually strives to neutralize White's advantage and achieve equality, or to develop dynamic counterplay in an unbalanced position." (WK 2010)

*Middlegame*

The word "middlegame" in chess refers to "the part of the game when most pieces have been developed. Because the opening theory has ended, players have to assess the position, to form plans based on the features of the positions, and at the same time to take into account the tactical possibilities in the position." (WK 2010; T. Harding 2003)

During the middlegame, there are some important strategic themes. Although there can be different strategic themes, the following three suffice for the purpose of illustration: (WK 2010)

- "Minority Attack"—"is the attack of queenside pawns against an opponent who has more pawns on the queenside" and is "often appropriate just for some pawn structures, resulting from a specific group of openings. The study of openings should therefore be connected with the preparation of plans typical for resulting middlegames."
- "Combinations"—"are often connected with the attack against the opponent's king; some typical patterns have their own names, for example the Boden's Mate or the Lasker—Bauer combination."

- "Material"—is about "whether and how to reduce material and transform into an endgame (i.e., simplify). For example, minor material advantages can generally be transformed into victory only in an endgame, and therefore the stronger side must choose an appropriate way to achieve an ending. Not every reduction of material is good for this purpose; for example, if one side keeps a light-squared bishop and the opponent has a dark-squared one, the transformation into a bishops and pawns ending is usually advantageous for the weaker side only, because an endgame with bishops on opposite colors is likely to be a draw, even with an advantage of one or two pawns."

Again, there can be other strategic concerns, but these three suffice for the present illustration.

*Endgame*

The word "endgame" (or "end game" or "ending") refers to "the stage of the game when there are few pieces left on the board." (WK 2010)

In general, "[t]here are…main strategic differences between earlier stages of the game and endgame," and good examples are shown below. (WK 2010; T. Harding 2003)

- The Promotion of Pawns—"[d]uring the endgame, pawns become more important; endgames often revolve around attempting to promote a pawn by advancing it to the eighth rank."
- The Active Role of the King—unlike the earlier stages, the king "becomes a strong piece in the endgame. It is often brought to the center of the board where it can protect its own pawns, attack the pawns of opposite color, and hinder movement of the opponent's king."
- "Zugzwang"—during the endgame a player can be put in "a disadvantage because the player has to make a move."

In addition, there are different classifications of endgames: "For example, king and pawn endgames involve only kings and pawns on one or both sides and the task of the stronger side is to promote one of the pawns. Other more complicated endings are classified according to the pieces on board other than kings, e.g., 'rook and pawn versus rook endgame.'" (WK 2010)

## Tactics

By contrast, the term "tactics" in chess refers to "a sequence of moves which limits the opponent's options and may result in tangible gain," and these moves are therefore "short-term actions—so short-term that they can be calculated in advance by a human player or by a computer." (WK 2010 & 2010b)

In this sense, "[t]actics are usually contrasted with strategy, in which advantages take longer to be realized, and the opponent is less constrained in responding." (WK 2010b) The same contrast exists in military studies, where "tactics...are concerned with the conduct of an engagement, while strategy is concerned with how different engagements are linked." (WK 2010)

In any event, the "possible depth of calculation" in chess tactics "depends on the player's ability or the strength of the chess engine and the performance of the computer it is run on." (WK 2010)

This is true, especially "in 'tactical' positions with a limited number of forced variations where much less than the best move would lose quickly, strong players can calculate very long sequences of moves." (WK 2010)

Tactical moves can be simple or complicated. In simple tactical moves, there are only a few moves like one or two; but in complicated tactical moves, "[s]mple one-move or two-move tactical actions—threats, exchanges of material, double attacks etc.—can be combined into more complicated combinations, sequences of tactical maneuvers that are often forced from the point of view of one or both players." (WK 2010; T. Harding 2003)

Over time, "[t]heoreticians described many elementary tactical methods and typical maneuvers, for example pins, forks, skewers, batteries, discovered attacks (especially discovered checks), zwischenzugs, deflections, decoys, sacrifices, underminings, overloadings, and interferences." (WK 2010; T. Harding 2003)

In general, chess tactics can be classified into different concerns, and some good examples are, namely, (a) defense and offense, (b) the gain of material, (c) the use of sacrifices, (d) the manipulation of pawns, (e) the compulsion to move (or *zugzwang* in German), and (f) the intermediate move (or *zwischenzug* in German), as shown below.

### *Defense and Offense*

In this tactical move, "offense" refers to the situation when a "piece is...to attack (or threaten) an opponent's piece if, in the next move, it could capture that piece." (WK 2010b)

By contrast, "defense" in tactics refers to the opposite situation when a "piece is...to defend (or protect) a piece of the defender's color if, in case the defended piece were taken by the opponent, the defender could recapture right away." (WK 2010b)

A player, when attacked, has several options to choose from, and hereafter are some good examples. (WK 2010b)

- Ex: "Capture the attacking piece."
- Ex: "Move the attacked piece."
- Ex: "Interpose another piece in between the two."
- Ex: "Guard the attacked piece and permit an exchange."
- Ex: "Pin the attacking piece so the capture...[is] illegal or unprofitable."
- Ex: "Use a zwischenzug."
- Ex: "Create a counter-threat."

Again, although there can be other options, the seven ones above suffice for the present illustration.

*The Gain of Material*

This tactical move is to gain an opponent's piece, especially one of greater value, and it can be achieved in different ways (e.g., "discovered attack," "pin," "skewer," and "fork"). (WK 2010b)

For instance, in the chess tactics known as "fork"—a player "uses one piece to attack two or more of the opponent's pieces simultaneously, hoping to achieve material advantage, since the opponent can only counter one of the two threats. Knights are often used for forks, with their unique moving and jumping ability. A common situation is a knight jumping to c2 or c7, thereby threatening both the enemy rook and the king. Such 'king forks' are particularly effective, because the opponent is forced by the rules of the game to alleviate the threat to the king. The opponent cannot choose to defend the other piece, and thus cannot use a zwischenzug to complicate the situation." (WK 2010b)

And in the chess tactics known as "skewer"—a player "attacks two pieces in a line, similar to a pin, except that the enemy piece of greater value is in front of the piece of lesser value. After the more valuable piece moves away, the lesser piece can be captured." (WK 2010b)

*The Use of Sacrifices*

This tactical move refers to the move to sacrifice a piece in order to gain something else later, because "[s]acrificing some material is often necessary to throw the opponent's position out of balance. The sacrificed material is sometimes able to be regained with interest a couple of moves later. Pawn sacrifices in the opening are known as gambits; they are usually not intended for a short-term gain, but instead to achieve a more active position." (WK 2010b)

For instance, "[d]irect attacks against the enemy king are often started by sacrifices. A common example is sacrificing a bishop on h2 or h7, checking the king, who usually must take the bishop. This allows the queen and knight to develop a fulminate attack." (WK 2010b)

*The Manipulation of Pawns*

This tactical move is to manipulate pawns, such that they can be more powerful "than meets the eye." (Wk 2010b)

For illustration, "since pawns are the least valuable chess piece, they can be used to capture defended pieces. A single pawn approaching will force a more powerful piece, such as a rook or a knight, to retreat. A simple move of a pawn may reveal a hidden threat....Also, when pawns are arranged in a diagonal line, with the frontmost pawns guarded by the pawns behind, they form an almost impenetrable wall capable of protecting any pieces directly behind them. Furthermore, a pawn which has progressed all the way to the opposite side of the board may be promoted to any other piece except a king." (WK 2010b)

Of course, this tactical move assumes a strong pawn structure, but when it is weakly organized, "a weak pawn structure can be a big weakness" instead. (WK 2010b)

*The Compulsion to Move*

This tactical move (or *zugzwang* in German) refers to the situation "when a player is forced to make an undesirable move. The player is put at a disadvantage because he would prefer to pass and make no move, but a move has to be made, all of which weaken his position." (WK 2010b)

However, "[s]ituations involving zugzwang occur uncommonly, but when they do occur, it is almost always in the endgame, where there are fewer choices of available moves." (WK 2010b)

*The Intermediate Move*

This tactical move (or *zwischenzug* in German) refers to "a common tactic which, instead of countering a direct threat, a move is played which poses an even more devastating threat, often an attack against the queen or the king. The opponent has to counter that threat first, and this will ideally change the situation to a disadvantage for that player." (WK 2010b)

In fact, "a zwischenzug…might properly be called a counter-tactic instead. The effect of a zwischenzug is to change the status quo before a tactic can come to fruition. The near ubiquity of the zwischenzug makes long combinations all the more rare and impressive." (WK 2010b)

Again, there can be other tactical concerns, but these six (as illustrated above) suffice for the present illustration.

## The Theoretical Debate

This introduction to the relationship between strategy and tactics in chess thus leads us to a more important question concerning the extent to which chess is possible and desirable on the basis of understanding the importance of strategy and tactics.

For illustration, there are five main theoretical approaches in the debate to be summarized hereafter, and they can be called, in the absence of better words, namely, (a) the *mental* argument, (b) the *natural* argument, (c) the *social* argument, (d) the *cultural* argument, and (e) the *mediative-variative* argument (which is my original contribution to the debate)—to be elabroated hereafter, respectively.

Lest misunderstanding occurs, two clarifications should be made here, as this is something that I regularly emphasize in my previous books, whenever a theoretical debate is summarized for a topic, almost verbatim.

Firstly, there can be other variants within each argument (besides the ones to be summarized hereafter). The advantage to select them here has to do with their illustrative representation of the diverse theories in the literature (which are deemed sufficient for the purpose at hand in this book).

And secondly, the first four arguments are not necessarily mutually exclusive. For instance, those who advocate the "mental" argument can also consider the "cultural" argument, although they do not focus on the latter.

And the reverse is also true, in that those who make the "cultural" argument can also consider the "mental" argument, although they do not focus on the latter.

For instance, a version of the "mental" argument concerns the brilliant use of chess rules to win a game, but this concern overlaps with the consideration of aesthetics in the "cultural" argument (e.g., some tactical moves are made because they are "beautiful"), although it does not focus on the latter (as will be clear later).

In other words, their disagreement is often one in degree, not in kind.

## The Mental Argument

With these two clarifications in mind—the first major theoretical approach is called, in the absence of better words, the *mental* argument, which focuses, relatively speaking of course, more on the mental skills of the players in a chess game to compete with each other for winning the game.

Therefore, the word "mental" in the title of the argument is suggestive, because it refers to these mental skills of the players in question.

For illustration, consider two main positions in the mental argument, namely, (a) chess rules and (b) chess novelties—as explained below, respectively.

### *Chess Rules*

Arne Moll (2009), for instance, argued that "[c]hess, essentially, is not just a game of humans but a set of mathematical rules that obey basic laws of physics."

Thus, those who are mentally able to master all the mechanical rules of chess will have a better chance to win a chess game than someone else who is not so capable.

In fact, Moll (2009) went further and argued that a player can be much improved if they also learn all the rules of chess by "making use of computer analysis," especially after the chess engine famously known as "IBM Deep Blue" beat the world chess champion Gary Kasparov in a series of chess matches in 1997.

But for the critics, there are serious problems in the mental argument as suggested by people like Moll. Consider two of the problems below for illustration.

Firstly, one criticism of the "chess rules" variant of the mental argument is that it ignores other dimensions of chess in the context of the natural, social, and cultural arguments.

And secondly, another criticism of the "chess rules" variant of the mental argument is that it does not give sufficient attention to the other side of chess playing, namely, the subjective (or creative) side of the players in a chess game, which is not reduciable to the mechanical rules of chess.

*Chess Novelties*

Unlike the "chess rules" variant of the mental argument—there is another version of the mental argument, in that there is no lack of chess novelties to be developed over time.

A good illustration concerns the phenomenon of "chess composition," in that, just as there are music composers for new pieces of music, there are likewise "chess composers" for new games of chess.

In accordance to this view, chess is not a "science" as those in the "chess rules" variant would like us to believe; rather, chess is an "art," so that "[c]hess composition is the art of creating chess problems....A person who creates such problems is known as a chess composer." (WK 2010)

Chess composers often come up with new chess problems to challenge chess players in a novel way. For instance, new chess problems are proposed, "not...taken from an actual game" but "invented for the specific purpose of providing a problem," with such stipulations like "to checkmate black within a specified number of moves," "white to move first and checkmate black within a specified number of moves against any defense," "white to play must win or draw," and so on. (WK 2010; H. Weenink 1926; K. Howard 1961)

In fact, Robert James 'Bobby' Fischer (or simply Bobby Fischer), who "is widely considered one of the greatest chess players of all time," introduced in 1996 what he called "a variant of chess called Fischer Random Chess, also known as Chess960, that is intended to allow players to contest games based on their understanding of chess rather than their ability to memorize opening variations." (WK 2010c)

Fischer was highly critical of treating chess in terms of some mechanical rules and advocated instead the importance of creativity (for chess novelties) in playing chess.

His proposal for Chess960 (or "Fischer Random") "was designed to remove the importance of opening book memorization. Fischer complained in a 2006 phoned-in call with a television interviewer that talented celebrity players from long ago, if brought back from the dead to play today, would no longer be competitive, because of the progress in memorization of opening books. 'Some kid of fourteen today, or even younger, could get an opening advantage against Capablanca,' he said, merely because of opening-book memorization, which Fischer disdained. 'Now chess is completely dead. It is all just memorization and prearrangement. It's a terrible game now. Very uncreative.'" (WK 2010c; CBN 2006)

But for the critics, there are some problems with the "chess novelties" variant of the mental argument. Consider two of them below for illustration, as they are the mirror opposite of the "chess rules" variant.

Firstly, one criticism of the "chess novelties" variant of the mental argument is that it ignores other dimensions of chess in the context of the natural, social, and cultural arguments.

And secondly, another criticism of the "chess novelties" variant of the mental argument is that it does not give sufficient attention to the other side of chess playing, namely, the mechanical rules in chess, which allowed, for instance, the chess engine famously known as "IBM Deep Blue" to beat the world chess champion Gary Kasparov in 1997. It is interesting to note that Fischer introduced Chess960 prior to the historical victory of IBM Deep Blue over Kasparov.

## The Natural Argument

The second theoretical approach is called, in the absence of better words, the *natural* argument, which focuses, relatively speaking of course, more on relationship between the origin of chess and the understanding of the cosmos.

In this sense, the word "natural" in the title of the argument is suggestive, because it refers to the relevance of cosmological understanding to the game of chess.

### *Cosmology*

For instance, Rangachar Vasantha (2010) argued that "[t]he game of chess, which was being played in the ancient India as well as in the present times,...has its origin linked genetically to secret magical and religious rituals in those times," especially in the context of cosmology.

For Vasantha (2010), to understand this linkage between cosmology and chess requires a comprehension of the relationship between the outer world and the inner world: "A cosmos by definition presuppose an ordered universe. Cosmology is the study of the law and intelligence inherent in this ordered universe. There is an inner as well as an outer way of studying cosmology. The outer embraces sensible observation and the inner is expression of cosmological laws within one's own structure. The goal of spiritual disciplines is to unite the inner and outer, the greater and smaller into an inseparable integrity. The language of the archetype laws, which unite the inner and outer cosmos, is that of pattern and in particular number pattern. The game of chess is associated with the number pattern and thus to cosmology."

As an illustration, "[c]hess and other board games were derived from, and the moves of the pieces are being closely related to the movements of the celestial bodies and their numerical symbolism. By joining those cells that show the same number, certain geometrical diagrams are obtained.... These diagrams are linked to board games in so far as every move of a piece results in similar geometrical patterns." (R. Vasantha 2010)

Other chess scholars like Harold James and Ruthven Murray made the same argument, though in a different context of the Chinese chess known as "象棋 (Xiangqi)," because the character "象 sometimes means 'constellation' (i.e., a figure made of stars)" in an astronomical context, and "older Xiangqi simulated the movements of stars and other celestial objects in the sky." (WK 2010d)

But for the critics, there are some problems with the natural argument. Consider two of them below for illustration.

Firstly, one criticism of the natural argument is that it ignores other dimensions of chess in the context of the mental, social, and cultural arguments.

And secondly, another criticism of the natural argument is that it is not clear how much the apparent similarities between chess rules and "the movements of the celestial bodies and their numerical symbolism" are exaggerated and to what extent their differences are much neglected (in order to make a misleading linkage between the two, which may not exist otherwise).

## The Social Argument

The third theoretical approach is called, in the absence of better words, the *social* argument, which focuses, relatively speaking of course, more on the social origins of chess, say, for "military" and "entertainment" purposes.

In this sense, the word "social" in the title of the argument is suggestive, because it refers to the extent of the social functions of chess and their influence on chess development over time.

For illustration, therefore, consider the social argument in relation to (a) the military institution and (b) the entertainment industry, as shown below, respectively.

*The Military Institution*

The relationship between the military institution and the game of chess has a historic root, in that, in the history of Western chess, for instance, especially "in the Middle Ages and during the Renaissance, chess was a part of noble culture; it was used to teach war strategy and was dubbed the 'King's Game.'" (WK 2010; M. Vale 2001)

In this military tradition, "[g]entlemen are 'to be meanly seene in the play at Chestes,' says the overview at the beginning of Baldassare Castiglione's *The Book of the Courtier*" (which was translated by Sir Thomas Hoby in 1561). (WK 2010) Even today, "chess is…used in armies to train minds of cadets and officers." (WK 2010)

But this does not mean, of course, that gentlemen should learn nothing else (other than chess), as Castiglione thus explained: "And what say you to the game at chestes? It is truely an honest kynde of enterteynmente and wittie, quoth Syr Friderick. But me think it hath a fault, whiche is, that a man may be to couning at it, for who ever will be excellent in the playe of chestes, I beleave he must beestowe much tyme about it, and applie it with so much study, that a man may assoone learne some noble scyence, or compase any other matter of importaunce, and yet in the ende in beestowing all that laboure, he knoweth no more but a game. Therfore in this I beleave there happeneth a very rare thing, namely, that the meane is more commendable, then the excellency." (WK 2010; B. Castiglione 1900)

In the East, scholars like Cho-Yo (1905) studied Chinese and Japanese chess variants and treated them as "the Science and Art of War or Struggle," as Cho-Yo used the examples of Japanese chess (*sho-ngi*) and Chinese chess (*chong-kie* and *i-go*) for the study of the Russo-Japanese War during 1904-1905.

But for the critics, there are some problems with the "military institution" variant of the social argument. Consider two of them below for illustration.

Firstly, one criticism of the "military institution" variant of the social argument is that it ignores other dimensions of chess in the context of the mental, natural, and cultural arguments.

And secondly, another criticism of the "military institution" variant of the social argument is that it is not clear how much what happens in a chess game (in a comfortable living room) is really comparable to what happens in a battlefield (in the rough and more complex real world).

*The Entertainment Industry*

Another variant of the social argument has to do with the entertainment industry.

The market for chess in the entertainment industry can be quite vast, since it was estimated that, today, "[a]pproximately 600 million people worldwide know how to play chess, while 300 million people play intermittently." (WK 2010)

In popular culture, the entertainment industry wastes no time to use chess as one of its themes. For illustration, "J. K. Rowling's *Harry Potter* plays 'Wizard's Chess' while the characters of *Star Trek* prefer 'Tri-Dimensional Chess' and the hero of *Searching for Bobby Fischer* struggles against adopting the aggressive and misanthropic views of a real chess grandmaster. Chess has also been used as the core theme of a musical, *Chess*, by Tim Rice, Björn Ulvaeus and Benny Andersson." (WK 2010; E. O'Neill 2008)

Even on the internet, "ChessWorld (aka letsplaychess) is a correspondence chess server hosted by the Barnet Chess Club in England and with over 400,000 members in more than two hundred countries. The site appears to be a comprehensive online chess club with games, many forms of tournaments, user homepages, active chess and social forums, and even annual 'meet-ups' where members meet in person for chess and socializing. Non-members are permitted a certain number of games and are invited to play in special tournaments." (WK 2010e)

However, for the critics, there are some problems with the "entertainment industry" variant of the social argument. Consider two of them below for illustration.

Firstly, one criticism of the "entertainment industry" variant of the social argument is that it ignores other dimensions of chess in the context of the mental, natural, and cultural arguments.

And secondly, another criticism of the "entertainment industry" variant of the social argument is that it does not consider other social functions (say, for the military institution) other than its entertainment concern.

## The Cultural Argument

The fourth theoretical approach is called, in the absence of better words, the *cultural* argument, which focuses, relatively speaking of course, more on such cultural concerns with morality and aesthetics in chess.

In this sense, the word "cultural" in the title of the argument is suggestive, because it refers to the extent of moral and aesthetic concerns in chess development.

For illustration, consider the cultural argument in relation to (a) morality and (b) aesthetics, as shown below, respectively.

*Morality*

The moral dimension of chess should not be underestimated.

Even in the older days, for instance, as early as 1300 A.D., "chess was often used as a basis of sermons on morality. An example is *Liber de Moribus Hominum et Officiis Nobilium Sive Super Ludo Scacchorum* ('Book of the Customs of Men and the Duties of Nobles or the Book of Chess'), written by an Italian Dominican monk Jacobus de Cessolis circa 1300." (WK 2010)

This book by Jacobus de Cessolis "was one of the most popular of the Middle Ages. The work was translated into many other languages (first printed edition at Utrecht in 1473) and was the basis for William Caxton's *The Game and Playe of the Chesse* (1474), one of the first books printed in English. Different chess pieces were used as metaphors for different classes of people, and human duties were derived from the rules of the game or from visual properties of the chess pieces." (WK 2010; M. Olmert 1996; CHEAL 1907; J. Adams 2006)

Later, during "the Age of Enlightenment, chess appeared mainly for self-improvement"; Benjamin Franklin (2010), for instance, in his article "The Morals of Chess" in 1750 thus advocated the moral virtues of "foresight," "circumspection," and "caution" in playing chess: "The Game of Chess is not merely an idle amusement; several very valuable qualities of the mind, useful in the course of human life, are to be acquired and strengthened by it, so as to become habits ready on all occasions; for life is a kind of Chess, in which we have often points to gain, and competitors or adversaries to contend with, and in which there is a vast variety of good and ill events, that are, in some degree, the effect of prudence, or the want of it. By playing at Chess then, we may learn: I. Foresight, which looks a little into futurity, and considers the consequences that may attend an action....II. Circumspection, which surveys the whole Chess-board, or scene of action—the relation of the several Pieces, and their situations....III. Caution, not to make our moves too hastily...." (WK 2010)

But, for the critics, there are some problems with the "morality" variant of the cultural argument. Consider two of them below for illustration.

Firstly, one criticism of the "morality" variant of the cultural argument is that it does not consider other dimensions of chess in the context of the mental, natural, and social arguments.

And secondly, another criticism of the "morality" variant of the cultural argument is that it also does not focus on other cultural aspects of chess (like the issue of aesthetics).

*Aesthetics*

Another variant of the cultural argument concerns the issue of aesthetics in chess.

Indeed, "[t]he role of aesthetic evaluation in the appreciation of chess problems is very significant, and indeed most composers and solvers consider such compositions to be an art form." (WK 2010e)

However, "[t]here are no official standards by which to distinguish a beautiful problem from a poor one, and such judgments can vary from individual to individual as well as from generation to generation. Such variation is to be expected when it comes to aesthetic appraisal." (WK 2010e)

Vladimir Nabokov, for instance, offered his version of aesthetics in chess by focusing on such features as the "originality, invention, conciseness, harmony, complexity, and splendid insincerity" in chess playing. (WK 2010e)

Other chess scholars offer theirs in our times. For instance, "modern taste generally recognizes…[certain] elements to be important in the aesthetic evaluation of a problem," such as "The first move of the problem's solution (the key move or key) must be unique," "The key move of the solution should not be obvious," "The problem should be economical," etc. (WK 2010e)

It is thus no wonder that this aesthetic issue in chess becomes an important theme in the arts. For instance, "chess is often depicted in the arts; significant works, where chess plays a key role, range from Thomas Middleton's *A Game at Chess* and *Through the Looking-Glass* by Lewis Carroll to *The Royal Game* by Stefan Zweig or Vladimir Nabokov's *The Defense*. Chess is also important in films like Ingmar Bergman's *The Seventh Seal* or Satyajit Ray's *The Chess Players*." (WK 2010)

Yet, for the critics, there are some problems with the "aesthetics" variant of the cultural argument. Consider two of them below for illustration.

Firstly, one criticism of the "aesthetics" variant of the cultural argument is that it does not consider other dimensions of chess in the context of the mental, natural, and social arguments.

And secondly, another criticism of the "aesthetics" variant of the cultural argument is that it also does not include other cultural aspects of chess (like the issue of morality).

### The Mediative-Variative Argument

And fifthly, unlike the previous four arguments (viz., the mental, the natural, the social, and the cultural)—the fifth one to be introduced here can be called, in the absence of better words again, the *mediative-variative* argument, which learns from all other theoretical approaches (which include the four previous arguments and also other examples as will be introduced in the rest of this book), without, however, favoring any of them, nor trying to integrate them (as they are not necessarily compatible with each other, as the introduction of the theoretical debate in the previous sub-sections has shown).

## The Mediative-Variative Theory of Chess

My *mediative-variative* argument can be more precisely called *the mediative-variative theory of chess*, since it is to propose a new theory of chess.

There are two words here, *mediative* and *variative*, in the title of the argument.

My argument is "meditative" because it tries to mediate the competing theoretical approaches in the literature (say, on whether chess is a science or an art) by learning from them but without favoring any of them nor trying to integrate them (as they are not necessarily compatible with each other).

And my argument is also "variative," because it offers a visionary prediction of the different variants of chess that the post-humans will use as substitutes for the human ones (hitherto existing in human history) which will no longer be much useful in the post-human future.

With this clarification in mind—the mediative-variative theory of chess has five distinctive features to remember.

Firstly, my theory makes good use of all theoretical approaches in the literature on chess, be they about the mental argument, the natural argument, the social argument, or the cultural argument—especially from the most comprehensive combined perspectives of the mind, nature, society, and culture (as will be clear in Chapter Two and Chapter Three).

Secondly, just like many other theories of mine in my previous books, my theory here does not heavily favor any specific theory over others in the literature, nor trying to integrate them (as they are not necessarily compatible with each other).

Thirdly, my theory treats the issue of chess in the distinctive dialectic context of strategy and tactics in chess—especially in the larger dialectic context of no strategy without tactics, and vice versa, together with the subsequent transcendence of both.

Fourthly, my theory on strategy and tactics is useful not only in the context of chess but also in the context outside chess (e.g., of social organizations and social institutions). In other words, it is important in chess, in a small sense, and in life, in a broad sense (as summarized in *Table 1.1* and *Table 1.2*).

And fifthly, it contains fourteen major theses, namely, (a) the first thesis: the absoluteness-relativeness principle, (b) the second thesis: the predictability-unpredictability principle, (c) the third thesis: the explicability-inexplicability principle, (d) the fourth thesis: the preciseness-vagueness principle, (e) the fifth thesis: the simpleness-complicatedness principle, (f) the sixth thesis: the openness-hiddenness principle, (g) the seventh thesis: the denseness-emptiness principle, (h) the eighth thesis: the slowness-quickness principle, (i) the nineth thesis: the convention-novelty principle, (j) the tenth thesis: the evolution-transformation principle, (k) the eleventh thesis: the symmetry-asymmetry principle, (l) the twelveth thesis: the regression-progression principle, (m) the thirteenth thesis: the sameness-difference principle, and (n) the fourteenth thesis: the post-human variation, in relation to the three great future transformations of mind games—to be elaborated in the rest of the book, with a summary in the concluding chapter.

Of course, as this is something that I often emphasized in my previous books, other principles (besides the 13 as cited above) are also relevant, but these 13 are the most relevant in the current case study (in terms of the number of citations of each principle in the book).

Even then, in some cases, the difference between any two given principles, for instance, in terms of the number of citations in a book, is rather small, so the reason in those cases is more aesthetic (than otherwise), because it looks nicer to list only 13 theses for 13 principles (than 21 theses for all of the 21 principles) in the Table of Contents.

This is true, even if different studies of the same kind can yield different views about the degree of relevance for each principle, depending on the specific nature of a research in question, needless to say. So, if a different author analyzes the same subject matter in a different way, the relevance of the principles will be different.

With this clarification in mind—the seminal project here, if successful, will fundamentally change the way that we think about chess, from the combined perspectives of the mind, nature, society, and culture, with enormous implications for the human future and what I originally called its "post-human" fate.

# Theory and Meta-Theory

The *mediative-variative* argument, or *the mediative-variative theory of chess*, presupposes some kind of meta-theory, in special relation to methodology and ontology—just as all theories, either explicitly or implicitly, require their own versions of meta-theory.

In my case, as this is something that I often explained in my previous books, I already proposed my distinctive approach to methodology (known as "sophisticated methodological holism") and ontology (known as "existential dialectics").

The next two sections are to introduce my meta-theory, or existential dialectics and sophisticated methodological holism, respectively—to be repeated, *often verbatim* (with only a few updated revisions), from my previous works, as this is something that I often do in each new book of mine as an introduction to some background information for the convenience of the readers.

## A Unified Theory of Everything

In the end, my distinctive approach to meta-theory serves as a foundation to unify all domains of knowledge for an overarching understanding of all forms of reality (by way of some ontological principles at the meta-theoretical level and the comprehensive perspectives of the mind, nature, society, and culture at the theoretical level—as will be explained in the next two sections).

This is so, without commiting the sins of reductionism and reverse-reductionism (as often seen in many holistic approaches, with the fad of systems approach as a most recent notorious example, as already critically explained in *The Future of Post-Human Formal Science*, or *FPHFS*), since my distinctive approach makes good use of different schools of thought without favoring any of them nor trying to integrate them (as they are not necessarily compatible with each other), so as to include (or adjust for) heterogeneity, conflict, subjectivity, and complexity, for example.

In this sense, the word "unified" does not have to mean an integrative approach, in a narrow sense, nor a systems approach, in a broad one, both of which I have critically rebuked (in my previous books), while learning from them.

This overarching project (consisting of numerous theories of mine, as will be summarized in the next section), thus, constitutes my grand unifed theory of everything in the world and beyond.

## The Logic of Existential Dialectics

To start, the ontology of existential dialectics can be summarized in this section—*often verbatim* (with only a few updated revisions), from my previous works, as this is something that I regularly do in each new book of mine as an introduction to some background information for the convenience of the readers.

After all, this book, like all others of mine, is in conversation with all previous ones of mine, for the final aim to converge all of them into a larger grand project on the future of intelligent life, both here on earth and there in deep space unto multiverses.

The summary can be organized in relation to (a) the conception of existential dialectics (or its ontics), (b) the syntax of existential dialectics (or its ontomethodology and ontologic), (c) the semantics of existential dialectics (or its ontosemantics), and (d) the pragmatics of existential dialectics (or its ontopragmatics), respectively hereafter—again *often verbatim*.

This means that, for those readers who had read some of my previous books, the summary below is really not much new (*almost verbatim*, with only some updated revisions).

But for those who have never read my previous books, it is a good review of them.

### The Conception of Existential Dialectics (or Its Ontics)

This ontology was originally designated as "existential dialectics" in *Beyond Capitalism to Post-Capitalism* (herein abbreviated as *BCPC*), although it was already analyzed in other books of mine like the 2 volumes of *The Future of Human Civilization* (*FHC*), *The Future of Capitalism and Democracy* (*FCD*), *The Future of Post-Human Consciousness* (*FPHC*), and the 2 volumes of *Beyond Democracy to Post-Democracy* (*BDPD*).

Lest any confusion occurs, it is important to stress at the outset that the word 'existential' in "existential dialectics" has nothing to do with Existentialism, which I rebuked in *FHC*, *FCD*, and also *FPHC*. Instead, the word here merely refers to the existence of intelligent life (both primitive and advanced) in a broad sense.

*Some Basic Concepts*

The conception of existential dialectics (or its ontics) makes use of different concepts (like "sets," "elements," "relations," "operations," "functions," "truth values," "axioms," "postulates," and "principles"—as shown in *Table 4.3*), which are important for the understanding of any logic of ontology.

That clarified—existential dialectics, as a language of ontology, can be analyzed in different ways that a language, as an analogy, is often analyzed, namely, in relation to phonology ("the study of patterns of a language's sounds"), phonetics ("the study of the physical aspects of sounds of human language"), morphonology ("the study of the internal structure of words"), syntax ("the study of how words combine to form grammatical sentences"), semantics ("the study of the meaning of words [lexical semantics] and fixed word combinations [phraseology]"), and pragmatics ("the study of how utterances are used…literally, figuratively, or otherwise…in communicative acts"). (WK 2007)

But since my theory of existential dialectics makes use of the English language for communication here, it does not propose a new way to make the patterns of a language's sounds (as in phonology), to study the physical aspects of sounds (as in phonetics), or even to strive for a new organization of the internal structure of words for the English language (as in morphonology).

For this reason, the language of existential dialectics to be analyzed hereafter is not concerned with the phonological, phonetic, and morphonological aspects of the English language. Instead, the analysis will explore the syntax, semantics, and pragmatics of existential dialectics as a language of ontology to improve our understanding of reality.

*The Ontic Dispute*

But first, there is an ontic dispute to be addressed, which is related to what I called "the dilemma of ontology" as introduced earlier in *Sec. 1.4* of *The Future of Post-Human Knowledge* (*FPHK*).

A good point of departure concerns what I want to call *the ontic dispute*, on how to understand the very nature of ontology.

Traditionally, scholars in the field used to study the conception of ontology in terms of its entities (e.g., material objects, abstract numbers), properties (e.g., duration, plurality), and relationships (e.g., causation).

This is a good starting point, but it suffers from different problems. Consider three examples of bias below, which are quite common in the literature, and the recent version (of treating concepts and principles in ontology on the basis of "plurality," "dynamism," "duration," "interaction," "life," "consciousness," and "volition") by Reginald Firehammer (2005) only constitutes a latest addition to this old (impoverishing) tradition.

Firstly, the classification of entities, properties, and relationships is too material-centric, in often focusing more on the existence of material entities than on non-material ones.

Secondly, the classification is too anthropo-centric, in often favoring the factor of human agency in depicting reality.

And lastly, the classification is too confused among the three categories relating to each other, in often lumping them together without adequate consideration of their intricated interrelationships (both in relation to each and in relation to all).

A good alternative is precisely what I want to offer in this debate, that is, a new classification of ontological entities, properties, and their relationships in terms of *method*, *structure*, *process*, *agency*, and *outcome* instead. Let me explicate what these words mean below.

In relation to "method"—it concerns how ontology is to be studied. In this sense, the word 'method' here should not be confused with the different usage of the same term 'method' as a technique of investigation in the context of methodology (as already introduced in *Sec. 1.2* of *FPHK*).

In relation to "structure"—it concerns what the general patterns of entities, properties, and relationships are.

In relation to "process"—it concerns how these general patterns change over time.

In relation to "agency"—it concerns how agency affects any change of these general patterns over time (without putting, however, too much emphasis on the role of human agency).

And in relation to "outcome"—it concerns what the outputs of this change over time are.

## The Syntax of Existential Dialectics (or Its Ontomethodology and Ontologic)

The syntax of existential dialectics, analogously speaking, refers here to the structure of an ontology which can be studied by way of combining ideas into complex relationships like ontological principles to depict reality in the world—and can be analyzed in seven sub-sections, namely,

(a) the selection criteria, (b) further clarifications on classification, (c) ontomethodology, (d) ontologic, (e) the principles as short cuts, (f) the principles as family resemblances, and (g) the dialectic constraints imposed by the principles, respectively hereafter.

*The Selection Criteria*

To start, not any pair of relationship can be chosen as an ontological principle in existential dialectics; otherwise, there could be billions of them under the sky.

Four illustrations below suffice to clarify this point.

(a) Firstly, an ontological principle cannot *overlap* with other principles, in that it is too closely related with other ones.

Of course, one cannot totally rule out any relationship between two entities, especially in a complex world of everything being linked to everything else—but the comparison here is relative, not absolute.

In this sense, the flexibility-inflexibility pair cannot be an ontological principle, since it has something closely in common with other ones (like the simpleness-complicatedness principle, for instance).

As an illustration, something which is flexible in interpretation is more likely to allow multiple viewpoints and thus makes the overall picture more complicated than otherwise (simple).

(b) Secondly, an ontological principle cannot be *redundant* in relation to other principles, in that it somehow duplicates other principles.

In this sense, for instance, the directness-indirectness pair cannot be an ontological principle, because it somehow duplicates other ontological principles (like the preciseness-vagueness principle, for instance).

As an illustration, something which is direct means to be right to the point and does not need to go to an unnecessarily long loop—so it is less vague, in being clearer about the thing in question.

(c) Thirdly, an ontological principle cannot be *derived* from other ones—as if it is a child-parent relationship.

In this sense, the convenience-inconvenience pair cannot be an ontological principle, because it can be derived from other ones (like the slowness-quickness principle, for instance).

For instance, something which is convenient already implies that it can be obtained quickly (rather than slowly)—all things considered.

(d) And fourthly, an ontological principle cannot be *trivial* (or *parochial*)—as if it is only one component competing with other ones for inclusion in a set.

In this sense, the consistency-inconsistency pair cannot be an ontological principle, because its validity is limited (or parochial), as it can be easily challenged and replaced by opposing major (not minor) theories (e.g., the correspondence theory of truth, the pragmatic theory of truth, the semantic theory of truth, and whatnot), not just in accordance to the consistency theory of truth (for consistency).

In this sense, the consistency argument is only one among different opposing arguments in relation to the logic of what constitutes "truth."

Although these examples are not exhaustive, they illustrative the selectiveness of any ontological principle to be existentially dialectic.

*Further Clarifications on Classification*

Lest any misunderstanding carelessly occurs, some further clarifications on classification should be made here (and summarized in *Table 4.8*).

(a) Firstly, the total number of ontological principles is unknown, to be discovered later, as our knowledge of the world becomes more advanced.

Consequently, the principles as introduced in my books are not exhaustive, with new ones being added, whenever more of them are discovered in later research.

At least, future generations can pick up where I leave off and continue the discovery.

In this sense, my work should be treated as a pioneering effort for the development of a systematic, comprehensive analysis of a new general ontology for the future of knowledge.

(b) Secondly, the principles are not rigidly classified, as they can be reclassified in a different way.

For instance, the preciseness-vagueness principle is classified under the category of "structure" but can be reclassified under the category of "method," although in so doing, it has a different meaning in the context of method.

The same logic applies to the same-difference principle under the category of "outcome," which can be reclassified under the category of "structure," although in so doing, once more, it has a different meaning in the context of structure.

(c) Thirdly, the principles are to be understood together, not that each prinicple is to be treated independently of others. After all, the principles are mutually constraining, in that they work together as a whole.

(d) Fourthly, the principles are relevant to all subject matters, but some principles are, relatively speaking of course, more useful to some subject matters than others under certain conditions—as implied in the symmetry-asymmetry principle.

(e) And fifthly, the principles constitute only two levels of analysis, this time, at the ontological and methodological levels—while other levels of analysis (from the perspectives of the mind, nature, society, and culture) are also needed, in order to understand reality in its totality (as explained in the section on "sophisticated methodological holism," so as to avoid the dual dangers of reductionism and reverse-reductionism).

*Ontomethodology*

With these selection criteria and clarifications in mind—the syntax of existential dialectics can be divided into two main smaller sections for analysis, since they are related to each other in the context of some fundamental principles, that is, (a) ontomethodology and (b) ontologic, to be addressed hereafter, respectively.

In the context of ontomethodology, some ontological principles are proposed here in relation to the nature of methodology, which however is tied up with the general issue of ontology in the context of knowledge.

In this sense, ontomethodology (in the context of these ontological principles in the category of method) are related with ontologic in general (in the context of ontological principles in general).

Some good instances of ontological principles involving ontomethodology include the formalness-informalness principle, the predictability-unpredictability principle, the explicability-inexplicability principle, the partiality-totality principle, and the absoluteness-relativeness principle, which will be introduced in conjunction with ontological principles in the ontologic of existential dialectics in general hereafter.

*Ontologic*

With this clarification of the relationship between ontomethodology and ontologic in mind (which is comparable to the relationship between methodology and ontology as described in *Sec. 1.3* of *FPHK*)—the first three principles in existential dialectics were introduced in *BCPC*. And more principles were later added to the logic in my subsequent books (like *FC*, *FAE*, and others).

More specifically, in *BCPC*, I proposed three principles (based on previous works of mine, not just something out of the blue) for the ontological logic, namely, (a) the regression-progression principle on the direction of history, (b) the symmetry-asymmetry principle on the relationships among existents, and (c) the change-constancy principle on the alteration of things—or in a more elegant term, the dynamics of space-time (in *FC*).

In *The Future of Complexity* (*FC*), I added three more principles, on top of the three aforementioned, namely, (d) the partiality-totality principle on the relationships between the parts and the whole, (e) the order-chaos principle on the pattern of things, and (f) the predictability-unpredictability principle on the occurrence of things—as part of the ontological logic of existential dialectics.

In *The Future of Aesthetic Experience* (or simply *FAE*), another principle is added, namely, (g) the evolution-transformation principle (on the multiple kinds of agency).

Also in *FAE*, I then reclassified the 7 principles into four categories, namely, (i) in relation to method, for the partiality-totality principle and the predictability-unpredictability principle, (ii) in relation to process, for the change-constancy principle and the order-chaos principle, (iii) in relation to agency, for the symmetry-asymmetry principle and the evolution-transformation principle, and (iv) in relation to outcome, for the regression-progression principle.

In *The Rise of Authoritarian Liberal Democracy* (*ALD*), I added two more principles, namely, (h) the softness-hardness principle on the force of agency (which is to be classified under the category about agency) and (i) the same-difference principle on the metamorphosis of change (which is to be classified under the category about outcome).

In *The Future of Information Architecture* (*FIA*), I introduce three additional principles, that is, (j) the simpleness-complicadness principle on the interconnection among things, (k) the preciseness-vagueness principle on the refinement of things, and (l) the slowness-quickness principle on the speed of change—with the first two in relation to structure and the third in relation to process.

In *The Future of Post-Human Unconsciousness* (*FPHU*), I added three more principles, that is, (m) the openness-hidden principle on the detection of things, in relation to structure, (n) the explicability-inexplicability principle on the underlying mechanisms of things, in relation to method, and (o) the expansion-contraction principle on the growth of things, in relation to process.

In *The Future of Post-Human Knowledge* (*FPHK*), I added one more principle, that is, (p) the absoluteness-relativeness principle on the multiplicity of entities, in relation to method.

In *The Future of Post-Human Mathematical Logic* (*FPHML*), I added one more principle, that is, (q) the formalness-informalness principle on the formal requirements of systems, in relation to method.

In *The Future of Post-Human Engineering* (*FPHE*), I added one more principle, that is, (r) the theory-praxis principle on the duality of knowledge, in relation to agency.

In *The Future of Post-Human Creative Thinking* (*FPHCT*), I added one more principle, that is, (s) the convention-novelty principle on the nature of creative thinking, in relation to agency.

In *The Future of Post-Human Geometry* (*FPHG*) here, I added one more principle, that is, (t) the finiteness-transfiniteness principle on the nature of numbers, in relation to structure.

And in *The Future of Post-Human Urban Planning* (*FPHUP*), I added one more principle, that is, (u) the denseness-emptiness principle on the distribution of entities in space, in relation to structure.

With this update, there are twenty-one principles so far in existential dialectics, that is, five principles for the category in relation to method (viz., the formalness-informalness principle, the partiality-totality principle, the predictability-unpredictability principle, the explicability-inexplicability principle, and the absoluteness-relativeness principle), five principles for the category in relation to structure (viz., the finiteness-transfiniteness principle, the simpleness-complicatedness principle, the preciseness-vagueness principle, the openness-hiddenness principle, and the denseness-emptiness principle), four principles for the category in relation to process (viz., the change-constancy principle, the slowness-quickness principle, the order-chaos principle, and the expansion-contraction principle), five principles for the category in relation to agency (viz., the theory-praxis principle, the convention-novelty principle, the symmetry-asymmetry principle, the evolution-transformation principle, and the softness-hardness principle), and two principles for the category in relation to outcome (viz., the regression-progression principle and the same-difference principle).

There are thus twenty-one principles in existential dialectics—so far (as summarized in *Table 4.4*).

### *The Principles as Short Cuts*

Yet, these principles should be treated with caution, lest misunderstanding occurs, since they do not constitute rigid dualities (or dichotomies).

The reason is that each pair in an ontological principle consists of two opposites, which are, however, merely short cuts both for multiple variations and degrees, as well as for different interactions with multiple other entities unlike them.

In this light, each pair can end up having hundred (or even thousand, if not more) different versions, which interact with hundred (or even thousand, if not more) other entities—as summarized in *Table 4.5*.

There are two clarifications here.

Firstly, this conception of shortcuts is not mutually exclusive nor absolute, in that the opposites can come in all shapes and sizes, with different degrees. For instance, by analogy, just as there are different degrees of the two colors "white" and "black"—there are likewise different degrees of the opposites in each ontological pair, to the extent that there can be multiple entities (not only two) interacting with each other in each pair.

And secondly, this conception of shortcuts do not ignore other possible entities in interacting with the two opposites (with their different versions) in each pair. For instance, by analogy, there are not only the two colors "white" and "black" as opposites, since there are other colors too besides them like "yellow," "green," "purple," or else. The same logic can be applied to each ontological pair, in that they also interact with other entities, not with only two of them (with their different versions).

Therefore, with these two clarifications in mind—each pair in an ontological principle serves only as an abbreviation for something more complicated and, therefore, although it contains two opposites, it should not be confused as a duality (dualism).

Existential dialectics rejects any dualism (or dichotomy) as too rigid and instead allows the multiplicity of entities, to the extent that between the two opposites in each pair exist many other alternatives to choose from. They are named in that short form for aesthetic elegance only, instead of listing all possible entities between the two opposites in the title (which would be too cumbersome and sound extremely awkward).

To be dialectic is to go beyond any rigid dichotomy and transcend into something different altogether in the long haul. One may be tempted to call the dialectic logic here with a different name like existential "multilectics" (instead of "dialectics"), but this naming is incorrect (or even misleading), for the two reasons aforestated.

After all, all the possible entities between any two opposites in an ontological pair are still about the two opposites, although each ontological pair come in all shapes and sizes (with different degrees) and do not exist by themselves but also interact with other entities unlike them. The virtue of revealing an ontological pair is to show how they relate within themselves (in multiple versions) and also interact with others unlike them (also in multiple versions).

This will be clear in a later sub-section titled "the dialectic constraints imposed by the principles."

*The Principles as Family Resemblances*

In addition, each ontological principle is generic, with some other comparable ontological pairs to be put in the same family (like a *family resemblance*)—as summarized in *Table 4.6*.

Ludwig Wittgenstein (1953) once suggested the idea of "family resemblance" in explaining different games classified under the same family called *games*.

Why should, for instance, playing football and chess as playing "games," when it is well understood that football is not the same as chess? The answer is that, although each game is different and has different rules, many of them (though not all) share, more or less, some commonalites (e.g., scoring as necessary for winning).

And this is so, even though some games share more than some others in any given selection of critieria, and no two games are exactly identical. So, Wittgenstein's point here is that there is no essential core which is common to all games, and the best that one can look for is some characteristics which are common to many (but not all) games. (A. Biletzki 2006)

(a) Appropriate Family Resemblances

In the context of existential dialectics, I want to distinguish two types of family resemblances, and, in the absence of better words, let me call them (a) *appropriate* family resemblances and (b) *inappropriate* family resemblances.

In regard to appropriate family resemblances—good examples include some of the pairs as already explained in an earlier section on selection criteria, like the flexibility-inflexibility pair, which can be put in the family resemblance of the simpleness-complicatedness principle, although the two pairs are not exactly identical. Likewise, the directness-indirectness pair can be put in the family resemblance of the preciseness-vagueness principle, although, again, the two pairs are not exactly identical.

In this sense, which specific pair in a family should be used to designate the name of the family can be at times a bit arbitrary, but with good reason.

For illustration, in the context of method, the partiality-totality principle can take the different form like individualisticness-holisticness, just as the explicability-inexplicability principle can take the different form like underlyingness-regularness—although each two pairs are not exactly identical and have slightly different meanings and usages.

A more comprehensive listing of this family resemblance for all other principles is shown in *Table 4.6* on the syntax of existential dialectics.

In fact, this section on family resemblances overlaps with the semantics of existential dialectics, which will be introduced later in the next section (and summarized in *Table 4.11*).

(b) Inappropriate Family Resemblances

However, there are some family resemblances which are not appropriate, and let me distinugish two of them by calling them, in the absence of better words, (b1) *inadequate* family resemblances and (b2) *compound* family resemblances (as summarized in *Table 4.10*).

(b1) In regard to "inadequate" family resemblances—there are some pairs which are not adequate, because they fail the test of selection criteria (as already pointed out in an earlier section), especially in relation to the criterion of not being *trivial* (or *parochial*)—as if it is only one component competing with other ones for inclusion in a set.

For instance, the *consistency-inconsistency* pair is inadequate, because it is only one component competing with other ones for inclusion in the larger set of the formalness-informalenss principle (like the *completeness-incompleteness* pair and the *soundness-unsoundness* pair, as discussed in my book titled *The Future of Post-Human Mathematical Logic*, or *FPHML* in short).

The same is true for the *cognition-noncognition* pair, which overlaps with both the completeness-incompleteness pair and the soundness-unsoundness pair in the larger set of the formalness-informalenss principle.

(b2) In regard to "compound" (or non-atomic) family resemblances—there are some pairs which are not appropriate for a different reason, because they fail the test of selection criteria (as already pointed out in an earlier section), especially in relation to the criterion of not *overlapping* with other principles, in that it is too closely related with other ones.

For instance, the *normality-abnormality* pair is compound (and thus inappropriate), because it overlaps with several other principles, like the symmetry-asymmetry principle, the density-emptiness principle, the convention-novelty principle, and the same-difference principle, for instance (as already discussed in *FPHP*).

The normality-abnormality pair overlaps with the symmetry-asymmetry principle, because being "abnormal" (say, for those with mental illnesses) is often treated not equally by others who are "normal." So, the relationship reveals the symmetry-asymmetry principle.

The normality-abnormality pair also overlaps with the density-emptiness principle, because being "abnormal" (say, for those with mental illnesses) is often not as dense in the overall (not local) distribution of a population as those who are "normal." So, the relationship reveals the density-emptiness principle.

The normality-abnormality pair also overlaps with the convention-novelty principle, because being "abnormal" (say, for those with mental illnesses) often exhibits non-conventional (non-conformist) ideas and behaviors when contrasted with those who are "normal." So, the relationship reveals the convention-novelty principle.

The normality-abnormality pair also overlaps with the same-difference principle, because being "abnormal" (say, for those with mental illnesses) shares some similarities in outcome when compared with those who are "normal" (in that both contribute to the control and regulation of human thoughts and behaviors in society, albeit in different ways). Yet, being "abnormal" also has something different from being "normal," in that abnormal thoughts and behaviors follow a different pattern when contrasted with those which are normal. So, the relationship reveals the same-difference principle.

These illustrations of indequate and compound family resemblances show how difficult it is to come up with an ontological principle in existential dialectics out of the blue, because many of them violate the selection criteria in one way or the other. It is all too easy to join two opposing words and simply make them into a pair, but it is all too difficult to pass the test of the selection criteria to be qualified as an ontological principle in existential dialectics.

Other examples of compound (and inadequate) family resemblances are shown in *Table 4.10* on "The Syntax of Existential Dialectics VII: Types of Inappropriate Family Resemblances."

### *The Dialectic Constraints Imposed by the Principles*

The principles, as they constitute the syntax of existential dialectics, are dialectic in character, such that, when they are applied, they impose dialectic constraints on how reality is to be understood.

Consider, say, the symmetry-asymmetry principle as an illustration here, in order to summarize two main characters of the dialectic constraints in question (as summarized in *Table 4.7*).

Firstly, to be dialectic here is to go beyond the narrow dichotomies (and, for that matter, any rigid multi-dimensional classificatory scheme), be they about "self" vs. "world," "freedom" vs. "unfreedom," "barbarity" vs. "civilization," "individuality" vs. "communality," and so on.

One way to do so (to go beyond) is to consider them all in terms of co-existence (without favoring one over the rest). For instance, my theory of "post-civilization" (to be summarized later in the section on the pragmatics of existential dialectics) is to go beyond barbarity and civilization in terms of understanding barbarity and civilization as being co-existent.

And the same logic can be said in relation to my theories of "post-democracy," "post-capitalsm," and others (also to be introduced later in the section on the pragmatics of existential dialectics), in regard to freedom vs. unfreedom, equality vs. inequality, communality vs. individuality, spirituality vs. materiality, and so on.

But to consider them all (in the dichotomies—and, for that matter, in any rigid multi-dimensional classifactory scheme) as co-existent is not the same as to imply that the opposing categories in any classificatory scheme are all equal, since, in accordance to the symmetry-asymmetry principle (as an illustration here), if they are equal in terms of being considered as co-existent, they are asymmetric in terms of being unequal in dialectic interaction (e.g., X can be more dominant than Y in case A, or Y is more dominant than X in case B), although in some special cases, they can be relatively equal (e.g., X and Y are relatively equal in case C).

For this reason, there are different versions of "post-democracy" and "post-capitalism" in my theories. As an illustration, in version I of the theory of post-democracy, freedom is more dominant than equality, whereas in version II of the theory of post-democracy, equality is more so than freedom.

But this "X more than Y" has to be understood in the context of dialectic logic (not in symbolic logic, as conventionally understood), in that both "X" and "Y" are important in post-democracy (in the context of dialectic logic), but in an asymmetry way.

By contrast, in symbolic logic, it often favors one over the other—be it in regard to privileging freedom over equality (in Fascism), favoring freedom relatively more than equality (in Liberal Democracy), or favoring equality relatively more than freedom (in Socialist Democracy). In regard to the latter two cases (about Liberal Democracy and Socialist Democracy), the difference between dialectic logic and symbolic logic can be one in degree, not in kind—in this sense, albeit not in other senses.

The same reasoning can be said about the relatonships between individuality and communality, between spirituality and materiality, and between formal legalism and informal legalism in the different versions of my theory of "post-capitalism."

Secondly, to be dialectic is to go beyond the narrow dichotomies (and, for that matter, any rigid multi-dimensional classificatory scheme) in another way, this time, in a transcendent way, that is, in exploring other possibilities or even other issues not considered within the narrow confines of narrow dichotomies (and, for that matter, any rigid multi-dimensional classificatory scheme).

As an analogy, in this second meaning, to go beyond the narrow dichotomy of "black" vs. "white" is not just to choose both "black" and "white" (as in the first meaning) but also to explore other options (e.g., "green," "purple," "blue," etc.—and, alternatively, "shade," "line," "curve," etc.).

By the same reasoning, to go beyond "democracy" is to transcend democracy (as in version III of my theory of "post-democracy") and to explore other possibilties of lifeforms (e.g., "floating consciousness," "hyper-spatial consciousness," etc., to live beyond the narrow obsession with freedom and equality).

This dialectic character of the principles in existential dialectics has important implications for the pragmatics of existential dialectics (as will be clear shortly, in the section on the pragmatics of existential dialectics).

## The Semantics of Existential Dialectics (or Its Ontosemantics)

The syntax of existential dialectics so understood in terms of ontological principles only gives us the structure of ontology in the world, in an abstract (general) sense. These principles by themselves do not tell us the specific meanings in a given context.

In order to grasp the specific meanings of the principles in a given context, it is necessary to study the semantics of existential dialectics (as summarized in *Table 4.11*).

And the analysis of the ontological principles as family resemblances in the previous section on syntax is only a starting point (and thus overlaps a bit with the semantics of existential dialectics here).

For this reason, I have often gone to great lengths in my previous books on different subjects to explain the specific meanings of the principles when applied in different contexts.

Consider two examples below for illustration, which are not exhaustive.

In *FPHST*, for instance, I used the first three principles (i.e., the change-constancy principle, the regression-progression principle, and the symmetry-asymmetry principle) to propose "the perspectival theory of space-time," for a better way to understand space and time—especially, though not exclusively, in relation to future post-human history (as summarized in *Table 1.3*, *Table 1.4*, *Table 1.5*, and *Table 1.6*).

In so doing, I had to introduce concepts and theories specific to the field of physics and other related fields (e.g., "absolute space" and "absolute time" in "classical mechanics" and "relative space-time" in "the theory of relativity").

And in *BNN*, I also exploited the three principles to propose the "transcendent" approach to the study of genes and memes as a new way to understand the interaction between nature and nurture.

In so doing, I had to explore concepts and theories in the world of evolutionary biology (e.g., "mutation," "variation," "adaptation," "selection," and "inheritance" in classical evolutionary theory) and neural biology (e.g., "chromosome," "gene," "DNA," "RNA," "protein," "neuron," "neural network," and "behavior").

Other examples, besdies these two here, are given in other books of mine on different topics.

Therefore, this semantic dimension of existential dialectics is important, since, in each of my books when the ontological principles are used, each principle becomes relevant in a different meaning when contextualized within a given case study in question.

In this book, for instance, the principles are relevant in relation to the subject matter of cosmology (as summarized in the concluding chapter), and they can be shown in all shapes and sizes (since they are also to be treated as "short cuts," as indicated before), without, however, being reduced to "specific ontology" (as opposed to "general ontology"), as explained in *Table 4.9*.

## The Pragmatics of Existential Dialectics (or Its Ontopragmatics)

Even the semantics of existential dialectics is not enough to understand reality, since semantics by itself does not tell us the nature of intentionality and interests in relation to the matrix of power formations and conflicting interests when existential dialectics as an ontology is applied.

Surely, one cannot completely separate the pragmatics of existential dialectics from its semantics (and for that matter, its syntax), but the point here is to give each a more elaborated analysis.

For this reason, the pragmatics of existential dialectics needs to be studied too, in special relation to the inclusion of power formation and conflicting interests in the application of the principles of existential dialectics (as summarized in *Table 4.12*).

It is thus no wonder that all my previous books have shown in different ways how and why these ontological principles and their theoretical applications can reveal the future world to come in a way that humans have never known, in a totally different battleground for competing human (and later post-human) interests by myriad groups fighting for their own versions of hegemony.

Perhaps a summary of my previous works in relation to existential dialectics and their contributions to my new theories as proposed over the years is deemed revealing here, for the further understanding of the pragmatics of existential dialectics (in close relation to the semantics of existential dialectics too).

That stressed—the following summary in four sub-sections is something that I often do in my previous books too, as a way to introduce the intellectual background of my ideas.

So, for those readers who had read my previous works, the following sub-sections serve as a reminder, as they are repeated hereafter *often verbatim* (with only some updated changes, with the rest kept intact).

These sub-sections are, namely, (a) the two-way street connecting theory with meta-theory, (b) the theoretical application of existential dialectics, (c) direct and indirect applications of existential dialectics, and (d) multiple applications of existential dialectics—to be summaried herefafter, respectively.

*The Two-Way Street Connecting Theory and Meta-Theory*

However, lest any misunderstanding occurs, it is important to remember that the pragmatics of existential dialectics is not a one-way street (that is, using the ontological principles for theoretical insights in praxis) but a two-way one, namely, (a) from meta-theory to theory, and (b) from theory to meta-theory.

(a) On one side of the street, the ontological principles can inspire some theoretical insights in praxis, that is, in relation to some specific fields.

(b) On the other (opposing) side of the street, however, the study of a subject matter in the specific fields in turn reveals some more hitherto unknown ontological principles to be discovered and identified.

For this reason, three new principles were added in *FC*, one in *FAE*, two in *ALD*, and three in *FIA*, on top of the original three in *BCPC*, for example (as there are more)—after some research on the specific subject matters.

Both theory and meta-theory enrich each other in all my works over the years.

*The Theoretical Application of Existential Dialectics*

This two-way street connecting theory and meta-theory has contributed to my original construction of 45 specific theories in 35 books (including this one here, up to now, in mid-2010) on numerous fields in all domains of knowledge, which can be classified in terms of specific "meta-theories" (about epistemology, methodology, logic, ontology, etc.) and specific "theories" (about the mind, nature, culture, society, and the rest—as will be further explained later in the sub-section on "the holistic organization of an inquiry" in the section about "sophisticated methodological holism").

The final aim of my numerous new specific theories (on both theory and meta-theory) is to converge all of them (as proposed in my numerous books) into a grand project about the future of intelligent life, both here on earth and in deep space.

In fact, all these theories of mine serve as a foundation to unify all domains of knowledge for an overarching understanding of all forms of reality (by way of some ontological principles and the comprehensive perspectives of the mind, nature, society, and culture), while allowing for heterogeneity, conflict, subjectivity, and complexity, for example.

With this grand project in mind—the summary of my specific theories on meta-theory are given in this section (on "existential dialectics") and the next (on "sophisticated methodological holism"), whereas the specific theories on theory will be summarized in the following categories of (a) society, (b) culture, (c) the mind, (d) nature, and (e) the rest, as shown below, in that order.

(a) Society

The category of "society" here can be divided into four sub-categories, i.e., (a1) social organizations, (a2) social institutions, (a3) social structure, and (a4) social systems.

(a1) Social Organizations

In relation to social organizations, I proposed a new thesis on the ambivalent nature of virtual organizations in the future, known as "the theory of e-civic alienation" in *FCD* (and summarized in *Table 4.19*).

And in *FPHO*, another theory called "the combinational theory of organization" is worked out, although it was first proposed in Ch.6 of *ALD*.

(a2) Social Institutions

There are different social institutions, be they about (a2.1) politics, (a2.2) economics, (a2.3) military affairs, and (a2.4) mass media, for instance.

(a2.1) In relation to politics, I offered different visions on (a2.1.1) political theory, (a2.1.2) comparative politics, and (a2.1.3) international relations, to cite three main ones.

(a2.1.1) In the context of political theory, my vision of a path-breaking political system in future times concerns the different forms of "post-democracy" to supersede democracy unto the post-human age.

My vision of post-democracy was called "the theory of post-democracy" (as summarized in *Table 4.22*, *Table 4.23*, and *Table 4.24*), whereas *Table 4.25* explains the distinctions among democracy, non-democracy, and post-democracy.

And *Table 4.26* stresses the multiple causes of the emergence of post-democracy, while *Table 4.27* clarifies some possible misunderstandings in regard to post-capitalism and post-democracy.

(a2.1.2) In the context of comparative politics, I also showed, in *ALD*, how and why authoritarianism can be an advanced form of liberal democracy (just as it has its own counterpart in non-liberal democracy), or what I originally called "authoritarian liberal democracy" (as summarized in *Table 4.21*), before it is to be superceded by "post-democracy."

(a2.1.3) In the context of international relations (both at the institutional and structural levels), I provided, in *BWT*, "the theory of cyclical progression of empire-building," for a better way to understand the logic of empire-building on earth and beyond—with the rise of what I originally envisioned as "hyper-empires" (like "the Chinese Union" and "the Indian Union"), "meso-empires" (like "the European Union" and "the North American Union") and "micro-empires" (like "the Middle Eastern Union" and "the Latin American Union") in the coming "post-post-Cold War era" as a visionary illustration, as summarized in *Table 4.20*.

In addition, in *FPHWP*, I proposed "the multifaceted theory of war and peace" to understand war and peace, in a small sense—or aggression and pacificity, in a broad sense.

(a2.2) In relation to economics, I also offered another vision, this time to understand the future forms of economic systems, that is, the different forms of "post-capitalism" to supersede capitalism unto the post-human age.

My vision of post-capitalism was referred to (in *FCD*, *BDPD* and *BCPC*) as "the theory of post-capitalism" (as summarized in *Table 4.28*, *Table 4.29*, *Table 4.30*, *Table 4.31*, and *Table 4.32*). And *Table 4.33* clarifies the differences among capitalism, non-capitalism, and post-capitalism, whereas *Table 4.34* shows multiple causes of the emergence of post-capitalism.

This vision on "post-capitalism" constitutes the third theoretical application of existential dialectics, this time, in that there is no wealth without poverty (or the wealth/poverty dialectics).

(a2.3) In relation to military affairs, I also proposed "the multifaceted theory of war and peace" (in *FCD*, although it was first so labeled in *FPHK*) for a new way to understand the causes of war—and, for that matter, the rise and fall of different forms of warfare over time.

This theory was further expanded in *FPHWP*, where I suggested a new way to understand war and peace, in a small sense—or aggression and pacificity, in a broad sense.

(a2.4) In relation to mass media, I also proposed "the totalistic theory of communication" in *The Future of Post-Human Mass Media* (or in short, *FPHMM*).

(a3) Social Structure

In relation to social structure, I proposed two additional theories, that is, "the theory of the cyclical progression of hegemony" in *FCD* (and also in *FPHC* and *BDPD*), and "the theory of the cyclical progression of empire-building" (in *BWT*, as already indicated above)—with important implications for the symmetry-asymmetry princple in existential dialectics, just to cite one example.

Also, in *BNN*, I also proposed "the theory of contrastive advantages" to understand the persistence of social structure by gender, race, ethnicity, or any other groupings.

(a4) Social Systems

In relation to social systems, there are several sub-categories for illustration, which can be about (a4.1) urbanization, (a4.2) technology, (a4.3) demography, (a4.4) the environment, and (a4.5) social trends as a whole.

(a4.1) In relation to urbanization, I proposed "the contingent theory of urban planning" in *FPHUP* for a new way to understand density and void.

(a4.2) In relation to technology, I worked out "the ambivalent theory of technology" in *FPHE* for a new way to understand the nature of technology.

Some other books (like *FPHMM*, *FAE*, *FCD*, and *FHC*) also deal with technology, and a good example concerns "the synthetic theory of information architecture" in *FIA* on taxonomy, network, and intelligence, in conjunction with the rise of technology.

(a4.3) In relation to demography, I proposed, first in *FHC* (and later in other books like *FCD, FPHC, FPHU,* etc.), "the theory of post-humanity," in relation to the "post-human" age at some distant point of "after-postmodernity," long after human extinction, to be eventually superseded by "post-humans" of various forms.

Some good candidates of the post-human lifeforms include, for instance, "thinking robots," "thinking machines," "cyborgs," "genetically altered superior beings," "floating consciousness," "hyper-spatial consciousness," "unfolding unconsciousness," "the hyper-sexual body," and "the hyper-martial body." This post-human vision of mine was first originally worked out in *FHC* and further elaborated in both *FCD, FPHC, FPHU, FPHS,* and *FPHMA*, for instance.

Lest any confusion occurs, I need to emphasize two clarifications here about the term "post-human" as a neologism in my works.

Firstly, the word "post-human" here should *not* be confused with another term which looks similar but has a totally different meaning in the literature of postmodernism, namely, "post-humanism"—which constitutes a critique of "humanism" as tradionally understood (especially, though not exclusively, in relation to the idea of progress in science and reason in the Enlightenment project). (WK 2008)

And secondly, the word "post-human" here should also *not* be confused with a similar term which is used to champion the ideology of technology for the future co-existence between humans and cyborgs in "trans-humanism." (WK 2008a)

Instead, my term "post-human" in relation to "posthuman-ism" also rejects "transhumanism" (especially, though not exclusively, in relation to the promises of technology) and refers to something else altogether, that is, the future extinction of humans and its post-human successors in deep space and beyond unto multiverses.

Later, in *BEPE*, I went further and provided a more comprehensive critique of "transhumanism," together with other issues.

With these two clarifications in mind (as summarized in *Table 1.31*)—in the end, a most fundamental question about intelligent life now has an answer, in that, if asked, "What is the future of human civilization?"—my answer in *FCD* (89) is thus: "As addressed in Ch.7 of *FHC*, a later epoch of the age of after-postmodernity (that is, at some point further away from after-postmodernity) will begin, as what I called the 'post-human' history (with the term 'post-human' originally used in my doctoral dissertation at M.I.T., which was finished in November 1995, under the title *After Postmodernity*, still available at M.I.T. library, and was later revised and published as *FHC*). The post-human history will be such that humans are nothing in the end, other than what culture, society, and nature (with some luck) have shaped them into, to be eventually superseded by post-humans (e.g., cyborgs, thinking machines, genetically altered superior beings, and others), if humans are not destroyed long before then."

(a4.4) In relation to the environment, I also addressed the issue of sustainability, especially in *FHC*, *FCD*, and *FPHUP*. For instance, in *FPHUP*, I specifically proposed a new theory, "the contingent theory of urban planning," on the need to go beyond the contemporoary obsession with sustainability, in the context of not only the environment but also other issues.

(a4.5) In relation to social trends as a whole, I proposed in Ch.9 of *FCD* (367-8), that "civilizational history will continue into the following cyclical progression of expansion, before it is to be superseded (solely as a high probability, since humans might be destroyed sooner either by themselves or in a gigantic natural calamity) by posthumans at some distant point in after-postmodernity (which I already discussed in *FHC*)" unto multiverses (different constellations of universes):

Local → Regional → Global →Solar →
Galactic → Clustery…→ Multiversal

In *BDPD*, this thesis of mine was referred to as "the theory of the cyclical progression of system integration and fragmentation," at the systemic level—in close relation to another thesis analyzed in Ch.9 of *FCD*, which was called, in the absence of better words, "the multifaceted theory of war and peace" in *FPHK* (as already indicated above).

In *FC*, I also proposed "the dialectic theory of complexity" to account for the factor of uncertainty, which can have a major role to play in the world.

(b) Culture

The category of "culture" here can also be divided into different sub-categories like (b1) aesthetics, (b2) religion, (b3) popular (mass) culture, (b4), logic, (b5) civilization as a whole, (b6) ethics, (b7) epistemology (methodology), (b8) metaphysics (ontology), and (b9) ideology.

However, (b7) epistemology (methodology), (b8) metaphysics (ontology), and (b9) ideology are already discussed elsewhere, be they about "meta-theory" (e.g., methodology and ontology under the categories of "existential dialectics" here and "sophisticated methodological holism" in the next section) or "theory" (e.g., ideology under the categories of "social institutions," as described earlier, and also of "civilization as a whole," as will be addressed shortly).

(b1) Aesthetics

In relation to aesthetics, I proposed "the transformative theory of aesthetic experience" in *FAE* about some great transformations of aesthetic experience in future history.

(b2) Religion

In relation to religion, I proposed "the comparative theory of religion" in *FHC* (and also in *FCD*, although it was first so labeled in *FPHK*).

(b3) Popular (Mass) Culture

In relation to popular (mass) culture, there are some sub-categories for illustration, which can be, for example, (b3.1) mass culture in general and (b3.2) sports (like chess).

(b3.1) In relation to mass culture in general, I proposed, in *FHC*, a new way to understand mass culture in terms of some fundamental dilemmas, which is called, in the absence of better words, "the dualistic theory of mass culture" in *FPHO*.

(b3.2) In relation to sports (like chess), I propose, in *FPHCHESS*, "the mediative-variative theory of chess" for a new way to understand tactics and strategy (in different traditions).

(b4) Logic

In relation to logic, I also worked out, in *FPHML*, "the contrastive theory of rationality" for a new way to understand rationality and knowledge (in the context of mathematical logic).

(b5) Civilization as a Whole

In relation to civilization as a whole, I also argued, in *FHC*, no freedom without unfreedom (as shown in *Table 4.13*) and no equality without inequality (as shown in *Table 4.14*), especially in relation to the seven dimensions of life existence first worked out in *FHC* (i.e., the technological, the everyday, the true, the holy, the sublime/beautiful, the good, and the just).

Even more interestingly, in *BDPD* and later in *BCIV*, I further revealed a theoretical refinement of existential dialectics, this time, in arguing that there is no civilization without barbarity, with human civilization to be eventually superseded by what I originally analyzed as "post-human post-civilization" (which should not be confused with "post-human civilization"), in the context of the freedom/unfreedom and equality/inequality dialectics.

In the final analysis, civilization cannot live without barbarity and has to learn to co-exist with it in ever new ways. Preposteorus as this may seem to many contemporaries—it is no more imperative to preserve civilization than necessary to destroy barbarity, and the ideal of civilization is essentially bankrupt, to be eventually superseded by "post-civilization" (as summarized in *Tables 4.35-4.43*).

This freedom/unfreedom and equality/inequality dialectics holds in pre-moderny, modernity, postmodernity, and what I orignally called "after-postmodernity" in *FHC* and *FCD*—especially, though not exclusively, in the context of the emergence of "post-humans."

There are two specific theories about "after-postmodernity" here. The first theory about "after-postmodernity" is about the trinity of pre-modernity, modernity, postmodernity, and after-postmodernity at the cultural level—and was called, in *BDPD*, "the theory of the trinity of modernity to its after-postmodern counterpart," although it was first analyzed in *FHC* and later in *FCD* and also *BCIV* (as summarized in *Table 1.34*, *Table 1.35*, *Table 1.36*, and *Table 1.37* and will be elaborated further in the later sub-section on "the rest" about future history).

The second theory about "after-postmodernity" is about pre-modernity, modernity, postmodernity, and after-postmodernity at the historical level. For instance, a conclusion in *FHC*, which is shocking indeed for many contemporaries, is none other than that "[t]he post-human history will therefore mark the end of human history as we know it and, for that matter, the end of human dominance and, practically speaking, the end of humans as well. The entire history of human civilization, from its beginning to the end, can be summarized by four words, linked by three arrows (as already discussed in *FHC*)":

Pre-Modernity → Modernity →
Post-Modernity → After-Postmodernity

In *BDPD*, this thesis of mine was specifically called "the theory of the evolution from pre-modernity to after-postmodernity," at the historical level—as shown in *Table 1.34*, *Table 1.35*, *Table 1.36*, and *Table 1.37* (and will be elaborated further in the later sub-section on "the rest" about future history).

Therefore, "[t]he end of humanity in the coming human extinction is the beginning of post-humanity. To say an untimely farewell to humanity is to foretell the future welcome of post-humanity." (P. Baofu 2002: 89) This thesis of mine was known in *BDPD* as "the theory of post-humanity," at the systemic level—as already indicated in (a4.3) above.

(b6) Ethics

In relation to ethics, I proposed, in *BEPE*, "the theory of post-ethics" for a new way to understand morality and immorality.

(c) The Mind

This important vision about "post-humans" (as summarized above) brings us to the category of "the mind" here, which, as is often understood in everyday's usage, refers to the "brain" and the "body" (together with their interactions) and has different levels of consciousness which can be both "spiritual" and non-spiritual.

Therefore, the "mind" already presupposes the "brain," "body," and "spirit" (or "soul"), so that one does not fall into the intellectual trap of the "mind-body dualism" as debunked in modern (and especially, postmodern) philosophy (which I already went to great lengths to explain in *The Future of Human Civilization*).

That said—the category of "the mind" can be divided into two sub-categories, namely, in relation to (c1) biology and (c2) psychology.

(c1) Biology

In relation to biology, there are two concerns here, namely, (c1.1) genes and intelligence and (c1.2) the human body.

(c1.1) In relation to genes and intelligence, I worked out "the theory of contrastive advantages" (which was originally proposed in *FCD* and further worked out in *BNN*, as summarized in *Table 1.24*), to show the nature-nurture interactions of multiple levels in action (e.g., the biological, the psychological, the structural, the systemic, the cultural, and others) for humans and post-humans on earth and beyond, to the extent that different groups, be they on the basis of race, gender, ethnicity, class, age, or else, are not equal, on average, and have never been, nor will they be, even if there can be exceptions.

(c1.2) In relation to the human body, for instance, in *FPHMA*, I suggested "the expansive-contractive theory of martial arts," in which two great future transformations of the martial body were envisioned, that is, both "with the hyper-martial body" and "without the body" (as summarized in *Table 1.32*)—although this issue can also be classified under the category of "culture."

And in *FPHS*, I proposed "the theory of virtual sexuality," in which three great future transformations of the sexual body were envisioned, that is, "virtual pleasure," "the hyper-sexual body," and "the post-sexual floating mind" (as summarized in *Table 1.33*).

(c2) Psychology

In relation to psypchology (which also overlaps with biology too, however), I also proposed some greater transformations of the mind to come in the future, especially though not exclusively in the post-human age.

These greater transformations of the mind in psychology can be further sub-divided into different categories, namely, (c2.1) consciousness, (c2.2) unconsciousness (together with subconsciousness), (c2.3) creativity, (c2.4) learning, (c2.5) aggression (and pacificity), and (c2.6) personality.

(c2.1) In relation to consciousness, I analyzed, in *FPHC*, the nature of consciousness in relation to the conceptual dimensions of consciousness (as shown in *Table 1.12*), the theoretical levels of consciousness (as shown in *Table 1.13*), the thematic issues of consciousness (as shown in *Table 1.14*), the different dimensions of human existence (viz., having-ness, belonging-ness, and being-ness) in relation to consciousness (as shown in *Table 1.15*, *Table 1.16*, *Table 1.17*, and *Table 1.18*).

In *BCPC*, the nature of consciousness was further explored in relation to cognition, emotion, and behavior, which can be in consciousness and other mental states (e.g., subconsciousness and unconsciousness), in the context of existential dialectics, as shown in *Table 1.19*, *Table 1.20*, *Table 1.21*, *Table 1.22,* and *Table 1.23*.

In fact, in *FCD*, I already proposed three different theories in relation to the limits of cognition, emotion, and behavior, namely, "the theory of cognitive partiality," "the theory of emotional non-neutrality," and "the theory of behavioral alteration."

Then, I argued that the existence of human consciousness will be superseded one day with "floating consciousness" (as first proposed in *FCD*) and "hyper-spatial consciousness" (as first proposed in *FPHC*) as a climax of evolution in consciousness, after the future extinction of human consciousness (as shown below):

Primordial consciousness → Human consciousness →
Post-human consciousness (with floating consciousness and hyper-spatial consciousness as a climax in the evolution of consciousness)

In *BDPD*, these latest theses of mine were called "the theory of floating consciousness" and "the theory of hyper-spatial consciousness," both at the cosmological and psychological levels (as summarized in *Table 1.26* and *Table 1.25*).

(c2.2) In relation to unconsciousness, for instance, in *FPHU*, the unconsious mind will undergo tremendous change, to the rise of what I originally called "unfolding unconsciousess" in relation to anomalous experience, as a climax of evolution in unconsciousness, in conjunction with the conscious mind (both human now and post-human later), especially at the psychological level, as shown below:

Primordial unconsciousness → Human unconsciousness →
Post-human unconsciousness (with unfolding unconsciousness
as a climax in the evolution of unconsciousness)

In *FPHU*, this thesis of mine is called "the theory of unfolding unconsciousness" in a small sense, or "the unfolding theory of anomalous experience," in a larger one, as shown in *Table 1.27* and *Table 1.28*.

(c2.3) In relation to creativity, for instance, in *FPHCT*, I also proposed "the comprehensive theory of creative thinking" to explore the role of creative thinking, in relation to technology and other factors, together with my original proposal of important creative techniques and traits (as shown in *Table 1.29*) and its double nature of desirability and undesirability (as shown in *Table 1.30*).

(c2.4) In relation to learning, for instance, in *FPHL*, I worked out "the multilogical theory of learning" for a new way to understand the nature of structure and context in language, in relation to learning and its future development in the context of the mind.

(c2.5) In relation to aggression (and pacificity), I proposed "the multifaceted theory of war and peace" in *FPHWP* (as mentioned above).

(c2.6) And in relation to personality, I worked out "the contrarian theory of personality" in *FPHP* for a new way to understand normality and abnormality.

(d) Nature

The category of "nature" here refers to the state of nature, which can be divided into some sub-categories like (d1) physics, (d2) cosmology, (d3) biology, (d4) geography, and (d5) geology.

But since biology in (d3) overlaps with biology in (c1)—only phyiscs, cosmology, geography, and geology are addressed hereafter.

(d1) Physics

In relation to physics, I also proposed "the perspectival theory of space-time" in *FPHST* (based on *FPHC*) for a new way to understand space and time, as shown in *Table 1.3*, *Table 1.4*, *Table 1.5*, and *Table 1.7*.

In *FPHG* (based on *FPHC* and *FPHST*), I proposed "the selective theory of geometry" for a new way to understand geometry, with implications for the understanding of space-time in the context of infinity, symmetry, and dimensionality for future lifeforms that our world has never known (as shown in *Table 1.6*, *Table 1.7*, *Table 1.8; Table 1.9*, *Table 1.10,* and *Table 1.11*).

And in *FC*, I proposed "the dialectic theory of complexity" for a new way to understand order and chaos in the state of nature and beyond.

(d2) Cosmology

In relation to cosmology, my theories on physics (as indicated above) already have important implications for understanding the nature and the future of the universe (which was also addressed in *FHC*, among other works aforcited).

I also proposed "the theory of hyper-spatial consciousness" in *FPHC* (and also in *FPHG*) for a new way to understand the future evolution of the mind to enter into different dimensions of space-time that our current world has never known.

In addition, in *BCOS*, I proposed "the theory of post-cosmology" for a new way to understand the beginnings and ends of the cosmos and beyond.

(d4) Geography

In relation to geography, I proposed "the theory of the geopower of nature" in *ALD* for a different way to understand the relationships between geography and other entities (like the mind, culture, and society).

(d5) Geology

And in relation to geology, I proposed "the resettlement theory of geology" in *FPHGEOL* for a different way to understand the relationships between statics and dynamics.

(e) The Rest

The category of "the rest" here refers to what cannot be strictly classified in any of the previous categories (i.e., society, culture, the mind, and nature). A good illustration of "the rest" refers to that which is, namely, (e1) historical.

(e1) Historical

Precisely here, I proposed "the theory of the evolution from pre-modernity to after-postmodernity" (as first worked out in *FHC* and later in *FCD* and *FPHC*), which overlaps—as a reminder from (b5)—with the issue concerning "civilization as a whole" in (b5).

For instance, in both *FCD* and *FPHC*, I worked out the structure of "post-human civilization" in terms of the trinity of after-postmodernity (i.e., "free-spirited after-postmodernity," "post-capitalist after-postmodernity," and "hegemonic after-postmodernity").

Both conceptually and theoretically, the trinity of after-postmodernity is a sequential extension to the trinity of modernity (i.e., "free-spirited modernity," "capitalist modernity,"and "hegemonic modernity") and the trinity of postmodernity (i.e., "free-spirited postmodernity," "capitalist postmodernity," and "hegemonic postmodernity") as first proposed in *FHC*.

And the trinity of pre-modernity (i.e., "pre-free-spirited pre-modernity," "pre-capitalist pre-modernity" and "hegemonic pre-modernity") was later conceived in *BCIV* to complete the historical set from pre-modernity to after-postmodernity.

In *BDPD*, this thesis about the trinity of pre-modernity, modernity, postmodernity, and after-postmodernity was collectively known as "the theory of the trinity of modernity to its after-postmodern counterpart," at the cultural level—as already so mentioned in (b5) earlier, and summarized in *Table 1.34*, *Table 1.35*, *Table 1.36*, and *Table 1.37*.

At the structural level, all these trinities are subject to the existential constraints (e.g., the freedom/unfreedom and equality/inequality dialectics in the context of "the cyclical progression of hegemony"), be the historical epoch in pre-modernity, modernity, postmodernity, or after-postmodernity in future times. Each of the historical epochs has its ever new ways of coming to terms with the ever new (different) mixtures of freedom/unfreedom and equality/inequality.

This is importantly so, not because, as is falsely assumed in conventional wisdom, one certain way is superior (or better) than another in terms of achieving more freedom and less unfreedom, or more equality with less inequality.

On the contrary, indeed, in each of the historical epochs, each increase of unfreedom greets each freedom achieved, just as each increase of inequality welcomes each equality achieved, albeit in ever new (different) ways. In *BDPD*, this thesis of mine was labeled as "the theory of the cyclical progression of hegemony," at the structural level, though it was first analyzed in *FCD*.

In *BDPD*, more theoretical applications of existential dialectics were further examined, in relation to five main features, in the context of the duality of oppression, namely, (a) that each freedom/equality achieved is also each unfreedom/inequality created, (b) that the subsequent oppressiveness is dualistic, both by the Same against the Others and itself and by the Others against the Same and themselves, (c) that both oppression and self-oppression can be achieved by way of downgrading differences (between the Same and the Others) and of accentuating them, (d) that the relationships are relatively asymmetric among them but relatively symmetric within them, even when the Same can be relatively asymmetric towards itself in self-oppression, and the Others can be likewise towards themselves, and (e) that symmetry and asymmetry change over time, with ever new players, new causes, and new forms, be the locality here on Earth or in deep space unto multiverses—as summarized in *Table.4.13*, *Table 4.14*, *Table 4.15*, and *Table 4.16*.

The same logic also holds both in relation to wealth and poverty (as addressed in *BCPC* and summarized in *Table 4.17* on the wealth/poverty dialectics) and in relation to civilization and barbarity (as addressed in *BCIV* and summarized in *Table 4.18* on the civilization/barbarity dialectics).

In *BDPD*, this thesis on existential dialectics was labeled as "the theory of existential dialectics."

*Direct and Indirect Applications of Existential Dialectics*

A different way to appreciate the usefulness of existential dialectics is by way of the analysis of its (a) direct and (b) indirect applications.

(a) In direct applications, on the one hand, the logic of existential dialectics can shed some theoretical insights on diverse phenomena in the world, and good instances are the usage of the principles of existential dialectics for the theoretical insights on the freedom/unfreedom dialectics, the equality/inequality dialectics, and the wealth/poverty dialectics—as introduced earlier in (b5) and (e) in the previous sub-section.

For instance, my books like *FPHST*, *BNN*, *FAE*, *FC*, *FIA*, *FPHU*, *FPHK*, *FPHML*, *FPHE*, *FPHMM*, *FPHCT*, *FPHG*, *FPHUP*, *FPHL*, *FPHO*, *FPHMA*, *FPHS*, *FPHLAW*, *FPHFS*, *FPHWP*, *FPHP*, *BEPE*, *BCOS*, *FPHGEOL*, and *FPHCHESS* use the principles to reveal some theoretical insights on the perspectives of space and time (as in *FPHST*), of nature and nurture (as in *BNN*), of beauty and ugliness (as in *FAE*), of order and chaos (as in *FC*), of taxonomy and network (as in *FIA*), of the unconscious mind (as in *FPHU*), of the nature of knowledge (as in *FPHK*), of mathematical logic (as in *FPHML*), of engineering (as in *FPHE*), of mass media (as in *FPHMM*), of creative thinking (as in *FPHCT*), of geometry (as in *FPHG*), of urban planning (as in *FPHUP*), of language (as in *FPHL*), of organization (as in *FPHO*), of martial arts (as in *FPHMA*), of sexuality (as in *FPHS*), of law (as in *FPHLAW*), of formal science (as in *FPHFS*), of aggression and pacificity (as in *FPHWP*), of normality and abnormality (as in *FPHP*). of morality and immorality (as in *BEPE*), of the beginnings and ends of the universe (as in *BOCS*), of statics and dynamics (as in *FPHGEOL*), and of tactics and strategy (as in *FPHCHESS*).

(b) In indirect applications, on the other hand, the theoretical insights can further be used to reveal other phenomena directly from them (viz., the theoretical insights) and therefore indirectly from the principles themselves.

A good illustration is of course the use of the theoretical insights on the freedom/unfreedom and equality/inequality dialectics for the understanding of the civilization/barbarity dialectics.

This distinction between direct and indirect applications may sound a bit academic, since even in indirect applications, the phenomena under study can still be directly related back to the principles themselves.

In the previous example, as an illustration, the civilization/barbarity dialectics can be directly related to the principles of existential dialectics without the intermediate role of the freedom/unfreedom and equality/inequality dialectics.

*Multiple Levels of Application*

There is another issue to be clarified, however. In other words, the theoretical insights can be applied to multiple levels of analysis—even though, in a given example, it may refer to one level or a few only.

For instance, in the example concerning the freedom/unfreedom dialectics, it can be used at the structural level (e.g., in relation to the theory of the cyclical progression of hegemony), but it can also be exploited for other levels (e.g., the theory of post-capitalism at the institutional level).

All these levels of application should not be misleadingly construed, as I stress this before, as a one-way street (that is, to use the ontological principles for theoretical application at multiple levels of specific analysis) but a two-way one, in which theoretical insights in praxis, when studied in more specific contexts, can in turn refine the nature of existential dialectics (for example, with the addition of new principles).

## Sophisticated Methodological Holism

The summary of my original meta-theory on methodology is provided below, again, to be repeated *often verbatim* from my previous works—as this is something I regularly do in each new book to introduce my past works, which are in conversation with this current one.

This means that, for those readers who had read some of my previous books, the summary below is really not much new (*almost verbatim*, with only some updated revisions).

That clarified—this original contribution of mine to the study of methodology takes the form of an approach known as "the theory of methodological holism" or "methodological holism" in short, as already worked out in *FPHC* (2004).

I have made good use of this methodological approach of mine for all of my previous works.

*The Confusion of Holism with Systems Approach*

Yet, it is imperative to stress at the outset, as I often emphasized this in all my previous books, that my approach of "methodological holism" does not oppose or exclude "methodological individualism" (as some readers may be tempted to assume, as is conventionally understood) but actually includes it.

The confusion here has to do with equating holism with systems approach in general (be it about systems theory, chaos theory, complexity theory, or else), which I debunked, while learning something from it, as already elaborated in *The Future of Complexity* (2007)—and *The Future of Post-Human Formal Science* (2010) is another addition to this critique of the intellectual fad of systems approach.

In other words, my meta-theory of holism, or sophisticated methodological holism in short, makes a distinction between doing holistic work and doing systems work, for two reasons.

The first reason is that holism, in my theory, does not imply systems approach, since it rejects systems approach because of the myriad problems that the literature on systems approach has been unable to resolve, while learning something from it.

And the second reason is that my theory is related to the other theory of mine, that is, existential dialectics, which does not exclude methodological individualism (but includes it, because it simply goes beyond both reductionism and reverse-reductionism in a dialectic way).

For these reasons (and others too, as summarized in *Table 4.1*), my version of methodological holism is *sophisticated*—not *vulgar* as sometimes used by inapt scholars employing the same term.

*The Ontological Constraints*

Sophisticated methodological holism is subject to some ontological constraints, and five good examples include "the partiality-totality principle," "the predictability-unpredictability principle," "the explicability-inexplicability principle," "the absoluteness-relativeness principle," and "the formalness-informalness principle" in existential dialectics.

Of course, there can be more than five ontological principles in relation to method, but the five examples here are sufficient to illustrate the point in question.

With this caveat in mind—the first two were already analyzed in previous books of mine like *FC* and *FAE*, whereas the third one was addressed in *FPHU*, and the last two were accounted for in *FPHK* and *FPHML*, respectively.

For now, it suffices to show that sophisticted methodological holism, because of these ontological constraints on methodology, targets two major sins of methodology, namely, what I call, in the absence of better words, (a) *reductionism* and (b) *reverse-reductionism*—both of which come in all shapes and sizes, to be summarized below (and also shown in *Table 4.2*).

### *Against the Varieties of Reductionism*

There are four versions of reductionism to be summarized here, which sophisticated methodological holism rejects, namely, (a1) conceptual, (a2) theoretical, (a3) methodological, and (a4) ontological forms of reductionism.

(a1) In conceptual reductionism, a good case in point concerns myriad dualities like mind vs. body, self vs. world, democracy vs non-democracy, and the like (as already addressed in *FHC*, *FPHC*, and *BDPD*, for instance).

(a2) In theoretical reductionism, an illuminating case study is best exemplified by what I originally called "the foundation fallacy" in *FPHST*, in any attempt to naively understand space-time from the physical perspective as the foundation and, consequently, to dangerously dismiss (or belittle) other perspectives.

In *FAE*, I elaborated further these versions of reductionism in the literature on aesthetics (e.g., form vs. content, representation vs. expression, critics vs. artists, and externalism vs. internalism).

In *FIA*, I revealed other forms of reductionism in the literature on information architecture (e.g., the constructivist argument and the representational argument).

In *FPHU*, I showed the persistent legacy of reductionism, this time, in the literature on anomalous experience (e.g., the obsession with physics, chemistry, and biology for explaining anomalous experience).

In *FPHE*, I examined another case of reductionism in action, in the context of engineering (e.g., technical constraints vs. normative constraints).

In *FPHMM*, I elaborated one more version of reductionism, in relation to the three domains of communication (e.g., the competing views on sending, connecting, and receiving).

In *FPHCT*, I explored another version of reductionism, in relation to invention and innovation (e.g., the bio-psychological argument vs. the socio-cultural arguments).

In *FPHG*, I identified another version of reductionism, in relation to infinity, symmetry, and dimensionality (e.g., the Euclidean argument vs. the non-Euclidean arguments).

In *FPHUP*, I analyzed another version of reductionism, in relation to density and void (e.g., the engineering argument and the ecology argument).

In *PFHL*, I examined another version of reductionism, in relation to structure and context (e.g., the structuralist argument and the contextualist argument).

In *PFHO*, I explored another version of reductionism, in relation to communcation, decision-making, and leadership (e.g., the rational-system argument vs. the natural-system argument vs. the open-system argument).

In *PFHMA*, I revealed another version of reductionism, in relation to the martial body and spirit (e.g., the spiritual argument vs. the materialist argument vs. the defensive argument).

In *PFHS*, I analyzed another version of reductionism, in relation to the sexual body and spirit (e.g., the naturalist argument vs. the constructivist argument).

In *PFHLAW*, I examined another version of reductionism, in relation to law (e.g., the necessity argument vs. the contengency argument).

In *FPHWP*, I show another version of reductionism, in relation to war and peace (e.g., the aggressivist argument vs. the pacifist argument).

In *BEPE*, I analyzed another version of reductionism, in relation to morality and immorality (e.g., the objectivist argument vs. the non-objectivist argument vs. the skeptical argument).

In *BCOS*, I examined another version of reductionism, in relation to the contested beginnings and speculative ends of the universe (e.g., the scientific argument vs. the religious argument vs. the esoteric argument vs. the metaphysical argument).

In *FPHP*, I showed another version of reductionism, in relation to normality and abnormality (e.g., the natural argument vs. the social argument vs. the cultural argument vs. the mental argument).

In *FPHGEOL*, I scrutinized another version of reductionism, in relation to statics and dynamics (e.g., the catastrophe argument vs. the uniformity argument vs. the revision argument).

And in *FPHCHESS* here, I show another version of reductionism, in relation to tactics and strategy (e.g., the natural argument vs. the social argument vs. the cultural argument vs. the mental argument).

(a3) In methodological reductionism, a good illustration can be the debate between different versions of qualitative and quantitative methods (as already analyzed in *FC* and also *FHC*).

In *FPHML*, I examined similar reductionism, this time, in the literature on mathematical logic (e.g., the obsession with consistency, soundness, and completeness).

And in *FPHFS*, I explored the problems of reductionism in the context of formal science (e.g., the analytical argument, as opposed to the synthetic argument).

(a4) In ontological reductionism, an excellent instance involves another debate, this time between emergentism and reductionism in complexity theory (as addressed in *FC*) and also in psychology (as elaborated in *FPHC*, in the context of Being and Belonging).

But to target against the varieties of reductionism constitutes only a side of the same coin, and the other side concerns the varieties of reverse-reductionism.

*Against the Varieties of Reverse-Reductionism*

The other side of the same coin is the reverse version of reductionism, which is what I want to call, in the absence of better words, *reverse-reductionism*.

My sophisticated version of methodological holism targets against the varieties of reverse-reductionism (just as it also rejects the ones of reductionism).

Perhaps nothing expresses better the popularity of reverse-reductionim than the "anything-goes" mentality in postmodernism of our time, as shown in the following four versions.

(b1) In conceptual reverse-reductionism, any concept of "art" (e.g., fine art, applied art, outsider art, junk art) is welcome in postmodernism (as already analyzed in Ch.4 of *FHC*—and also in *FAE*).

(b2) In theoretical reverse-reductionism, a variety of art and literary theories co-exist. Take the case of literary studies, as there are now Literary Structuralism, Marxist Literary Criticism, New Criticism, Phenomenology, Hermeneutics, Language-Game Literary Criticism, Feminist Literary Criticism, Reception Theory, Reader Response Criticism, Poststructuralism, Semiotics, Pyschoanalytic Literary Criticism, just to cite some well-known ones, with no one being said to be better than any others (as detailedly analyzed in Ch.4 of *FHC*). (S. Raman 1997)

In *BNN*, I even introduced "the compromise fallacy" as another good illustration of theoretical reverse-reductionism, in misleadingly treating both genetic and environmental approaches as equally valid.

(b3) In methodological reverse-reductionism, multiple methodologies are deemed as acceptable in postmodernism (e.g., doing art without praxis, doing art with praxis, and doing art by sublation), as analyzed in Ch.4 of *FHC*.

And in *FPHFS*, I exposed the problems of reverse-reductionism in the context of formal science in relation to systems theory (e.g., the synthetic argument).

(b4) In ontological reverse-reductionism, no privileged ontology is allowed, and the door is open practically for anything in postmodernism (e.g., the equal status of the ontology of Being vs. that of Becoming, as already addressed in Ch.4 of *FHC*—and also in *FPHC*).

In *FAE*, I also introduced another version of reverse-reductionism, that is, "the pluralist fallacy," in the context of understanding aesthetic experience, for instance—although this fallacy has been committed not exclusively in relation to the ontological level (but also at the conceptual, theoretical, and methodological ones).

These dual dangers against reverse-reductionism (in this sub-section) and reductionism (in the previous sub-section) are something that sophisticated methodological holism rejects. And in *FPHFS*, I addressed these two problems further.

This point about the dual dangers is important enough, since many scholars often favor one against the other, but it is vital to target both.

### *The Holistic Organization of an Inquiry*

With these dual dangers against reductionism and reverse-reductionism in mind—sophisticated methodological holism suggests that an inquiry of any given phenomenon is more complete, if treated in the context of a comprehensive analysis at all relevant levels, which challengingly encompass all the fields of human knowledge, ranging from the natural sciences and formal sciences through the social sciences to the humanities.

This section is something that I had already stressed in all my previous books and repeat (*often verbatim*) hereafter. But, for those readers who had read my previous books, this serves as a reminder.

With this reminder in mind—there are multiple ways to engage in a holistic inquiry with all relevent levels of analysis. Over the years, I have proposed different ways to fulfill this holistic methodological requirement, as repeated in my previous books.

Hereafter is a summary of four major ways, namely, (a) by discipline, (b) by domain, (c) by subject, and (d) by meta-analysis—all of which fulfill the holistic requirement of sophisticated methodological holism on a given issue.

(a) In a holistic organization by discipline, a good illustration is the multiple levels of analysis, namely, (a1) the micro-physical, (a2) the chemical, (a3) the biological, (a4) the psychological, (a5) the organizational, (a6) the institutional, (a7) the structural, (a8) the systemic, (a9) the cultural, (a10) the macro-physical (cosmological), and (a11) other relevant levels, which are either a combination of all these levels or the practical applications with a combination of them.

This holistic organization by discipline is used in *FCD* and *BCPC*, for instance.

(b) In a holistic organization by domain, some good candidates include the classification of different perspectives of analysis in relation to (b1) nature, (b2) the mind, (b3) society, and (b4) culture, as already worked out in *BCIV*, *FPHST*, *FC*, *FAE*, *FPHU*, *FPHK*, *FPHE*, and *FPHFS*.

Culture in (b4), in this re-classification, is the same as culture in (a9) and can be further divided into different sub-categories, with good examples like (a9i) epistemology (methodology), (a9ii) aesthetics, (a9iii) ethics, (a9iv) metaphysics (ontology), and (a9v) religion. For clarification, epistemology (methodology) in (a9i) and metaphysics (ontology) in (a9iv) overlap with the category of "meta-theory."

Society in (b3) includes the organizational in (a5), the institutional in (a6), the structural in (a7), and the systemic in (a8).

The mind (b2) has more to do with the chemical in (a2), the biological in (a3), and the psychological in (a4), although it can overlap with (a1), (b3) and (b4), for instance.

It should be stressed that the "mind" here also includes the categories like the "brain," "body," and the "spirit" (or the "soul")—whenever needed in a given context, so that one does not fall into the intellectual trap of the "mind-body dualism" in the history of modern (and especially, postmodern) philosophy, which I already went to great lengths to explain in *The Future of Human Civilization*. After all, in everyday's usage, the "mind" already implies the "brain" and the "body" (together with their interactions) and has different levels of consciousness which can be both "spiritual" and non-spiritual.

And nature in (b1) refers to the micro-physical in (a1) and the macro-physical (cosmological) in (a10), although it can also overlap with (a2), (a3) and (a4).

(c) In a holistic organization by subject, some excellent examples concern the analysis of consciousness in *FPHC* (i.e., on Having,

Belonging, and Being); of civilization in *FHC* (i.e., the True, the Holy, the Everyday, the Technological, the Beautiful/Sublime, the Good, and the Just); of international politics in *BWT* (i.e., Hyper-Empires, Meso-Empires, and Micro-Empires); of nature and nurture in *BNN* (i.e., genes and memes); of information architecture in *FIA* (i.e., taxonomy and network); of mathematical logic in *FPHML* (i.e., consistency, soundness, and completeness); of communication in *FPHMM* (i.e., sending, connecting, and receiving); of creative thinking in *FPHCT* (i.e., invention and innovation); of geometry in *FPHG* (i.e., infinity, symmetry, and dimensionality); of urban planning in *FPHUP* (i.e., density and void), of language in *FPHL* (i.e., structure and context), and of organization in *FPHO* (i.e., communication, decision-making, and leadership), in *FPHMA* (i.e., the martial body and the martial spirit), in *FPHS* (i.e., the sexual body and the sexual spirit), in *FPHLAW* (i.e., necessity and contingency), in *FPHWP* (i.e., aggression and pacificity), in *FPHP* (i.e., normality and abnormality), in *BEPE* (i.e., morality and immorality), in *BCOS* (i.e., the beginnings and ends of the universe), in *FPHGEOL* (i.e., statics and dynamics), and in *FPHCHESS* (i.e., tactics and strategy)—by incorporating all the relevent levels of analysis as cited above in each.

(d) And in a holistic organization by meta-analysis, two good cases in point involve the works on post-capitalism in *BCPC* and on authoritarian liberal democracy in *ALD*, with the classification of analysis in theory and meta-theory.

The essential point to remember here, as I thus stress in all my previous books, is that the multiple levels of analysis can be reorganized in many different ways, insofar as none of the levels (if relevant to an inquiry in question) is ignored or dismissed, to avoid the danger of reductionism (and for that matter, the one of reverse-reductionism, in the opposite direction).

### *Three Clarifications*

Three clarifications are needed here, however, to avoid any hasty misunderstanding.

Firstly, the entities in each classification are not mutually exclusive. For instance, in the holisitc organization by domain, nature in (b1) can also be linked to the chemical in (a2), the biological in (a3), and the systemic in (a8). And the mind (b3) can alternatively be related to the micro-physical in (a1) and the chemical in (a2), for example.

Secondly, there is also the important factor of luck (or uncertainty in conventional usage), but it is already implied or allowed in each of the organizations of an inquiry—especially in relation to the predictability-unpredictability principle and the order-chaos principle in existential dialectics. This topic about uncertainty was already addressed in *FHC*—and more extensively, in *FC*, in the context of order and chaos.

And lastly, the comparison in a classification is not absolute, but relative, as there are often some shades of gray, not exactly black or white (figuratively speaking).

Some further clarifications and qualifications of sophisticated methodological holism are shown in *Table 4.2*.

*Some Distinctive Usefulness*

In the end, sophisticated methodological holism—when applied, especially though not exclusively, as illustrated in my numerous works—can enrich the understanding of reality in some distinctive ways. Here are four examples (as revealed in each of my books).

Firstly, it provides a comprehensive analysis of a subject matter, from which much can be learned about reality, both in relation to the perspectives of the mind, nature, society, and culture.

Secondly, it offers a new classification of the subject matter.

Thirdly, it suggests some visions of the future in relation to the subject matter in question.

And fourthly, it proposes some insights on meta-theory (e.g., methodology and ontology) in general—with the clear understanding, however, of the dilemma of specific vs. general ontology (as shown in *Table 4.9* on the syntax of existential dialectics in the context of the dilemma of ontology).

For this very reason, all of these ways are important, without reducing one into the analysis of another. (*FPHCT*)

# Chapter Outline

With this summary of my two original meta-theories (on methodology and ontology, *almost verbatim* from my previous works) in mind—this book is organized in four main parts, namely, (a) Part I on the introduction to this book, (b) Part II on strategy, (c) Part III on tactics, and (d) Part IV on the conclusion of the book.

The book is thus organized in four chapters.

The introductory chapter here, that is, Chapter One titled "Introduction—The Fun of Chess," begins by first describing the didactic tale of two chess sensibilities and then clarifying the variants of chess and the relationship between strategy and tactics in chess—together with the theoretical debate in the literature, my mediative-variative theory of chess, the relationship between theory and meta-theory, existential dialectics, sophisticated methodological holism, and some further clarifications.

Chapter Two titled "Strategy and Its Inadequacy" explores the nature of strategy in chess from the four perspectives of the mind, nature, society, and culture, with a deconstructive analysis of each.

Chapter Three titled "Tactics and Its Narrowness" examines the nature of tactics in chess from the four perspectives of the mind, nature, society, and culture, with a constructive critique of each.

The last chapter titled "Conclusion—The Future of Chess" summarizes the analysis in the book in the dialectic context of strategy and tactics in chess, with fourteen major theses, namely, (a) the first thesis: the absoluteness-relativeness principle, (b) the second thesis: the predictability-unpredictability principle, (c) the third thesis: the explicability-inexplicability principle, (d) the fourth thesis: the preciseness-vagueness principle, (e) the fifth thesis: the simpleness-complicatedness principle, (f) the sixth thesis: the openness-hiddenness principle, (g) the seventh thesis: the denseness-emptiness principle, (h) the eighth thesis: the slowness-quickness principle, (i) the nineth thesis: the convention-novelty principle, (j) the tenth thesis: the evolution-transformation principle, (k) the eleventh thesis: the symmetry-asymmetry principle, (l) the twelveth thesis: the regression-progression principle, (m) the thirteenth thesis: the sameness-difference principle, and (n) the fourteenth thesis: the post-human variation, in relation to the three great future transformations of mind games—to be elaborated in the rest of the book, with a summary in the concluding chapter.

This seminal project is to fundamentally alter the way that we think about chess, from the combined perspectives of the mind, nature, society, and culture, with enormous implications for the human future and what I originally called its "post-human" fate.

## Some Clarifications

But some additional clarifications are needed, before a formal analysis of the nature of chess is to begin in Chapter Two.

These clarifications, lest any potential misunderstanding by the readers occurs, refer to (a) the conversation with my previous books, (b) the illustrative selection of case studies and examples, (c) detailed analysis versus overall analysis, (d) two distinctive features of using quotations, and (e) the use of neologisms.

These clarifications are something that I also regularly repeat in my previous books too, so these points are repeated hereafter, *often verbatim*—albeit in a different context this time.

## In Conversation with My Previous Books

Firstly, my theory of post-ethics is constructed on the theoretical foundation of my previous books (i.e., *FHC*, *FCD*, *FPHC*, *BDPD*, *BCPC*, *BCIV*, *FPHST*, *BNN*, *BWT*, *FC*, *FAE*, *ALD*, *FIA*, *FPHU*, *FPHK*, *FPHML*, *FPHE*, *FPHMM*, *FPHCT*, *FPHG*, *FPHUP*, *FPHL*, *FPHO*, *FPHMA*, *FPHS*, *FPHLAW*, *FPHFS*, *FPHWP*, *FPHP*, *BCOS*, *BEPE*, and *FPHGEOL*—as explicated in "The List of Abbreviations").

Just consider the following two illustrations, namely, (a) in relation to other related theoretical debates and (b) in relation to different visions as worked out in my previous books.

(a) The first illustration is that the theoretical debate on the nature of morality and immorality also involves a few other related theoretical issues.

These other related theoretical issues have to do with some deeply contested theoretical debates, which, however, I had already gone to great lengths to analyze in my numerous (previous) books, as summarized below. It should also be stressed, lest misunderstanding occurs, that each of the theoretical debates (as listed below) does not exhaust the issues covered in the books, as each book covers more issues than what is listed below (for illustration only).

- On Sports and Mass Culture
  - –Ex: 2 volumes, *The Future of Human Civilization*
  - –Ex: *The Future of Post-Human Martial Arts*
  - –Ex: *The Future of Post-Human Mass Media*
- On Creativity
  - –Ex: *The Future of Post-Human Creative Thinking*
  - –Ex: *The Future of Aesthetic Experience*
- On science
  - –Ex: 2 volumes, *The Future of Human Civilization*
  - –Ex: *The Future of Complexity*
  - –Ex: *The Future of Post-Human Knowledge*
  - –Ex: *The Future of Post-Human Space-Time*
  - –Ex: *Beyond Cosmology to Post-Cosmology*
  - –Ex: *The Future of Post-Human Geometry*

- On Technology and Communication
  - –Ex: 2 volumes, *The Future of Human Civilization*
  - –Ex: *The Future of Post-Human Engineering*
  - –Ex: *The Future of Post-Human Mass Media*
- On Morality
  - –Ex: 2 volumes, *The Future of Human Civilization*
  - –Ex: *Beyond Ethics to Post-Ethics*
  - –Ex: *The Future of Capitalism and Democracy*
  - –Ex: *Beyond Democracy to Post-Democracy*
- On Aesthetics
  - –Ex: 2 volumes, *The Future of Human Civilization*
  - –Ex: *The Future of Aesthetic Experience*
- On Warfare
  - –Ex: 2 volumes, *The Future of Human Civilization*
  - –Ex: *The Future of Post-Human War and Peace*
  - –Ex: *The Future of Post-Human Martial Arts*
  - –Ex: *The Future of Capitalism and Democracy*
  - –Ex: *Beyond Democracy to Post-Democracy*
- On Spirituality
  - –Ex: 2 volumes, *The Future of Human Civilization*
  - –Ex: *The Future of Capitalism and Democracy*
  - –Ex: *Beyond Capitalism to Post-Capitalism*
  - –Ex: *The Future of Post-Human Sexuality*
  - –Ex: *The Future of Post-Human Martial Arts*
  - –Ex: *The Future of Post-Human Unconsciousness*
  - –Ex: *The Future of Post-Human Consciousness*
  - –Ex: *Beyond Ethics to Post-Ethics*
- On Being and Belonging
  - –Ex: *The Future of Post-Human Consciousness*
  - –Ex: 2 volumes, *The Future of Human Civilization*
- On Different Dimensions of Human Existence
  - –Ex: 2 volumes, *The Future of Human Civilization*
- On Violence
  - –Ex: *The Future of Capitalism and Democracy*
  - –Ex: *The Future of Post-Human Law*
  - –Ex: *The Future of Post-Human War and Peace*
  - –Ex: *The Future of Post-Human Sexuality*
  - –Ex: *The Future of Post-Human Martial Arts*
- On Conflict Resolution
  - –Ex: *The Future of Capitalism and Democracy*
  - –Ex: *The Future of Post-Human War and Peace*

• On Modernity, Post-Modernity, and After-Postmodernity
  –Ex: 2 volumes, *The Future of Human Civilization*
  –Ex: *Beyond Civilization to Post-Civilization*
• On Pre-Modernity
  –Ex: *Beyond Civilization to Post-Civilization*
• On Uncertainty
  –Ex: *The Future of Complexity*
  –Ex: 2 volumes, *The Future of Human Civilization*
• On Cognition
  –Ex: *The Future of Post-Human Unconsciousness*
  –Ex: *The Future of Post-Human Consciousness*
  –Ex: *The Future of Capitalism and Democracy*
  –Ex: *The Future of Post-Human Creative Thinking*
  –Ex: *The Future of Post-Human Personality*
• On Intuition
  –Ex: *The Future of Post-Human Unconsciousness*
  –Ex: *The Future of Post-Human Consciousness*
  –Ex: *Beyond Capitalism to Post-Capitalism*
  –Ex: *The Future of Capitalism and Democracy*
• On Emotion
  –Ex: *The Future of Post-Human Unconsciousness*
  –Ex: *The Future of Post-Human Consciousness*
  –Ex: *The Future of Aesthetic Experience*
  –Ex: *Beyond Capitalism to Post-Capitalism*
  –Ex: *The Future of Capitalism and Democracy*
  –Ex: *The Future of Post-Human Personality*
• On Behavior
  –Ex: *The Future of Post-Human Language*
  –Ex: *The Future of Post-Human Unconsciousness*
  –Ex: *The Future of Post-Human Consciousness*
  –Ex: *Beyond Capitalism to Post-Capitalism*
  –Ex: *The Future of Capitalism and Democracy*
  –Ex: *The Future of Post-Human Sexuality*
  –Ex: *The Future of Post-Human Personality*

Therefore, those who are interested in these other theoretical issues (and debates) which are related to the current context can consult these other books of mine for more details.

(b) And the second illustration is that one of the major theses of this book concerns the future, be it human and later post-human, which is closely linked with the different visions that I had worked out in my previous books, in relation to the mind, nature, society, and culture.

Of course, for the convenience of the readers, a summary of some of these visions will be provided whenever necessary, as shown in the two sections on meta-theory above (*often verbatim*) and in different tables at the end of this chapter and the concluding chapter, for instance.

Yet, I still expect the readers to read my previous books directly for more details—as this book is written in conversation with my previous ones.

## Case Studies and Examples

Secondly, the case studies and examples in the chapters are not exhaustive but solely illustrative. Nor are they necessarily mutually exclusive, as they can be reclassified in a different way. And exceptions are allowed.

One is tempted to assume, however, that, since the case studies and their examples are not exhaustive, more case studies and examples are needed. But the problem here is that just adding more of them does not necessarily change the conclusions to be drawn and may even lead to redundancy.

After all, the case studies and examples used in this book are chosen with care and deemed sufficient for the purpose at hand, even though they are solely illustrative.

## Detailed Analysis versus Overall Synthesis

Thirdly, there is a distinction between detailed analysis and overall synthesis, in that the former is, especially though not exclusively, more for the specialized viewpoint of a specific discipline, whereas the latter is, especially though not exclusively, more for the broad horizon of an interdisciplinary approach—relatively speaking, of course.

Or, by analogy, there is a distinction between depicting a "tree" (with tiny details) in a "forest" and viewing the entire "forest" itself (with the broader horizon).

This book, like all previous ones of mine, tries to strike a balance between the two forms of understanding, although this does not imply that the two forms should be exactly equally offered, in an exact 50-50 balance.

Instead, the book, more often than not, does not allow the tiny details to obscure the larger picture in a given case study, because the entire book is inter-disciplinary in nature and tries to look at the entire "forest," not being bogged down with the detailed analysis of an individual "tree" in the forest, by analogy.

In this sense, although detailed analysis can be provided whenever needed for a given case study, the important point to remember is the overall horizon with its broadness of scope.

## Two Distinctive Features of Using Quotations

Fourthly, the way that quotes are used in this book (and others of mine) has two distinctive features, as explained below.

The first distinctive feature is that, at the end of each quotation, sometimes there are more than one reference—in which case the first reference is for the original citation, and the rest of the references are for the convenience of the reader for more information about the issue under quotation.

And the second feature is that quotations are used as often as possible, so as to let others speak for themselves, without the nuisance of putting my own words into their mouths. Even more importantly, this allows me to use the original text as an evidence for critical analysis at the end of each section.

## The Use of Neologisms

And finally, I use different neologisms in my books, mostly to introduce my original concepts and theories, and good instances include those here or elsewhere in my previous books (e.g., "the mediative-variative theory of chess," "post-democracy," "hyper-spatial consciousness," "post-capitalism," and whatnot). Needless to say, they are used here solely for our current intellectual convenience, as they will be renamed differently in different ways in future history.

As I thus wrote in *FCD* (508-9), "all these terms 'post-capitalism,' 'post-democracy'...and other ones as introduced in...[the] project (e.g.,...'posthuman elitists,' and 'posthuman counter-elitists,' just to cite a few of them) are more for our current intellectual convenience than to the liking of future humans and post-humans, who will surely invent more tasteful neologisms to call their own eras, entities, and everything else, for that matter. But the didactic point here is to use the terms to foretell what the future might be like, not that its eras and entities must be called so exactly and permanently. After all, William Shakespeare (1995: Act II,

Scene II, Line 47) well said long ago: 'What is in a name? That which we call a rose by any other name would smell as sweet.'"

As I had also stressed time and again before, each of the neologisms can be re-written as a different "X," only to be re-named differently by the powers that be in different eras of future history.

With these clarifications for thought in mind—I now proceed to Chapter Two on strategy and its indequacy.

**Table 1.1. Strategy and Tactics in Chess**

- **Strategy**
  - —Opening
    - Ex: development
    - Ex: control of the center
    - Ex: king safety
    - Ex: pawn structure
  - —Middle Game
    - Ex: kinds of attacks
    - Ex: combinations of attacks
    - Ex: preparation for endgame
  - —Endgame
    - Ex: importance of pawns
    - Ex: the stronger role of the king
    - Ex: zugswang

- **Tactics**
  - —Ex: defense and offense
  - —Ex: the gain of material
  - —Ex: the use of sacrifices
  - —Ex: the manipulation of pawns
  - —Ex: the compulsion to move
  - —Ex: the intermediate move

*Notes*: The examples in the categories are solely illustrative (not exhaustive), and the comparison is relative (not absolute), nor are they necessarily mutually exclusive. And some can be easily re-classified elsewhere. As generalities, they allow exceptions.
*Source*: From *Sec. 1.3* of *FPHCHESS*. See book for citations.

**Table 1.2. Strategy and Tactics in General**

- **Strategy**
  - —Goal Setting
    - Ex: communicating to others about the goal
      - –Ex: why it is desirable
      - –Ex: how it is achievable
    - Ex: dealing with others in regard to
      - –Ex: the support of allies
      - –Ex: the resistance of opponents
  - —Goal Implementation
    - Ex: relying on the support of allies
      - –Ex: delivering short-term benefits while holding long-term goals
    - Ex: fighting the resistance of opponents
    - Ex: dividing the process in different stages (or time frames)
    - Ex: making good use of different search strategies
  - —Goal Assessment
    - Ex: preparing for
      - –Ex: an exit strategy (Plan B) in case of failure
      - –Ex: a victory arrangement in case of success
    - Ex: being ready to
      - –Ex: resign or give up if failing
      - –Ex: reward the allies if winning
    - Ex: being alert about
      - –Ex: the too early declartion of victory
      - –Ex: the too late acceptance of defeat

- **Tactics**
  - —Ex: be creative
  - —Ex: be flexible
  - —Ex: be tolerant of uncertainty
  - —Ex: be multifaceted

---

*Notes*: The examples in the categories are solely illustrative (not exhaustive), and the comparison is relative (not absolute), nor are they necessarily mutually exclusive. And some can be easily re-classified elsewhere. As generalities, they allow exceptions.
*Source*: From Chapter 4 of *FPHCHESS*—but also *FPHCT* and *FPHO*. See books for citations.

**Table 1.3. The Theoretical Debate on Space-Time (Part I)**

- **Isaac Newton's Absolutist (Substantivist) Theory of Space-Time**
  —space and time are independent from each other. The structure of space-time is $E3 \times E1$ (with the structure of space, *P*, as the set of spatial locations in a three-dimensional Euclidean space, *E3*, and the structure of time as the set of temporal moments, *T*, in the one-dimensional real time, *E1*).
  —space and time are also independent from the effects of matter and events. The existence of space and time is possible even in a world absent of matter (and, for that matter, even in a world absent of events), as if they were material objects but with their total unchangingness thorough time.

- **Albert Einstein's Relativist Theory of Space-Time**
  —space and time are interchangeable (not absolute), just as matter and energy are equivalent (not independent) with the famous equation, $E = mc^2$ (as in the special theory of relativity in 1905).
  —space-time and matter-energy are also relative in a grand union (as in the general theory of relativity in 1915). Thus, each pair affects the other pair, as "matter 'tells' spacetime how to curve [in a non-Euclidean geometry] and curved spacetime 'tells' matter how to behave....Space contracts near mass and dilates away from it. Time dilates near mass and contracts away from it. Clocks positioned farther away from the mass of the earth run faster than clocks closer to the earth." (L. Shlain 1991: 328-330)

*(continued on next page)*

## Table 1.3. The Theoretical Debate on Space-Time (Part II)

- **Peter Baofu's Perspectival Theory of Space-Time**
  —space and time can be understood from multiple perspectives, be they in relation to culture, society, nature, and the mind, with each perspective revealing something about the nature of space-time and simultaneously delimiting its view. This is subject to "the regression-progression principle" in existential dialectics.
  —each perspective of space and time exists in society and culture with good reasons, with some being more successful and hegemonic (dominant) than others. This is subject to "the symmetry-asymmetry principle" in existential dialectics.
  —space and time will not last, to be eventually superseded (altered) by post-humans in different forms (e.g.,stretching/shrinking space-time, engineering more dimensions of space-time, and manipulating multiverses), be they here in this universe or in multiverses. Thus, even the physical existence of space-time cannot last forever, with ever more transformations in the process. This is subject to "the change-constancy principle" in existential dialectics.
  —the conventional wisdom (especially by physicists) of treating the physical perspective of space and time as the foundation of all other perspectives (of space and time) and of regarding them as much less important is a form of reductionism,committing what I call *the foundation fallacy*, in misleadingly dismissing the multiple perspectives of space and time in relation to culture, society, nature, and the mind.

*Notes*: The examples in each category are solely illustrative (not exhaustive), and the comparison is relative (not absolute), nor are they necessarily mutually exclusive. Some can be easily re-classified elsewhere. As generalities, they allow exceptions.
*Source*: A summary of *Sec.1.1*, *Sec.1.2*, and *Sec.1.3* of *FPHST*—and, for that matter, the rest of *FPHST*

**Table 1.4. Main Reasons for Altering Space-Time**

- **The Need to Make New Energy-Matter**
  —Ex: manipulating molecular bonds for new materials
  —Ex: creating nanotechnologies on the atomic scale
  —Ex: engineering the atomic nucleus
  —Ex: restructuring most elementary particles
  —Ex: inventing new forms of matter and energy

- **The Need to Create New Space-Time**
  —Ex: creating "warp drive" (as in science fiction) for space travel
  —Ex: creating "pocket universes"

- **The Need to Conquer the Cosmos unto Multiverses**
  —Ex: spreading floating consciousness and hyper-spatial consciousness, besides other forms that humans have never known, in the cosmos and beyond unto multiverses for ultimate conquest

*Notes*: The examples in each category are solely illustrative (not exhaustive) nor necessarily mutually exclusive, and the comparison is relative (not absolute). As generalities, they allow exceptions. Also, it should be stressed that the three reasons are all related, in that they all contribute to the evolution of intelligent life in the cosmos unto multiverses in the most distant future beyond our current knowledge.
*Sources*: A summary of *Sec.6.2* of *FPHST*. See also *FHC*, *FCD*, *FPHC*, *FC*, *FPHU*, and *FPHG*, for example.

**Table 1.5. The Technological Frontiers of the Micro-World**

- **Type I-Minus**
  —Ex: building structures and mining

- **Type II-Minus**
  —Ex: playing with the genetic makeups of living things

- **Type III-Minus**
  —Ex: manipulating molecular bonds for new materials

- **Type IV-Minus**
  —Ex: creating nanotechnologies on the atomic scale

- **Type V-Minus**
  —Ex: engineering the atomic nucleus

- **Type VI-Minus**
  —Ex: restructuring most elementary particles

- **Type Ω-Minus**
  —Ex: altering the structure of space-time

*Notes*: As already indicated in *Sec.4.4.2.2* of *FPHC*, the problem with this micro-classification (from Barrow's work) is that the civilization types (with the exception of Type Ω-Minus, for example) are not quite distinct, since many of them can be achieved more or less in a civilization, to the extent that Type II-minus and Type III-minus, just to cite two plausible types, can be historically contemporaneous, relatively speaking, unlike the vast historical distance between, say, Type O and Type I (or Type I and Type II) civilizations. In other words, the micro-classification here is not very useful to understand civilization types but is revealing to see the technological frontiers of the micro-world.
*Sources*: A reconstruction from J.Barrow (1998:133), as originally shown in *Table 4.7* of *FPHC*. See *FPHC* for more info.

**Table 1.6. Finity, Transfinity, and Infinity (Part I)**

- **Infinity**
  —The term "infinity," which is "symbolically represented with ∞," derives from the Latin *infinitas* (or "unboundedness"), to refer to "several distinct concepts—usually linked to the idea of 'without end'—which arise in philosophy, mathematics, and theology." (WK 2008c)
  —In the context of mathematics, "infinity" is often "treated as if it were a number (i.e., it counts or measures things: 'an infinite number of terms') but it is a different type of 'number' from the real numbers. Infinity is related to limits,...large cardinals,...projective geometry, extended real numbers and the absolute Infinite," for instance. (WK 2008c)

- **Transfinity**
  —To avoid confusion, my usage of the word "transfinite" (as used here in the ontological principle) is not the same as the one which was originally coined by Georg Cantor to refer to "cardinal numbers or ordinal numbers that are larger than all finite numbers, yet not necessarily absolutely infinite" (in the mind of God), as opposed to "relative" infinity (in the mind of Man). (WK 2008d) In other words, for Cantor, there are two kinds of infinity, "absolute" and "relative"—with "relative" infinity or "transfinity" (in the mind of Man) being dependent on "absolute" infinity (in the mind of God) for its existence.
  —By contrast, my usage of the word "transfinite" here differs radically from the one used by Cantor (and other mathematicians) and is more limited, in light of the problems confronting any attempt to understand the idea of infinity, be it by intuition, imagination, and conception (as already shown in the prevous three sub-sections).
  —With this clarification in mind, I allow numbers which can be many times larger—or smaller, for that matter—than the finite things that we encounter in daily life, but they do not have to be related to the idea of infinity at all (which may not exist).

*(continued on next page)*

**Table 1.6. Finity, Transfinity, and Infinity (Part II)**

• **Transfinity *(cont'd)***

—Of course, there may be some *borderline* cases, in which it is not clear whether the number in question is transfinite (in my usage) or simply a mathematical convenience. A good example of a borderline case is the Planck unit of length for "the smallest space possibly measured in nature," which is "less than billionths of trillionths of trillionths of an inch" (or something like $1.6 \times 10^{-35}$ meters). (P. Baofu 2006a; N. McAleer 1987: 219; WKV 2008; D. Corbett 2008)

—The ideas concerning symmetry and dimensions in the examples above were addressed in Chapter Three and Chapter Four of *FPHG*. For now, it suffices to further add that it is mind-boggling to even conceive or imagine extreme numbers like the Planck unit of length (or others). Does it really exist at all? Or is it merely a mathematical convenience to describe a physical situation that the human mind does not understand properly?

—In any event, "transfinity" (in my usage) can be used as a further extension (of number) to the finite numbers in daily life—albeit with some borderline cases in mind.

• **Finity**

—It refers to numbers which are bounded (that is, with an end), especially (though not exclusively) in relation to things in everyday life.

*Source*: From *Sec. 1.4* and *Sec. 2.2.3* of *FPHG*

**Table 1.7. Theoretical Speculations of Multiverses**

- **"Baby Universes" (Ex: Andre Linde and others)**
  —Ex: In a flat universe theory, "even if our part of it eventually collapses,...some spots in the cosmos would suddenly start inflating on their own, creating brand-new 'baby universes.'" (P. Baofu 2000: 623)

- **"Parallel Universes" (Ex: Stephen Hawking and others)**
  —Ex: In quantum cosmology, there allows the existence of infinite numbers of parallel universes, with tunneling among them. (M. Kaku 1994: 256) Hawking later revised his views on this.

- **"Pocket Universes" (Ex: Alan Guth)**
  —Ex: "As the pocket universes live out their lives and recollapse or dwindle away, new universes are generated to take their place....While life in our pocket universe will presumably die out, life in the universe as a whole will thrive for eternity." (A. Guth 1997: 248; P. Baofu 2002: 482)

- **"Brane Worlds" (Ex: Warren Siegel, Lisa Randall, and others)**
  —Ex: Our universe is stuck on a membrane of space-time embedded in a larger cosmos, with different brane worlds connecting and/or colliding with each other.

*Notes*: These examples are solely illustrative (not exhaustive), and some of the items can be reclassified somewhere else. Nor are they always mutually exclusive. Since they are generalities, exceptions are expected.
*Source*: From *Table 4.8* of *FPHC*

**Table 1.8. The Confusion between "Many Worlds" and "Multiverse" (Part I)**

- **First Problem**
  —The first problem in this second confusion concerns a debate between the "many-worlds" interpretation of quantum mechanics by Hugh Everett (1957) and others, and the Copenhagen interpretation by Niels Bohr and Werner Heisenberg on the effects of quantum states after interacting with an external environment (e.g., by a measuring observer).
  —On the one hand, the Copenhagen interpretation of quantum mechanics is more indeterministic in its concern with the probabilities of the outcomes of a wave function after interacting with an external environment (e.g., by a measuring observer using light to detect, for example, "an electron in a particular region around the nucleus at a particular time." (WK 2008e) Each wave function for the state of a particle refers to "a mathematical representation used to calculate the probability for it to be found in a location, or state of motion." (WK 2008f) And the "wavefunction collapse" means that "the act of measurement causes the calculated set of probabilities to 'collapse' to the value defined by the measurement." (WK 2008f)
  —On the other hand, the Everett's interrpetation is more deterministic, in rejecting "the objective reality of wavefunction collapse" and "instead explaining the subjective appearance of wavefunction collapse with the mechanism of quantum decoherence." (WK 2008g) The words "quantum decoherence" here refers to "the mechanism by which quantum systems interact with their environments to exhibit probabilistically additive behavior," in such a way that "the quantum nature of the system is simply 'leaked' into the environment." (WK 2008g) For example, "[a]s a result of an interaction, the wave functions of the system and the measuring device become entangled with each other. Decoherence happens when different portions of the system's wavefunction become entangled *in different ways* [italic added] with the measuring device...." (WK 2008g) Thus, for Everett, the appearance of "wavefunction collapse" can be explained by "quantum decoherence" for quantum states to branch out in different independent worlds after interacting with an external environment.

*(continued on next page)*

**Table 1.8. The Confusion between "Many Worlds" and "Multiverse" (Part II)**

- **First Problem** ***(cont'd)***
  —Yet, a major criticism against Everett's interpretation is that these different independent worlds in Everett's sense "will never be accessible to us," because of the interference by the measuring observer in an external environment. (WK 2008g)
  —For illustration, "once a measurement is done, the measured system becomes entangled with both the physicist who measured it and a huge number of other particles, some of which are photons flying away towards the other end of the universe; in order to prove that the wave function did not collapse one would have to bring all these particles back and measure them again, together with the system that was measured originally. This is completely impractical, but even if one can theoretically do this, it would destroy any evidence that the original measurement took place (including the physicist's memory)." (WK 2008g)
  —But then, the critics ask, If these independent worlds in Everett's sense are inaccessible to us, how do we know much of anything really about them?

- **Second Problem**
  —What is relevant at the quantum mechanical level may not necessarily apply to the world at the physicsl level.
  —In fact, the persistent inability among contemporoary physicists to unifiy the theory of relativity for large objects at the physical level and quantum mechanics for tiny objects at the quantum mechanical level should warn us about the danger of reductionism, by reducing the understanding of different universes ("parallel universes") at the physical level from a generalization of "many worlds" at the quantum mechanical level.

- **Third Problem**
  —There is no empirical evidence to suggest that there are infinite (or extremely many) possible worlds in the cosmos as the many-worlds model by Everett would have us to believe.

*Source*: From *Sec.4.4.1* of *FPHG*

**Table 1.9. Hyperspace and Its Challenge (Part I)**

• **First Problem**

—Hyperspace for intra- and inter-universal travel is assumed to be possible, on the basis of four main assumptions; although there may be others, these are the often cited ones.

—The first assumption concerns the existence of "entry" and "exit" points in hyperspace, which is easier said than done to enter, go through, and then exit hyperspace.

—For instance, using hyperspace "requires complex calculations…of single or multiple hyperspace jumps and the control of the jump….Peturbations such as those experienced by ship in space from the gravitional field around an object such as a planet or even a star are exacerbated in hyperspatial travel, since mass in real space distorts hyperspace in an equal measure. 'Jumping' near to a gravitational mass is likely to make resulting exit from hyperspace to be highly uncertain, with the level of improbability *i* increasing with the square of the distance to the nearest gravitional 'well.'" (WK 2008h)

—Besides, "[o]nly spaceships equipped with a special force field can enter hyperspace, because exposure to hyperspace even for short period of time is hazardous to unprotected humans." (WK 2008h)

• **Second Problem**

—The second assumption concerns the faster-than-light speed. It is well known that Einstein argued for the speed of light as the ultimate limit, but nowadays phyiscists do allow the violation of this limit under certain conditions (like the hypothetical existence of tachyons, as will be explained further in a later sub-section on time travel).

—Even then, there is a challenge to explain in science fiction "why ships can travel faster than light in hyperspace," and two good arguments are that "hyperspace may be smaller than real space and therefore a star ship's propulsion seems to be greatly multiplied, or else the speed of light in hyperspace is not a barrier as it is in real space. Whatever the reasoning, the general effect is that ships traveling in hyperspace seem to have broken the speed of light, appearing at their destinations much more quickly and without the shift in time that the Theory of Relativity would suggest." (WK 2008h)

*(continued on next page)*

**Table 1.9. Hyperspace and Its Challenge (Part II)**

- **Third Problem**
  —The third assumption concerns the speed of time (to be measured, say, by a clock), but this is elaborated in a different section on the related issue of "time travel" (as shown in a different table on time travel).

- **Fourth Problem**
  —The fourth assumption concerns spatial shortcut, as it is possible, because of the curvature of space, in that "the manifold of ordinary three-dimensional space is curved in four or more 'higher' spacial dimensions (a 'hyperspace' in the geometric sense…). This curvature causes certain widely separated points in three-dimensional space to nonetheless be 'adjacent' to each other four-dimensionally. Creating an aperture in 4D space (a wormhole) between these locations can allow instantaneous transit between the two locations; a common comparison is that of a folded piece of paper, where a hole punched through two folded sections is more direct than a line drawn between them on the sheet. This idea probably arose out of certain popular descriptions of General Relativity and/or Riemannian manifolds, and may be the original form from which later concepts of hyperspace arose." (WK 2008h)
  —But this does not mean that hyperpsace is always safe. In fact, "in some science fiction, the danger of hyperspace travel is due to the chance that the route through hyperspace may take a ship too close to a celestial body with a large gravitational field, such as a star. In such scenarios, if a starship passes too close to a large gravitational field while in hyperspace, the ship is forcibly pulled out of hyperspace and reverts to normal space. Therefore, certain hyperspace 'routes' may be mapped out that are safe, not passing too close to stars or other dangers." (WJ 2008h)

*Source*: From *Sec.4.4.1* of *FPHG*

**Table 1.10. The Problems of Time Travel into the Future (Part I)**

• **First Problem**

—The first problem concerns "time dilation under the Theory of Special Relativity" (which separates space-time from mass-energy as special cases). Suppose you are on a moving train and an observer outside the train is watching you along the tracks. As your train moves along at the speed of light (or something close to it), "time, as measured by your watch, ticked along at a slower pace than time measured by the observer. Not only that, distance changed, too. For the observer, a one-foot ruler whizzing by on the train would have measured less than a foot." (NO 2000)

—In fact, you may not even notice that this slower clicking of the clock; in fact, everything looks normal to you inside the train as it was before: "The weird thing is that, for you on the train, time wouldn't seem to be moving slower and your ruler wouldn't be shorter—all would appear normal." (NO 2008) By contrast, you instead think that "time on the rest of the Earth would appear to be ticking along slower and its rulers would be shorter." (NO 2000) So, both of you insist that the other clock (not his) is clicking more slowly—assuming, of course, an inertial frame of reference.

—Now, change the story a bit, in that, this time, instead of an "inertial reference frame," acceleration is allowed, and you are on your way to board a spaceship. As an illustration, suppose "[y]ou board a spaceship and take off for deep space. The ship approaches the speed of light. Time for you seems to pass as it always has. It takes you about five seconds to tie your shoe. But to an observer on Earth (assuming he or she could watch you), you are moving at a snail's pace. It takes hours to tie your shoe." (NO 2000) In any event, "you continue on your journey. You slow down, stop, and accelerate back to Earth. You arrive home. You have aged two years during your flight. Two hundred years have passed on Earth. You have successfully travelled forward through time." (NO 2000) In this case, both the observer (if he is still alive) and you agree that you have aged less.

*(continued on next page)*

**Table 1.10. The Problems of Time Travel into the Future (Part II)**

- **First Problem *(cont'd)***

  —In other words, when you are back to Earth, a lot of things have changed since, and you may find rather strange to be in a new environment which you hardly recognize, when compared with what you used to think as your city, or the Earth (two years ago, well, in accordance to your time of measurement). This means that you are in the future (two hundred years later), even though you are still relatively young (only two years older). But, can you go back? Well, no. The reason is that "[a]ccording to relativity, you can only move through time in one direction." (NO 2000)

- **Second Problem**

  —The second problem concerns "time dilation under the Theory of General Relativity" (which unifies space-time with mass-energy in a general theory), the same result of time dilation occurs.

  —For instance, when "one clock is deeper in a gravity well than the other," a result is that "the clock deeper in the well" will "tick…more slowly; this effect must be taken into account when calibrating the clocks on the satellites of the Global Positioning System, and it could lead to significant differences in rates of aging for observers at different distances from a black hole." (WK 2008i)

  —More precisely, by way of another example, "it has been calculated that, under general relativity, a person could travel forward in time at a rate four times that of distant observers by residing inside a spherical shell with a diameter of 5 meters and the mass of Jupiter. For such a person, every one second of their 'personal' time would correspond to four seconds for distant observers. Of course, squeezing the mass of a large planet into such a structure is not expected to be within our technological capabilities in the near future." (WK 2008i; R. Gott 2002)

  —The point here, however, is that time travel into the future, in this sense, is allowed, theoretically speaking, but in a very limited way, from the theory of relativity. But one can also ask, What if the theory of relativity were wrong? Well, in that case, there would be a need to find a better theory to explain time travel into the future.

*Source*: From *Sec.4.4.2* of *FPHG*

**Table 1.11. The Problems of Time Travel into the Past (Part I)**

- **First Problem**
  - —The first problem concerns the faster-than-light-speed, since, in accordance to the theory of special relativity, when an object goes faster than the speed of light, something weird can happen, in that, instead of having the clock clicking more slowly, it is moving backward, in relation to some inertial frame of reference—which then raisies the issue of whether time travel into the past is possible at all.
  - —But the problem here is about the violation of the law of casuality. In everyday language, an excellent illustration of causality violation is that, suppose you travel back into the past and then kill your parents, this then means that you could not have been born. But this is not true, since you are alive.
  - —Yet, "in the case of a hypothetical signal moving faster than light, there would always be some frames in which the signal was received before it was sent, so that the signal could be said to have moved backwards in time." The violation of casuality exists in this case.
  - —The first part of the problem is that, in accordance to the theory of special relativity, "it would take an infinite amount of energy to accelerate a slower-than-light object to the speed of light"—let alone the energy needed to propel an object to go faster than the speed of light. (WK 2008i) Although some suggest the possibility of "negative energy," it remains to be seen to what extent this is true.
  - —And the second part of the problem is that, "although relativity does not forbid the theoretical possibility of tachyons which move faster than light at all times, when analyzed using quantum field theory it seems that it would not actually be possible to use them to transmit information faster than light and there is no evidence for their existence." (WK 2008i; S. Chase 1993)

*(continued on next page)*

## Table 1.11. The Problems of Time Travel into the Past (Part II)

• **Second Problem**

—The second problem concerns cosmic string and black holes, since it is the theory of general relativity which "extends the special theory to cover gravity, illustrating it in terms of curvature in spacetime caused by mass-energy and the flow of momentum." (WK 2008i)

—On the other hand, "[g]eneral relativity describes the universe under a system of field equations…and there exist solutions to these equations that permit what are called 'closed time-like curves,' and hence time travel into the past….The first of these was proposed by Kurt Gödel, a solution known as the Gödel metric, but his (and many others') example requires the universe to have physical characteristics that it does not appear to have." (WK 2008i; K. Thorne 1994)

—But then, the problem now becomes: "Whether general relativity forbids closed time-like curves for all realistic conditions is unknown." (WK 2008i)

• **Third Problem**

—The third problem concerns wormholes and related ones, and "[w]ormholes are a hypothetical warped spacetime which are also permitted by the Einstein field equations of general relativity, although it would be impossible to travel through a wormhole unless it was what is known as a traversable wormhole." (WK 2008i; M. Visser 1995) In other words, "a wormhole is a hypothetical topological feature of spacetime that is fundamentally a 'shortcut' through space and time. Spacetime can be viewed as a 2D surface, and when 'folded' over, a wormhole bridge can be formed. A wormhole has at least two mouths which are connected to a single throat or tube. If the wormhole is traversable, matter can 'travel' from one mouth to the other by passing through the throat." (WK 2008j)

—But the problem here is three-fold, as shown in what follows.

*(continued on next page)*

**Table 1.11. The Problems of Time Travel into the Past (Part III)**

• **Third Problem *(cont'd)***
—The first part of the problem is that "there is no observational evidence for wormholes." (WK 2008j)
—The second part of the problem is that it is not clear how exactly wormholes (and black holes, for that matter) could allow time travel into the past at all.
—And the third part of the problem is that if using wormholes (and black holes, for that matter) to travel into the past requires something to go faster than light too, then this raises another issue concerning the two problems as previously indicated in the sub-section on the faster-than-light-speed.

• **Fourth Problem**
—The fourth problem is that some physicists have argued that, even should time travel into the past be possible, by way of these three methods, the person still would not return to the same past history that he exactly experienced before, but in a different one. (WK 2008i)
—In other words, "strictly speaking," time travel into the past is not "really" possible, in light of these criticisms, so there is no need to worry that one could travel into the past and kill one's parent, as a way to violate the law of casuality.

*Source*: From *Sec.4.4.2* of *FPHG*

**Table 1.13. The Theoretical Levels of Consciousness (and Other Mental States) (Part I)**

- **At the Micro-Physical Theoretical Level**
  —Quantum-Mechanics
  —Electromagnetism

- **At the Chemical Theoretical Level**
  —Biochemistry

- **At the Biological Theoretical Level**
  —Evolutionary Biology
  —Neuroscience
  —Artificial Intelligence

- **At the Psychological Theoretical Level**
  —Psychodynamic Psychology
    - Psychoanalytical Psychology
    - Analytical Psychology
    - Socially Oriented Psychology

  —Experimental (Behavioral) Psychology
    - Operant Reinforcement Theory
    - Stimulus-Response Theory
    - Social Learning Theory

  —Cognitive (Gestalt) Psychology
    - Humanist Psychology
    - Existential Psychology
    - Field Theory

  —Psychometric Psychology
    - Idiographics
    - Constitutional Psychology

  —Social Psychology
    - Symbolic Interactive Theory
    - Social Exchange Theory

*(continued on next page)*

**Table 1.13. The Theoretical Levels of Consciousness (and Other Mental States) (Part II)**

- **At the Organizational Theoretical Level**
  —Managerial-Bureaucratic Theory
  —Oligarchic Theory
  —Network Theory

- **At the Institutional Theoretical Level**
  —Functionalist Theory
  —Anomic Theory

- **At the Structural Theoretical Level**
  —Conflict Theory
    - Marxian Theory
    - Critical Theory
    - Weberian Theory

  —Games Theory (in Formal Theory)
  —Feminist Theory
    - Feminist Concerns
    - Feminist Hopes

- **At the Systemic Theoretical Level**
  —Equilibrium Theory
  —System Theory
  —Chaos Theory

- **At the Cosmological Theoretical Level**
  —Superluminal Model
  —The Theory of Floating Consciousness

*(continued on next page)*

**Table 1.13. The Theoretical Levels of Consciousness (and Other Mental States) (Part III)**

- **At the Cultural Theoretical Level**
  —Substantive Theories
  - Structuralist Theory
  - Post-Structuralist Theory (in Postmodernism)

  —Meta-Theories
  - Epistemic Objectivism vs. Epistemic Historicism
  - Epistemic Subjectivism vs. Epistemic Non-Subjectivism
    –Phenomenology
    –Ethnomethodology
    –Hermeneutics
  - Epistemic Relativism vs. Epistemic Absolutism
  - Epistemic Reductionism vs. Epistemic Emergencism

- **At Other Levels**
  —Structuration Theory
  —Reflexive Socioanalysis

*Notes*: These examples are solely illustrative (not exhaustive), and some of the items can be reclassified somewhere else. Nor are they always mutually exclusive. Since they are generalities, exceptions are expected.
*Source*: From many different sources as indicated in *FPHC*

**Table 1.14. The Thematic Issues of Consciousness (and Other Mental States)**

- **The Factor of History**
  —Pre-Modernity
  —Modernity
  —Postmodernity
  —After-Postmodernity

- **The Importance of Needs**
  —Having (e.g., the everyday, the technological)
  —Belonging (e.g., the just and the good)
  —Being (e.g., the true, the holy, the sublime/beautiful, and the good)

*Notes*: These examples are solely illustrative (not exhaustive). Nor are they always mutually exclusive. Since they are generalities, exceptions are expected. Also, some categories overlap with others; for instance, the good is also related with being-ness, just as the everyday can be connected with belong-ness.
*Sources*: From *FPHC*, based on *FHC* and *FCD*

**Table 1.15. Having, Belonging, and Being in Consciousness**

- **Having (e.g., About the Everyday and Technological)**
  —Ex: Physiological needs (e.g., thirst, hunger, sex)
  —Ex: Safety (e.g., stability, freedom from fear and anxiety)

- **Belonging (e.g., About the Good and Just)**
  —Ex: Social interaction
  —Ex: Friendship, acquaintance
  —Ex: Love, family
  —Ex: Self-respect, respect from others

- **Being (e.g., About the True, Holy, Beautiful/Sublime, and Good)**
  —Ex: Understanding of reality about self and world
  —Ex: Spiritual quest for holiness, beauty, sublimity, and goodness

*Notes*: These examples are solely illustrative (not exhaustive), and some of the items can be reclassified somewhere else. Nor are they always mutually exclusive. Since they are generalities, exceptions are expected.
*Sources*: From *FPHC*, based on *FHC* and *FCD*

**Table 1.16. The Having-Ness of Consciousness (and Other Mental States)**

- **At the Micro-Physical Level**
  —Ex: The finer physics of nutrition

- **At the Chemical Level**
  —Ex: Food guide pyramid

- **At the Bio-Psychological Level**
  —Ex: The hierarchy of needs

- **At the Institutional Level**
  —Ex: The production of commodity fetishism

- **At the Organizational Level**
  —Ex: Mass standardization (and customization); bureaucratic rationality

- **At the Structural Level**
  —Ex: The divide between the haves and the have-nots

- **At the Cultural Level**
  —Ex: The bondage of greater expectations

- **At the Systemic Level**
  —Ex: The technological transformation of poverty with transvaluations

- **At the Cosmological Level**
  —Ex: Floating consciousness in deep space without the human physiological needs

*Notes*: These examples are solely illustrative (not exhaustive), and some of the items can be reclassified somewhere else. Nor are they always mutually exclusive. Since they are generalities, exceptions are expected.
*Sources*: From *FPHC*. See also *FCD* and *FHC*.

**Table 1.17. The Belonging-Ness of Consciousness (and Other Mental States)**

- **At the Micro-Physical Level**
  —Ex: Statistical physics and social networks

- **At the Chemical Level**
  —Ex: Olfaction and social relations

- **At the Bio-Psychological Level**
  —Ex: Dual human nature on relationships

- **At the Institutional Level**
  —Ex: Private property and social cleavage

- **At the Organizational Level**
  —Ex: Legal formalism and iron cage

- **At the Structural Level**
  —Ex: Class conflicts and social alienation

- **At the Cultural Level**
  —Ex: Ruthless society and the culture of law

- **At the Systemic Level**
  —Ex: Human extinction and post-human relations

- **At the Cosmological Level**
  —Ex: The transcendence of equality in the cosmos and beyond

*Notes*: These examples are solely illustrative (not exhaustive), and some of the items can be reclassified somewhere else. Nor are they always mutually exclusive. Since they are generalities, exceptions are expected.
*Sources*: From *FPHC*. See also *FCD* and *FHC*.

**Table 1.18. The Being-Ness of Consciousness (and Other Mental States) (Part I)**

- **At the Micro-Physical Level**
  —Ex: Space-time and matter-energy in classical mechanics, quantum mechanics, and the theory of relativity

- **At the Chemical Level**
  —Ex: Drugs and well being

- **At the Biological Level**
  —Ex: Neuroscience and mystical experiences

- **At the Psychological Level**
  —Ex: Reductionism and emergencism

- **At the Institutional Level**
  —Ex: The proliferation of cults and sects, and the corporate-art industry

- **At the Organizational Level**
  —Ex: The organization of ideas in the scheme of things

- **At the Structural Level**
  —Ex: Disciplinary power and control in the order of things

- **At the Systemic Level**
  —Ex: Mutualism and ecological psychology

- **At the Cultural Level**
  —Ex: The recurrence of competing spirits, and floating consciousness as a climax of evolution

*(continued on next page)*

**Table 1.18. The Being-Ness of Consciousness (and Other Mental States) (Part II)**

- **At the Cosmological Level**
  —Ex: The search for elsewherewhen, and the evolution of other consciousnesses (e.g., the hyper-spatial consciousness)

*Notes*: These examples are solely illustrative (not exhaustive), and some of the items can be reclassified somewhere else. Nor are they always mutually exclusive. Since they are generalities, exceptions are expected.
*Sources*: From *FPHC*. See also *FCD* and *FHC*.

**Table 1.19. Cognitive Partiality in Different Mental States**

- **The Conscious**
  - —*Biased*
    - Ex: Self-overratingness
    - Ex: Self-modesty
    - Ex: Self-fulfilling prophesy
  - —*Shallow*
    - Ex: Attribution error
    - Ex: Heuristics
    - Ex: Illusory thinking
  - —*Short-term*
    - Ex: Deindividuation
  - —*Materialistic*
    - Ex: Hierarchy of needs

- **The Subconscious**
    - Ex: mental reconstruction
    - Ex: stereotyping

- **The Unconscious**
    - Ex: automatic processing
    - Ex: cognitive intuition

*Notes*: These examples are solely illustrative (not exhaustive), and some of the items can be reclassified somewhere else. Nor are they always mutually exclusive. Since they are generalities, exceptions are expected. And the comparison is relative, not absolute.
*Source*: From *BCPC*

**Table 1.20. Emotional Non-Neutrality and Behavioral Alteration in Different Mental States**

- **The Emotional**
  - —*Conscious*
    - Ex: Feelings
  - —*Subconscious*
    - Ex: Moods
  - —*Unconscious*
    - Ex: Empathy
    - Ex: Instincts

- **The Behavioral**
  - —*Conscious*
    - Ex: Foot-in-the-door
    - Ex: Role playing
    - Ex: Saying-is-believing
  - —*Subconscious*
    - Ex: Discrimination
  - —*Unconscious*
    - Ex: Non-Verbal Communication

*Notes*: These examples are solely illustrative (not exhaustive), and some of the items can be reclassified somewhere else. Nor are they always mutually exclusive. Since they are generalities, exceptions are expected. And the comparison is relative, not absolute.
*Source*: From *BCPC*

**Table 1.21. The Limits of Intuition in Unconsciousness**

- **Cognitive Intuition (or Intuitive Thinking)**
  - —*Features*
    - Ex: mental schemas
    - Ex: expertise
    - Ex: blindsight
  - —*Problems*
    - Ex: inaccurate
    - Ex: non-explanatory

- **Emotional Intuition**
  - —*Features*
    - Ex: empathy
    - Ex: instincts
  - —*Problems*
    - Ex: over-generalized
    - Ex: non-explanatory

- **Behavioral Intuition**
  - —*Features*
    - Ex: prosemics
    - Ex: kinesics
  - —*Problems*
    - Ex: unreliable
    - Ex: non-explanatory

*Notes*: These examples are solely illustrative (not exhaustive), and some of the items can be reclassified somewhere else. Nor are they always mutually exclusive. Since they are generalities, exceptions are expected. And the comparison is relative, not absolute.
*Source*: From *BCPC*

**Table 1.22. The Wealth/Poverty Dialectics in Different Mental States: The Case of Cognition**

- **The Conscious**
  - —*Biased*
    - Ex: If richer in self-serving bias, then poorer in understanding others
  - —*Shallow*
    - Ex: If richer in heuristic thinking, then poorer in comprehending reality
  - —*Short-term*
    - Ex: If richer in short-term tendency, then poorer in long-term planning
  - —*Materialistic*
    - Ex: If richer in physiological obsession, then poorer in spiritual enlightenment

- **The Subconscious**
  - Ex: If richer in mental reconstruction, then poorer in accuracy of recall
  - Ex: If richer in stereotyping, then poorer in understanding groups

- **The Unconscious**
  - Ex: If richer in automatic processing, then poorer in depth of analysis
  - Ex: If richer in cognitive intuition, then poorer in reliability of judgments

*Notes*: These examples are solely illustrative (not exhaustive), and some of the items can be reclassified somewhere else. Nor are they always mutually exclusive. Since they are generalities, exceptions are expected. And the comparison is relative, not absolute.
*Source*: From *BCPC*

**Table 1.23. The Wealth/Poverty Dialectics in Different Mental States: The Case of Emotion and Behavior**

- **Emotion**
  - —*Conscious*
    - Ex: If richer in relying on familiarity and similarity for social attraction, then poorer in having diverse social relations
  - —*Subconscious*
    - Ex: If richer in moodiness, then poorer in sober thinking
  - —*Unconscious*
    - Ex: If richer in empathy, then poorer in task-oriented efficiency
    - Ex: If richer in instincts, then poorer in critical thinking

- **Behavior**
  - —*Conscious*
    - Ex: If richer in foot-in-the-door technique, then poorer in instant Gratification
    - Ex: If richer in role playing, then poorer in identity stability
  - —*Subconscious*
    - Ex: If richer in discriminating, then poorer in social harmony
  - —*Unconscious*
    - Ex: If richer in kinesics and proxemics, then poorer in verbal skills, relatively speaking.

*Notes*: These examples are solely illustrative (not exhaustive), and some of the items can be reclassified somewhere else. Nor are they always mutually exclusive. Since they are generalities, exceptions are expected. And the comparison is relative, not absolute.
*Source*: From *BCPC*

**Table 1.24. The Theoretical Debate on Nature vs. Nurture (Part I)**

- **The Environmental Approach**
  —*Thesis*: It focuses, relatively speaking, more on the environment (culture and society) in explaining the achievement gap among individuals and for that matter, countries or regions, when contrasted with the natural factors.
  —*Discourse*: Examples include Jose Ortega y Gasset ("Man has no nature; what he has is history"), Ashley Montagu ("Man is man because he has no instincts, because everything he is and has become he has learned from his culture, from the man-made part of the environment, from other human beings"), Stephen Jay Gould ("[The] brain [is] capable of a full range of behaviors and predisposed to none"), and Jesse Jackson (who blames white racism for the failure of blacks to close the achievement gap between whites and blacks in America). The works on dependency theory in international political economy (with a Marxian influence) and on the Protestant work ethic (by Max Weber) also point to this environmental direction.

- **The Genetic Approach**
  —*Thesis*: It focuses instead, relatively speaking again, on hereditory factors (e.g., genes and evolution) in explaining the achievement gap among individuals and for that matter, countries or regions, when contrasted with the envrionmental factors.
  —*Discourse*: Examples are Hans Eysenck and William Sheldon (in constitutional psychology), Konrad Lorenz (in his work on innate aggressive human nature), Gary Marcus (on the complexities of human thought by a tiny number of genes) and Robert Plomin, together with Michael Owen and Peter McGuffin (on the genetic basis of complex human behaviors).

*(continued on next page)*

**Table 1.24. The Theoretical Debate on Nature vs. Nurture (Part II)**

---

- **The Compromise Approach**
  —*Thesis*: It seeks the middle-of-the-road argument in regard to nature and nurture and regards all differences among individuals and groups as the result of the mixture of both nature and nurture, more or less equally.
  —*Discourse*: Examples include C. Murray and R. Herrnstein ("It seems highly likely to us that both genes and environment have something to do with this issue") and Dan Dennett ("Surely 'everyone knows' that the nature-nurture debate was resolved long ago, and neither side wins since everything – is – a – mixture – of – both – and - it's – all – very -complicated…."

- **The Transcendent Approach**
  —*Thesis*: It goes beyond both nature and nurture (without, however, committing the compromise fallacy) in showing their closely intertwined interactions in producing the behavioral differences as often seen in individual human endeavors on the micro scale, and for that matter, in country (or regional) endeavors on the macro one—in the context of my five theses, namely, (a) the compromise fallacy, (b) no oppression without self-oppression, (c) no success without failure, (d) the factor of randomness, and (e) the post-human vision, to be elaborated in Chapter Six.
  —*Discourse*: Peter Baofu proposed this approach on the basis of his "theory of contrastive advantages" (as an original theoretical contribution to the debate, which was first proposed and analyzed in *The Future of Capitalism and Democracy*). In the end, the human genes will not last, to be eventually superseded by post-human life forms, so the debate between genes and memes has obscured something profoundly important about the future that the world has never known. And the debate is also misleading and faulty in its dichotomy.

---

*Source*: From *BNN*. See the book for citations and details.

**Table 1.25. Physical Challenges to Hyper-Spatial Consciousness**

- **The Understanding of a Higher-Dimensional World of Space-Time**
  —Ex: 4 for traditional aspects of space-time (e.g., length, width, breadth and time) plus 6 more new dimensions in theory of hyper-space, with profound implications for practical applications to new forms of consciousness.

- **The Mastering of Dark Matter and Dark Energy**
  —Ex: "ordinary matter" (e.g., atoms, molecules) as a mere 4.4% of the universe, with 23% made of "cold dark matter" and the rest (about 73%) of mysterious "dark energy," with fundamental significance to questions about the limit of the speed of energy (or info), the availability of energy for use, and the nature of space-time, just to cite some examples.

- **The Exploration of Multiverses**
  —Ex: theoretical speculation of other universes (e.g., "baby universes," "gateways" in black holes, "wave function of the universe," "many worlds," "brane worlds"), with potentially seminary discoveries of different physical laws in relation to matter-energy and space-time, and vital differences to the future of post-human conquest of other universes (for the emergence of new forms of consciousness).

*Notes*: These examples are solely illustrative (not exhaustive), and some of the items can be reclassified somewhere else. Nor are they always mutually exclusive. Since they are generalities, exceptions are expected. The point here is to give a rough picture of the evolution of consciousness to the hyper-spatial consciousness and others totally unknown to current earthlings. As a note of clarification, it makes no difference to my argument as to whether or not the hyper-spatial consciousness may emerge before, during, and after floating consciousness.
*Source*: From *Table 4.5* of *FPHC*

**Table 1.26. The Theory of Floating Consciousness (Part I)**

---

• **At the Micro-Physical Level**
—Ex: intelligent life without the human physical-chemical system

• **At the Chemical Level**
—Ex: space radiation and toxins

• **At the Bio-Psychological Level**
—Ex: exo-biological evolution in deep space
—Ex: genetic engineering of new beings

• **At the Institutional Level**
—Ex: post-capitalism
—Ex: post-democracy

• **At the Organizational Level**
—Ex: less legal-formalistic routines

• **At the Structural Level**
—Ex: alien forms of violence

• **At the Cultural Level**
—Ex: transcending freedom
—Ex: transcending equality

---

*(continued on next page)*

**Table 1.26. The Theory of Floating Consciousness (Part II)**

- **At the Cosmological Level**
  —Ex: parallel universes
  —Ex: pocket universes

- **At the Systemic Level**
  —Ex: space habitats (in zero-gravity)

*Notes*: Each example draws from the works of different scholars in the field. For instance, at the cosmological level, the idea of parallel universes is from the theoretical speculation in quantum cosmology by Stephen Hawking and others, while the one of pocket universes comes from the theoretical work of Allan Guth at MIT. And at the institutional level, I proposed post-capitalism and post-democracy in *FCD* (and later, from *BDPD* and *BCPC*). In addition, the examples are solely illustrative (not exhaustive), and some of the items can be reclassified somewhere else. Nor are they always mutually exclusive. Since they are generalities, exceptions are expected.
*Sources*: From *FPHC*—and, originally, from *FCD*

**Table 1.12. The Conceptual Dimensions of Consciousness (and Other Mental States)**

- **On Heredity and Time**
  - —Heredity and the Environment
  - —The Past and the Present
- **On Layers of Mental States and Abnormality**
  - —Consciousness, Unconsciousness, and Preconsciousness
  - —Normality and Paranormality
- **On Organicity and Motivation**
  - —Mechanicity and Organicity
  - —Primary Motivations and Multiple Motivations
- **On Other (Mostly Epistemic) Considerations**
  - —Synthesis and Analysis
  - —Situation and the Subject
  - —Process and Outcome
  - —Reasoning and Other Modes of Thinking
  - —Meta-Conceptual Nominalism and Realism

*Notes*:: These examples are solely illustrative (not exhaustive), and some of the items can be reclassified somewhere else. Nor are they always mutually exclusive. Since they are generalities, exceptions are expected.
*Sources*: From *FPHC*. A re-construction, but with my own contribution, originally from G.Lindzey & C.Hall, *Introduction to Theories of Personality* (NY: John Wiley & Sons, 1985).

**Table 1.27. The Potential of Unfolding Unconsciousness**

- **Superior Senses**
  - —*Through space*
    - Ex: clair-sentience (in feeling)
    - Ex: clair-voyance (in seeing)
    - Ex: clair-audience (in hearing)
    - Ex: clair-austance (in tasting)
    - Ex: clair-alience (in smelling)
    - Ex: clair-cognizance (in knowing)
  - —*Through time*
    - Ex: pre-cognition (in knowing), pre-sentiment (in feeling)
    - Ex: retro-cognition (in knowing), retro-sentiment (in feeling)

- **Superior Contacts**
  - —*With living minds*
    - Ex: telepathy
  - —*With the dead*
    - Ex: mediumship
  - —*With other-living entities*
    - Ex: telekinesis

*Notes*: The categories and examples are solely illustrative (not exhaustive), and the comparison is relative (not absolute), nor are they necessarily mutually exclusive. And some can be easily re-classified elsewhere. As generalities, they allow exceptions.
*Source*: A summary of Ch.5 of *FPHU*

**Table 1.28. The Future Exploration of Unfolding Unconsciousness**

- **In the Context of Nature**
  —Ex: the higher-dimensional space-time in hyperspace theory
  —Ex: the particle-wave interactions in quantum mechanics
  —Ex: the study of brain waves in electromagnetic theory

- **In the Context of the Mind**
  —Ex: the unconscious fantasies in psychoanalysis
  —Ex: the mind-altering drugs in biochemistry
  —Ex: the manipulation of neural activity in neurobiology

- **In the Context of Culture**
  —Ex: the correlation between anomalous experience and moral/religious/aesthetic interests in culture studies
  —Ex: the role of myths and language in epistemology

- **In the Context of Society**
  —Ex: different organizational agendas in anomalous research
  —Ex: the scholarly divide on anomalous study
  —Ex: interests and propaganda in institutioanl research
  —Ex: the transformative power of technology in research on social systems
  —Ex: the natural and non-natural selection of other worlds in exo-biology
  —Ex: the role of the post-human conscious mind in qualitative demography

*Notes*: The examples in each category are solely illustrative (not exhaustive), and the comparison is relative (not absolute), nor are they necessarily mutually exclusive. And some can be easily re-classified elsewhere. As generalities, they allow exceptions.
*Source*: A summary of Chs.2-5 of *FPHU*, in conjunction with the rest of the book

**Table 1.29. Creative Techniques and Traits**

- **Creative Techniques**
  - —*Comprehensiveness*
    - Ex: Broadness of Scope
    - Ex: In-Depth Analysis
    - Ex: Multifaceted Taxonomy
    - Ex: Numerous Theories
  - —*Arrangement*
    - Ex: Separation (or Subtraction)
    - Ex: Combination (or Addition)
    - Ex: Replacement
    - Ex: Reversal
    - Ex: Accommodation
    - Ex: Stretching
  - —*Serendipity*
    - Ex: Play
    - Ex: Chance

- **Creative Traits**
  - —*Openness / Risk*
    - Ex: Ability to welcome new ideas or to do new things, with risk
  - —*Discipline*
    - Ex: Ability to sacrifice whatever it takes to succeed
  - —*Resilience / Confidence*
    - Ex: Ability to take failures to mind, not to heart
  - —*Trust*
    - Ex: Ability to cooperate with others for team work, if necessary

*Notes*: The examples in the categories are solely illustrative (not exhaustive), and the comparison is relative (not absolute), nor are they necessarily mutually exclusive. And some can be easily re-classified elsewhere. As generalities, they allow exceptions.
*Source*: From *Sec.4.2* of *FPHCT*. For more details, also read the rest of the book.

**Table 1.30. The Desirability of Creativity, and Its Dark Sides (Part I)**

- **In Relation to Instrumental Rationality**
  - —*Desirability*
    - Ex: For the sake of efficiency and effectiveness in this capitalist era of our time.
  - —*Undesirability*
    - Ex: What Max Weber (1930) called the "iron cage" of capitalism, where live the "sensualists without heart, specialists without spirit"—or something which the Frankfurt School has forcefully asked us to be freed from.

- **In Relation to Substantive Rationality**
  - —*Desirability*
    - Ex: For the sake of God, the King, Motherland, or other comparable ideals.
  - —*Undesirability*
    - Ex: Domination and oppression of various forms—something which the moderns had struggled to free themselves from in the first place, since the dawn of modernity.

- **In Relation to Autonomous Rationality**
  - —*Desirability*
    - Ex: For the sake of itself, the autonomy of creative endeavor.
  - —*Undesirability*
    - Ex: The decadence of its own degeneration (e.g., the donwsides of postmodernism in our postmodern times, or what Nietzsche once called the unbearable "unreality" and "falseness" of the autonomous artist's "innermost existence"—and, in other cases, the painful suffering from different mental illnesses for those highly creative individuals).

*(continued on next page)*

**Table 1.30. The Desirability of Creativity, and Its Dark Sides (Part II)**

*Notes*: The examples in the categories are solely illustrative (not exhaustive), and the comparison is relative (not absolute), nor are they necessarily mutually exclusive. And some can be easily re-classified elsewhere. As generalities, they allow exceptions.
Also, I already went to great lenghts to explain the desirability and its dark sides of all these three rationalities (although there may be others too) in the two volumes of *FHC*—and, for that matter, *FCD*.
*Source*: From *Sec.4.9* of *FPHCT*. For more details, also read the rest of the book.

**Table 1.31. Posthuman-Ism, Post-Humanism, and Trans-Humanism**

- **Post-Humanism**
  —The neologism "post-human" used in my books should *not* be confused with another term which looks similar but has a totally different meaning in the literature of postmodernism, namely, "post-humanism"—which constitutes a critique of "humanism" as tradionally understood (especially, though not exclusively, in relation to the idea of progress in science and reason in the Enlightenment project). (WK 2008)
  —My works reject the project of "postmodernism" and propose the future world of what I orignally called "after-postmodernity" in *FHC* and *FCD*, for instance.

- **Trans-Humanism**
  —Also, the neologism "post-human" used in my books should *not* be confused with a similar term which is used to champion the ideology of technology for the fturue co-existence between humans and cyborgs in "trans-humanism." (WK 2008a)
  —Instead, my term "post-human" in relation to "posthuman-ism" also rejects "trans-humanism" (especially, though not exclusively, in relation to the promises of technology) and refers to something else altogether, that is, the future extinction of humans and its post-human successors in deep space and beyond unto multiverses.
  —Chapter Two of my book on ethics (*BEPE*) provides a more comprehensive critique of trans-humanism.

- **Posthuman-Ism**
  —With these two clarifications in mind, the "post-humans" (as envisioned in my books) can take different forms, and I proposed, in my different books, some of them, such as "unfolding unconsciousness," "floating consciousness," "hyper-spatial consciousness," "thinking machines," "thinking robots," "genetically altered superior beings," "cyborgs," and others (as already analyzed in my previous works).
  —They will evolve in conjunction with other visions of mine in relation to nature, society, and culture.

*Sources*: From *Sec.1.6* of *FPHE* (and also *BEPE*). For more details, also read the rest of the books (and other books of mine).

**Table 1.32. Three Great Future Transformations of the Martial Body (Part I)**

- **Virtual Battle**
  - —Ex: virtual fight
  - —Ex: virtual memory

- **The Hyper-Martial Body**
  - —*In relation to the physical dimension*
    - Ex: striking (e.g., punching, kicking, trapping, acupressure-striking, and so on in an amazing way)
    - Ex: grapping (e.g., throwing, pinning, joint-locking, and so on in an amazing way)
    - Ex: running and walking (e.g., on different surfaces)
    - Ex: jumping (e.g., through walls, trees, etc.)
    - Ex: flying (e.g., through space)
    - Ex: enduring pain (e.g., absorbing punches, kicks, and so on in an amazing way)
    - Ex: using extremely advanced hi-tech weapons
    - Ex: healing (e.g., with special energy, etc.)
    - Ex: entering (e.g., into different dimensions of space)
    - Ex: sensing (e.g., in extraordinary ways through space and time, etc.)
    - Ex: wearing hi-tech suits for defensive and offensive purposes (e.g., like a hi-tech armor)
  - —*In relation to the mental dimension*
    - Ex: the ability to relax the martial body in a way that current humans cannot.
    - Ex: the ability to concentrate the martial mind in a way that current humans are not accustomed to.
    - Ex: the ability to breathe and be timely in preparing the martial body for fighting in a way that current humans are not good at.

- **The Post-Martial Floating Mind**
  - —Ex: the mind without the human body
  - —Ex: the quest for spirituality beyond martial arts

*(continued on next page)*

**Table 1.32. Three Great Future Transformations of the Martial Body (Part II)**

*Notes*: The examples in the categories are solely illustrative (not exhaustive), and the comparison is relative (not absolute), nor are they necessarily mutually exclusive. And some can be easily re-classified elsewhere. As generalities, they allow exceptions.
*Source*: From *Sec. 2.5.1* of *FPHMA*. See text for citations.

**Table 1.33. Three Great Future Transformations of the Sexual Body**

- **Virtual Pleasure**
  —Ex: virtual Sex
  —Ex: virtual Recall

- **The Hyper-Sexual Body**
  —*In relation to the physical dimension*
  - Ex: the ability to perform differenet sexual positions without the various physical limitations that humans currently have.
  - Ex: the ability to produce unusual amount of sexual energy to endure in sexual acts in a way that humans currently cannot.
  - Ex: the ability to engage in multiple sexual experiences comparable to (but different from) non-human sexuality (e.g., hermaphrodite, monoecious, sex-switching, and many other types) that current humans cannot.

  —*In relation to the mental dimension*
  - Ex: the ability to relax the sexual body in a way that current humans are not accustomed to.
  - Ex: the ability to concentrate the sexual mind in a way that current humans are not good at.
  - Ex: the ability to breathe and be patient in preparing the sexual body for sexuality in a way that current humans are not able to.

- **The Post-Sexual Floating Mind**
  —Ex: the mind without the human body
  —Ex: the quest for spirituality beyond sexuality

*Notes*: The examples in the categories are solely illustrative (not exhaustive), and the comparison is relative (not absolute), nor are they necessarily mutually exclusive. And some can be easily re-classified elsewhere. As generalities, they allow exceptions.
*Source*: From *Sec. 3.4.2* of *FPHS*. See text for more info.

**Table 1.34. The Trinity of Pre-Modernity**

- **Pre-Free-Spirited Pre-Modernity (Pre-Modernism) and Its Internal Split**
  —Competing worldviews and values both within and between linear centric (e.g., Islamic, Christian, Judaic, Imperial Roman) and cyclical-centric (e.g., Confucian, Taoist, Hindu, and Buddhist) orientations
  —Compare modernism with pre-modernism here in relation to the seven dimensions of human existence like the true and the holy (e.g., different versions of epistemic dogmas and religious superstitions), the everyday and the technological (e.g.,different versions of non-technophilism and non-consumerism), the beautiful/sublime (e.g., different versions of aesthetic non-autonomy), and the good and the just (e.g., different versions of moral particularism).

- **Pre-Capitalist Pre-Modernity (Pre-Modernization) and Its Own Discontents**
  —Competing versions of societal arrangements (e.g., feudalism, monarchism, and the holy order)

- **Hegemonic Pre-Modernity and Its Countervailing Forces**
  —Different power centers and their enemies (e.g., the Roman Empire and the "barbarian hordes," the "Holy Crusades" and the Muslims, the Middle Kingdom and the invading tribes, different social castes in India, and warring Greek city-states)

*Notes*: The examples in each category are solely illustrative (not exhaustive) nor necessarily mutually exclusive, and the comparison is relative (not absolute). As generalities, they allow exceptions. Also, it does not matter what the "base" era is in the analysis of any trinity. And in the present context, the "base" era is modernity (for instance, with its "free-spirited modernity" and the other two parts). So, for pre-modernity, the trinity takes the form of, say, "pre-free-spirited pre-modernity," together with the other two parts.
*Sources*: From Ch.2 of *BCIV* and also the 2 volumes of *FHC*

**Table 1.35. The Trinity of Modernity (Part I)**

- **Free-Spirited Modernity (Modernism) and Its Internal Split**
  - —*On the True and the Holy*
    - The freedom from the dogmas of the past to the better understanding of, and union with, the world and self (Ch.3 of *FHC*)
    - Alternative discourses: about the true (e.g., anti-science discourses) and the holy (non-mainstream theologies) (Ch.3 of *FHC*)
  - —*On the Technological and the Everyday*
    - The freedom from life harshness to the higher technophilic, consumeristic lifeform (Ch.2 of *FHC*)
    - Alternative discourses: about the everyday (e.g., transcendental mindsets) and the technological (e.g., Arcadianism) (Ch.2 of *FHC*)
  - —*On the Good and the Just*
    - The freedom from the theo-aristocratic tyranny to the moral universality for a just society (Ch.5 of *FHC*)
    - Alternative discourses: about the just (e.g., Communism, Anarchism) and the good (e.g., Nazism/Fascism, and Zarathustrianism) (Chs.5-6 of *FHC*)
  - —*On the Beautiful and the Sublime*
    - The freedom from the external distortion of aesthetic pleasure to the boundless infinity of totality in artistic autonomy (Ch.4 of *FHC*)
    - Alternative discourses: about the beautiful/sublime (e.g., kitsch and historical avant-gardism) (Ch.4 of *FHC*)

- **Capitalist Modernity (Modernization) and Its Own Discontents**
  - —*During the Industrial Revolution*
    - Ex: Marx on the institution of inequality (Ch.1 of *FHC*)
  - —*During the Modern Rational-Instrumental Epoch*
    - Ex: Weber on the politics of soft liberal institutions (Ch.5 of *FHC*)
  - —*During the Great Depression*
    - Ex: Keynes on the myth of the free market (Chs.1-3 of *FHC*)
  - —*During the Cold War*
    - Ex: Lasch on the narcissistic culture industry (Chs.2-3 of *FHC*)

*(continued on next page)*

**Table 1.35. The Trinity of Modernity (Part II)**

- **Hegemonic Modernity and Its Countervailing Forces**
  —*The Legacies of Colonialism and Imperialism*
  - Ex: European colonization of most of the modern world (Ch.1 of *FHC*)

  —*The Struggle for Decolonialization*
  - Ex: The countervailing forces of resentment, rechantment, and regionalism (Chs.1 & 6 of *FHC*)

*Notes*: The examples in each category are solely illustrative (not exhaustive) nor necessarily mutually exclusive, and the comparison is relative (not absolute). As generalities, they allow exceptions.
*Sources*: From the 2 volumes of *FHC*—and also from *FCD*

**Table 1.36. The Trinity of Post-Modernity**

- **Free-Spirited Postmodernity (Postmodernism) and Its Internal Split**
  —*On the True and the Holy*
  - Postmodern performative turn for knowing and its enemies (Ch.3 of *FHC*)
  - Postmodern comparative theology and its opponents (Ch.3 of *FHC*)

  —*On the Technological and the Everyday*
  - Postmodern corporate technological mindset and its adversaries (Ch.2 of *FHC*)
  - Postmodern postmaterialism and its critics (Ch.2 of *FHC*)

  —*On the Good and the Just*
  - Postmodern politics of difference and its foes (Ch.5 of *FHC*)

  —*On the Beautiful and the Sublime*
  - Postmodern deconstruction and its dissenters (Ch.4 of *FHC*)

- **Capitalist Postmodernity (Postmodernization) and Its Own Discontents**
  —*During the Post-Cold War and Beyond*
  - Ex: post-Fordism and its shortcomings (Ch.6 of *FHC*; Chs.6-7 of *FCD*)

- **Hegemonic Postmodernity and Its Countervailing Forces**
  —*The Debate on the Global Village*
  - Ex: uni-civilizationalism vs. multi-civilizationalism (Ch.6 of *FHC*)

  —*The Resistance Movement*
  - Ex: rechantment and the politics of civilizational claims (e.g., Islamic, Confucian and other ethos in relation to the Same) (Ch.6 of *FHC*; Ch.10 of *FCD*)
  - Ex: resentment and the politics of resurgence (e.g., the rising Chinese superpower, the growing EU, and other players in relation to the U.S. and her allies) (Ch.6 of *FHC*; Ch.8 of *FCD*)
  - Ex: regionalism and the politics of inequality (e.g., trans- or international blocs, the North-South divide, NGO's) (Ch.6 of *FHC*; Ch.5 of *FCD*)

*Notes*: The examples in each category are solely illustrative (not exhaustive) nor necessarily mutually exclusive, and the comparison is relative (not absolute). As generalities, they allow exceptions.
*Sources*: From *FCD* and the 2 volumes of *FHC*

**Table 1.37. The Trinity of After-Postmodernity**

- **Free-Spirited After-Postmodernity (After-Postmodernism) and Its Internal Split**
  —The discourse of naked contingency (Ch.10 of *FCD*; Ch.4 of *FPHC*)

- **Post-Capitalist After-Postmodernity (After-Postmodernization) and Its Own Discontents**
  —Different versions of post-capitalism and post-democracy, and their enemies (Ch.10 of *FCD*; Chs.3-4 of *FPHC*)

- **Hegemonic After-Postmodernity and Its Countervailing Forces**
  —The Cyclical Progression of Hegemony in Multiverses (Chs.9-10 of *FCD*; Ch.4 of *FPHC*)

*Notes*: The examples in each category are solely illustrative (not exhaustive) nor necessarily mutually exclusive, and the comparison is relative (not absolute). As generalities, they allow exceptions.
*Sources*: From *FCD* and also *FHC*

# • PART TWO •

## *Strategy*

# CHAPTER 2
## *STRATEGY AND ITS INADEQUACY*

---

> Strategy requires thought; tactics requires observation.
> —Max Euwe (WK 2010)

## The Foresight of Strategy

The game of chess requires both strategy and tactics. As already summarized in *Sec. 1.3*, strategy and tactics are closely related.

For instance, "chess strategy consists of setting and achieving long-term goals during the game" (like "where to place different pieces"), but "tactics concentrate on immediate manoeuvre. These two parts of chess thinking cannot be completely separated, because strategic goals are mostly achieved by the means of tactics, while the tactical opportunities are based on the previous strategy of play." (WK 2010)

Thus, any study of chess requires an examination of the nature of both strategy and tactics, as one cannot be completely separated from the other.

With this close relationship between strategy and tactics in mind, this book offers a dialectic treatment of strategy and tactics, in that, if strategy has its foresight, it has its uncertainty too—just as, if tactics has its usefulness, it likewise has its limitation.

The inquiry concernng strategy is the subject matter of this chapter (Chapter Two), whereas that of tactics will be addressed in the next chapter (Chapter Three).

However, it should be stressed that, because of this close relationship between strategy and tactics, there is a scholarly challenge here, in that one cannot analyze strategy (or tactics, for that matter) separately.

This then means that the issues concerning strategy in this chapter (Chapter Two) and tactics in the next chapter (Chapter Three) often overlap. Thus, it is a challenge for this book to devote a chapter on strategy and another chapter on tactics.

With this reminder in mind—a good way to analyze strategy and its inadequacy (in this chapter) is by way of the evaluation of the extent to which strategy is in fact both possible and desirabe.

This chapter thus takes the challenging task to provide a comprehensive analysis of strategy in the context of chess from the four perspectives of the mind, nature, society, and culture (in accordance to my sophisticated methodological holism, as explained in *Sec. 1.8*).

So, this chapter is organized in four main sections, in relation to (2.2) strategy and the mind, (2.3) strategy and nature, (2.4) strategy and society, and (2.5) strategy and culture—to be examined hereafter, in that order (and summarized in *Table 2.1*).

# Strategy and the Mind

Strategy, when analyzed from the perspective of the mind, can reveal some distinctive insights on its possibility and desirability, and this can be shown by way of two case studies, namely, (2.2.1) chess expertise, and psychological explanations, and (2.2.2) chess games, and the development of mental skills—to be addressed in what follows, in that order.

## Chess Expertise, and Psychological Explanations

A commonly asked question in chess concerns why some players are so brilliant. What exactly is that something unusual in their minds which allows them to beat others in chess games so brilliantly?

A good answer, as this will be explained below, is that a brilliant strategy has much to do with such mental qualities as memory, perception, imagination, and practice, among others. And this is also relevant to tactics, for that matter.

### *The Phenemenon of Chess Prodigy*

Consider, for instance, the phenomenon of "chess prodigy," who are, on average, "children who play chess so well that they are able to beat Masters and even Grandmasters, often at a very young age. Chess is one of the few sports where children can compete with adults on equal ground; it is thus one of the few skills in which true child prodigies exist. Expectations can be high for chess prodigies; while some become World Champions, others fail to make progress in adulthood." (WK 2010h)

In general, "[o]ne measure of chess prodigies is the age at which they gain the International Grandmaster title. It should be noted that this title has only existed since 1950, and also that the title has become less difficult to obtain in recent years....Nevertheless, there have been more strong prodigies in recent years." (WK 2010h)

So, before 1950, good examples of early chess prodigies include "Paul Morphy (1837–1884) and José Raúl Capablanca (1888–1942), both of whom won matches against strong adult opponents at the age of 12; and Samuel Reshevsky (1911–1992), who was giving simultaneous exhibitions at the age of six. Morphy went on to be unofficial World Champion (before the official title existed), Capablanca became World Champion, and Reshevsky—while never attaining the title—was in the top few players in the world for many years." (WK 2010h; CBN 2006a)

After 1950, good examples of child prodigies (who have held the "International Grandmaster" title are shown below. (WK 2010h)

- "1950 David Bronstein USSR 26 years"
- "1952 Tigran Petrosian USSR 23 years"
- "1955 Boris Spassky USSR 18 years"
- "1958 Bobby Fischer USA 15 years, 6 months, 1 day"
- "1991 Judit Polgár Hungary 15 years, 4 months, 28 days"
- "1994 Péter Lékó Hungary 14 years, 4 months, 22 days"
- "1997 Etienne Bacrot France 14 years, 2 months, 0 days"
- "1997 Ruslan Ponomariov Ukraine 14 years, 0 months, 17 days"
- "1999 Bu Xiangzhi China 13 years, 10 months, 13 days"
- "2002 Sergey Karjakin Ukraine 12 years, 7 months, 0 days"

Of course, there can be other criteria to classify chess prodigies, such as the criterion of someone "to become Grandmasters before their fifteenth birthday," but the examples above suffice for illustration in the current context. (WK 2010h)

*The Varieties of Psychological Explanations*

Over the years, there are different explanations for the success of these unusual chess players, and a good illustration concerns the field of chess psychology, where "[t]here is an extensive scientific literature." (WK 2010; R. Grabner 2007; A. De Groot 1996; F. Gobet 2004; D. Holding 1985)

Consider some examples below, for illustration.

For instance, back in 1894, "[Alfred] Binet conducted one of the first psychological studies into chess. It investigated the cognitive facilities of chess masters. Binet hypothesized that chess depends upon the phenomenological qualities of visual memory but after studying the reports by master participants, it was concluded that memory was only part of the chain of cognition involved in the game process. The players were blindfolded and required to play the game from memory. It was found that only masters were able to play successfully without seeing the board for a second time and that amateur or intermediate players found it to be an impossible task. It was further concluded that experience, imagination, and memories of abstract and concrete varieties were required in grand master chess." (WK 2010j) T. Robbins (1996) and his colleagues also found in their research the importance of memory and other comparable abilities in chess players.

Another scholar, Adriaan de Groot, "concurred with Alfred Binet that visual memory and visual perception are important attributors and that problem-solving ability is of paramount importance." (WK 2010i) Specifically, de Groot (1965), "in his doctoral thesis, showed that chess masters can rapidly perceive the key features of a position. According to de Groot, this perception, made possible by years of practice and study, is more important than the sheer ability to anticipate moves." (WK 2010)

This is so, because good memorization, while also important, is not that unique. In fact, "de Groot...showed that chess masters can memorize positions shown for a few seconds almost perfectly," but "[t]he ability to memorize does not, alone, account for this skill, since masters and novices, when faced with random arrangements of chess pieces, had equivalent recall (about half a dozen positions in each case). Rather, it is the ability to recognize patterns, which are then memorized, which distinguished the skilled players from the novices. When the positions of the pieces were taken from an actual game, the masters had almost total positional recall." (WK 2010; R. Heuer 1999)

Other scholars like "[K.] Ericsson [1993] and colleagues have argued that deliberate practice is sufficient for reaching high levels of expertise, like master in chess." (WK 2010) It is thus no wonder that this kind of finding is instrumental for "local government, schools, and student organizations all over the world" to "implement...chess programs." (WK 2010i)

For instance, "New York based Chess-In-The-Schools, Inc. has been active in the public school system in the city since 1986. It currently reaches more than 30,000 students annually. The American Foundation for Chess has initiated programs in partnership with local school districts in several U.S. cities, including Seattle, San Diego, Philadelphia, and Tampa. The Chess'n Math Association promotes chess at the scholastic level in Canada. Chess for Success is a program for at-risk schools in Oregon. Since 1991, the U.S. Chess Center in Washington, D.C. teaches chess to children, especially those in the inner city." (WK 2010i; CITS 2010; AFC 2010; CFS 2010)

However, more recent research indicates that factors other than practice are important. For example, "[F.] Gobet [2007] and colleagues have shown that stronger players start playing chess earlier, that they are more likely to be left-handed, and that they are more likely to be born in late winter and early spring." (WK 2010)

*The Criticisms of Psychological Explanations*

However, these varieties of psychological explanations (aforecited) have their weaknesses.

Consider a few of them below, for illustration.

Firstly, one criticism of the varieties of psychological explanation is that the very existence of the varieties in question already shows that there is no consensus to convince every chess scholars about the reasons for brilliant chess performance. Surely, such factors as perception, imagination, memory, and practice are all important, but the qualifications about them vary from case to case, from group to group, etc.

Secondly, another criticism of the varieties of psychological explanation is that they do not take into account of the cultural, social, and natural arguments (as introduced in *Sec. 1.4*).

Thirdly, still another criticism of the varieties of psychological explanation is that it is also not clear how desirable each of the factors in question has really been proved to be. For instance, the benefits of chess practice are still open to question, such that, for the critics, "the results of research studies have failed to produce unambiguous support for the intellectual benefits of playing chess." (WK 2010i; T. Redman 2006)

*The Importance of the Lesson*

Even then, these criticisms of the varieties of psychological explanation should not be misinterpreted as a dismissal of their usefulness. Instead, they are shown here for us to learn something from the two opposing sides of the debate, such that the possibility and desirability of strategy (from the perspective of the mind with chess psychology as a case study here) are not to the extent that its respective defenders would like us to believe.

More importantly, this analysis of chess pscyhology can teach us something valuable about the ontological principles in existential dialectics, and good examples include the absoluteness-relativeness principle, the predictability-unpredictability principle, the explicability-inexplicability principle, the preciseness-vagueness principle, the simpleness-complicatedness principle, the openness-hiddenness principle, the denseness-emptiness principle, the slowness-quickness principle, the expansion-contraction principle, the theory-praxis principle, the convention-novelty principle, the evolution-transformation principle, the symmetry-asymmetry principle, the regression-progression principle, and the same-difference principle.

For instance, in relation to the absoluteness-relativeness principle, if there is absolute viewpoint (e.g., the absolute view of de Groot that perception and memory are important for brilliant chess playing), there is relative counterpart (e.g. what is important for de Groot in regard to perception and memory in chess playing is not necessary so for Ericsson and his colleagues, who emphasize more "deliberate practice"). And the reverse direction also holds true.

In relation to the predictability-unpredictability principle, if there is predictability (e.g., the predictable tendency of a chess prodigy to "play chess so well" that he can "beat Masters and even Grandmasters, often at a very young age"), there is unpredictability (e.g., the more difficult task to predict which particular chess prodigy will end up making a major contribution to the world of chess, because "while some become World Champions, others fail to make progress in adulthood"). And the reverse direction also holds true.

In relation to the explicability-inexplicability principle, if there is explicability (e.g, the explanation by scholars like Ericsson and his colleagues that "deliberate practice is sufficient for reaching high levels of expertise, like master in chess"), there is inexplicability (e.g, the lack of sufficient explanation, if looked within the worldview of Ericsson and his colleagues, of why chess prodigies succeed in playing chess brilliantly "at a very young age," even without the long history of many years of practice that other players may have, by comparison). And the reverse direction also holds true.

In relation to the preciseness-vagueness principle, if there is preciseness (e.g., the precise identification of "a very young age" as one major measurement of chess prodigies), there is vagueness (e.g., the vagueness of this criterion itself, since "a very young age" can be 12 years, or 13 years, or 14 years, etc., as shown in the list of child prodigies). And the reverse direction also holds true.

In relation to the simpleness-complicatedness principle, if there is simpleness (e.g., the relatively simple view of Ericsson and his colleagues that "deliberate practice is sufficient for reaching high levels of expertise, like master in chess"), there is complicatedness (e.g., the relatively complicated view of Alfred Binet and his colleagues that brilliant chess performance requires not only memory, but also imagination and experience, for instance). And the reverse direction also holds true.

In relation to the openness-hiddenness principle, if there is openness (e.g., the open exploration of the relationship between practice and good performance in the research on chess psychology by Ericsson and his colleagues), there is hiddenness (e.g., the hidden bias in the research on chess psychology by Ericsson and his colleagues, in that for the critics, "the results of research studies have failed to produce unambiguous support for the intellectual benefits of playing chess"). And the reverse direction also holds true.

In relation to the denseness-emptiness principle, if there is denseness (e.g., the relatively denser concentration of chess prodigies in Eastern countries like the former Soviet republics and China, as shown in the list of modern International Grandmasters), there is emptiness (e.g., the relatively less dense, or more empty, concentration of chess prodigies in Africa, as shown in the list of modern International Grandmasters). And the reverse direction also holds true.

In relation to the slowness-quickness principle, if there is slowness (e.g., the relatively slower speed of an ordinary chess player to win in a competitive game of chess), there is quickness (e.g., the relativley faster speed of a chess prodigy to win in a competitive game of chess). And the reverse direction also holds true.

In relation to the expansion-contraction principle, if there is expansion (e.g., the relatively more developed ability of a chess prodigy to rely on his memory, imagination, and perception to play chess), there is contraction (e.g., the relatively less developed ability of a chess prodigy to rely on his long history of many years of chess practice to play chess, which he lacks). And the reverse direction also holds true.

In relation to the theory-praxis principle, if there is theory (e.g., the theoretical construction of the importance of "deliberate practice" for playing chess well), there is praxis (e.g., the practical application of the theory of "deliberate practice" by such organizations like Chess-In-The-Schools, Inc., the American Foundation for Chess, and others to encourage children to play chess). And the reverse direction also holds true.

In relation to the convention-novelty principle, if there is convention (e.g., the conventional wisdom about chess prodigies, in regard to their unusual memory, perception, and imagination in playing chess), there is novelty (e.g., the alternative novel challenge to the conventional wisdom by Ericsson and his colleagues, in that "deliberate practice" is very important for brilliant chess performance instead). And the reverse direction also holds true.

In relation to the evolution-transformation principle, if there is evolution (e.g, the natural evolution of humans to play with each other for fun in the state of nature), there is transformation (e.g., the technical transformation of human playing with each other by the invention of chess). And the reverse direction also holds true.

In relation to the symmetry-asymmetry principle, if there is symmetry (e.g., the co-existence of different chess players in society), there is asymmetry (e.g., the high regard towards chess prodigies, but the low regard towards mediocre chess players). And the reverse direction also holds true.

In relation to the regression-progression principle, if there is regression (e.g., the regression made in the practice of chess, in that "the results of research studies have failed to produce unambiguous support for the intellectual benefits of playing chess"), there is progression (e.g., the progress made in the practice of chess for better performance, as shown by the work of Ericsson and his colleagues). And the reverse direction also holds true.

And in relation to the same-difference principle, if there is similarity in outcome (e.g., the contribution to the molding and regulation of beliefs and behaviors, regardless of whether this be achieved by way of playing chess or by way of making war), there is difference in outcome (e.g., the contribution to the molding and regulation of beliefs and behaviors by way of playing chess, when one side fights with the other side only on the chess board—but the contribution to the molding and regulation of beliefs and behaviors by way of making war, when one side fights with the other side in the battlefield). And the reverse direction also holds true.

## Chess Games, and the Development of Mental Skills

The disccusion of chess psychology (in the previous section) points to a related but controversial issue concerning the mental benefits of playing chess.

For instance, a good question to ask is, Is playing chess beneficial to the mind?

A good answer, as this will be summarized below, is that brilliant strategy can be benefitted from such mental skills as logical thinking, planning ability, memory, imagination, concentration, self-confidence, and patience. And this is also relevant to tactics, for that matter.

*The Mental Benefits of Playing Chess*

For the advocates of playing chess, the answer to the above question is affirmative enough, period.

For illustration, consider the follow seven ways that playing chess can benefit the mind, as claimed by the advocates (and summarized in *Table 2.2*), namely, (a) "improving memory," (b) "improving concentration," (c) "stoking imagination and creativity," (d) "developing logical thinking," (e) "building self-confidence," (f) "improving ability to plan," and (g) cultivating patience (and will power). (CT 2009)

(a) Improving Memory

The first benefit of playing chess, as claimed by its advocates, is that it can improve memory: "Chess players will always strive to improve their games so that they can win more games. This incentive of satisfaction in working out a win provides an urge to learn the opening theories, the variations, tactics in middle game, end game theories and so on. Through your effort to memorize such things, you will find an improvement in your memory. Like body muscles, memory also needs exercise to develop. Playing chess automatically provides this exercise. The improved memory has another positive effect. It helps to guard against the onset of Alzheimer's disease that tends to affect a good number of people at their old age." (CT 2009)

(b) Improving Concentration

The second benefit of playing chess, as claimed by its advocates, is that it can also improve concentration: "The very nature of chess requires a player to concentrate on his own moves and his opponent's. This continues from move to move and you try to be alert so that you do not make a wrong move or fall into some trap laid by the opponent. This type of focused thinking helps the concentration power of your mind." (CT 2009)

(c) Stoking Imagination and Creativity

The third benefit of playing chess, as claimed by its advocates, is that it can stoke imagination and creativity: "Because of continued research by experts to find the best openings for both the players, you will often find the same set of initial moves. But have you ever found any research on the best set of middle game moves? No, because the possibilities are so numerous that no one can foresee how the game will move. In middle game, you are on your own and you see the complex positions in your own way. You also try to visualize the position after your own set of planned moves and the opponent's expected responses. Thus, you get to be creative and your imagination provides the lead. Your ability in other fields gets a boost through the development of these aspects of your mind." (CT 2009)

(d) Developing Logical Thinking

The fourth benefit of playing chess, as claimed by its advocates, is that it can develop logical thinking: "In chess, neither you nor your opponent makes a move randomly. Each has certain plans and ideas, and moves are made to give shape to it. This is an automatic logical process and over a time, this kind of thinking becomes a habit in your other activities also." (CT 2009)

(e) Building Self-Confidence

The fifth benefit of playing chess, as claimed by its advocates, is that it can build self-confidence: "When you are playing, you are the master of your decisions as there is nobody else to help you out. You take those decisions after analyzing the game situations to the best of your capability. This builds up the mental strength to take decisions in other situations in life and that develops your self-confidence also. Even when you lose games, you will not get deterred if you are in the habit of post-play analysis to find out where you went wrong and this knowledge will add to your confidence." (CT 2009)

(f) Improving the Ability to Plan

The sixth benefit of playing chess, as claimed by its advocates, is that it can improve the ability to Plan: "In chess, you are always planning your actions and trying to foresee what results will be achieved through those actions. This planning ability will soon become your second nature." (CT 2009)

(g) Cultivating Patience (and Will Power)

And the seventh benefit of playing chess, as claimed by its advocates, is that it can cultivate patience and will-power: "Any chess game will try your patience and nerves and test your will-power to prevail upon your opponent. Your mind will benefit from these demands of the game." (CT 2009)

*The Response from the Critics*

Yet, there is no lack of criticisms concerning the claims about the mental benefits of playing chess. Consider a few of them below, for illustration, as summarized in *Table 2.2*.

Firstly, one criticism of the claim about the mental benefits of playing chess is that it remains to be conclusively proved, because "the results of research studies have failed to produce unambiguous support for the intellectual benefits of playing chess." (WK 2010i; T. Redman 2006)

Secondly, another criticism of the claim about the mental benefits of playing chess is that it fails to take into account of other factors, be they social or cultural, which can contribute to its development or retardation. For instance, if there are these obvious mental benefits of playing chess as claimed by its adovcates, how can one explain the intriguing discrepancy that chess playing is not very popular in black Africa but is highly popular in the former Soviet republics and East Asian countries (like China, Korea, and Japan)? In fact, most of the holders of the "International Grandmaster" title are from the former Soviet republics and China, for instance.

And thirdly, still another criticism of the claim about the mental benefits of playing chess is that it does not consider chess-related deaths as the dark side of playing chess.

In fact, many people do not know that "chess has been associated with a number of anecdotes, and some relate to games that have resulted in the murder of one of the players involved. The reliability of many of these anecdotes is suspect, but some appear to be based in fact." (WK 2010k)

For instance, "the anecdote with the most supporting evidence is given in the book *Chess or the King's game* (1616) by Augustus, Duke of Lüneburg….The anecdote states that Okarius (also spelt Okar or Otkar), the prince of Bavaria, had a son of great promise residing at the Court of King Pippin. One day Pippin's son was playing chess with the young Prince of Bavaria, and became so enraged at repeatedly losing that he hit the prince on the temple with one of his rooks and killed him on the spot. This anecdote is repeated in another Bavarian Chronicle, and in a work by Metellus of Tegernsee about Saint Quirin and other documents refer to his death while at Pippin's court." (WK 2010k)

Another famous incident involves "King Canute (c. 994–1035) of Denmark, England and Norway," who "ordered an earl killed after a disagreement about a chess game. By one account, the king made an illegal move that angered Earl Ulf, who knocked over the board and stormed off, after which the king sent someone to kill him." (WK 2010k; GC 2010)

Later on, "two separate incidents in 13th century London involving men of Essex resorting to violence resulting in death as an outcome of playing chess further caused sensation and alarm. The growing popularity of the game—now associated with revelry and violence—alarmed the Church." (WK 2010r; I. Riddler 1998)

It is no wonder, in this context, that "Saint Peter Damian [once] denounced the bishop of Florence...for playing chess even when aware of its evil effects on the society." (WK 2010r; I. Riddler 1998)

In addition, over time, "[t]he practice of playing chess for money became so widespread during the 13th century that Louis IX of France issued an ordinance against gambling in 1254. This ordinance turned out to be unenforceable and was largely neglected by the common public, and even the courtly society, which continued to enjoy the now prohibited chess tournaments uninterrupted." (WK 2010r; M. Vale 2001)

Surely, there can be other incidents besides these chess-related deaths and corrupted lifestyles, but the examples aforecited are deemed sufficient to illustrate the important point here about the critique of the claim about the mental benefits of playing chess.

*The Didactic Nature of the Story*

But this does not mean that chess playing is therefore useless. On the contrary, the introduction to these criticisms concerning the claim about the mental benefits of playing chess allows us to see the two opposing sides of the debate, such that the possibility and desirability of strategy (from the perspective of the mind with the controversy about the mental benefits of playing chess as a case study here) are not to the extent that its respective defenders would like us to believe.

Moreover, this introduction to the controversy can be didactic enough in revealing the relevance of the ontological principles in existential dialectics, and good examples include the absoluteness-relativeness principle, the partiality-totality principle, the predictability-unpredictability principle, the explicability-inexplicability principle, the finiteness-transfiniteness principle, the preciseness-vagueness principle, the simpleness-complicatedness principle, the openness-hiddenness principle, the denseness-emptiness principle, the slowness-quickness principle, the expansion-contraction principle, the theory-praxis principle, the convention-novelty principle, the evolution-transformation principle, the symmetry-asymmetry principle, the softness-hardness principle, the regression-progression principle, and the same-difference principle.

For instance, in relation to the absoluteness-relativeness principle, if there is absoluteness (e.g., the absolute view of the advocates of chess playing in regard to the mental benefits of playing chess), there is relativeness (e.g., what is true for the advocates of chess playing in regard to the mental benefits is not necessarily sure for the critics, who pointed out that "the results of research studies have failed to produce unambiguous support for the intellectual benefits of playing chess"). And the reverse direction also holds true.

In relation to the partiality-totality principle, if there is partiality (e.g., the partial thinking of each player to make moves which would help him beat the other side in a game of chess), there is totalty (e.g., the more totalistic view of the entire game of chess, in that "the possibilities [of moves] are so numerous that no one can foresee how the game will move," such that the whole is not the sum of its parts, that is, not the sum of the individualist views of the players obsessed with winning). And the reverse direction also holds true.

In relation to the predictability-unpredictability principle, if there is predictability (e.g, the predictable tendency of a chess player to try his best to make moves which, in his judgement, would help him beat his opponent in a chess game), there is unpredictability (e.g., the more difficult task to predict which particular chess player would behave to his opponent in a particular way after the end of a particular game, in light of the history of chess-related deaths or violence). And the reverse direction also holds true.

In relation to the explicability-inexplicability principle, if there is explicability (e.g., the explanation of the importance of chess playing on the basis of the mental benefits, as claimed by the adovcates), there is inexplicabiltiy (e.g,. the lack of sufficient explanation of the intriguing discrepancy that chess playing is not very popular in black Africa but is highly popular in the former Soviet republics and East Asian countries, if the mental benefits of playing chess are so obvious as its advocates often claim). And the reverse direction also holds true.

In relation to the finiteness-transfiniteness principle, if there is finiteness (e.g., the finite number of the players, that is, 2 in a game, and of the pieces in Western chess, that is, 32 on the chess board), there is transfiniteness (e.g., the transfinite number of possible moves in chess, because "the possibilities [of moves] are so numerous that no one can foresee how the game will move"). And the reverse direction also holds true.

In relation to the preciseness-vagueness principle, if there is preciseness (e.g., the precise identification of the seven ways that playing chess can benefit the mind, by the advocates), there is vagueness (e.g., the vagueness in the seven beneficial ways of chess playing to the mind, because they can mean different things to different scholars and players, like how exactly imagination or creativity can help a chess player beat his opponent in a particular game). And the reverse direction also holds true.

In relation to the simpleness-complicatedness principle, if there is simpleness (e.g., the relatively simple analysis of the benefits of playing chess in seven ways, as claimed by the advocates), there is compicatedness (e.g., the relatively more complicated analysis of the relationship between chess playing and its mental impact, as shown in the criticisms by the critics). And the reverse direction also holds true.

In relation to the openness-hiddenness principle, if there is openness (e.g., the open exploration of the relationship between chess playing and its mental benefits), there is hiddenness (e.g., the hidden bias in the claim by the advocates of chess playing in regard to the mental benefits, because they do not focus on other factors of chess development and the dark side of chess playing, like the incidents of chess-related deaths). And the reverse direction also holds true.

In relation to the denseness-emptiness principle, if there is denseness (e.g.., the relatively denser concentration of individuals who play chess and also believe in the mental benefits of chess playing), there is emptiness (e.g., the relatively less dense, or more empty, concentration of individuals who play chess but seriously question the claim about the mental benefits of chess playing). And the reverse direction also holds true.

In relation to the slowness-quickness principle, if there is slowness (e.g., the relatively slower readiness of poets to plan logically, think calculatively, and wait strategically both in poetry and other life activities), there is quickness (e.g., the relatively quicker readiness of chess players to to plan logically, think calculatively, and wait patiently both in chess playing and other life activities). And the reverse direction also holds true.

In relation to the expansion-contraction principle, if there is expansion (e.g., the relatively more developed ability of chess players to think logically and creatively), there is contraction (e.g., the relatively less developed ability of chess players to think emotionally and mystically). And the reverse direction also holds true.

In relation to the theory-praxis principle, if there is theory (e.g., the theoretical construction of the mental benefits of chess playing), there is praxis (e.g., the practical application of the view concerning the mental benefits of chess playing by its advocates to encourage others to play more chess). And the reverse direction also holds true.

In relation to the convention-novelty principle, if there is convention (e.g., the conventional wisdom about the mental benefits of chess playing), there is novelty (e.g., the alternative novel challenge to the conventional wisdom about chess playing by the critics, who pointed out the other side of chess playing like the incidents of chess-related deaths). And the reverse direction also holds true.

In relation to the evolution-transformation principle, if there is evolution (e.g., the natural evolution of humans in the state of nature to fight with each other from time to time), there is transformation (e.g., the technical transformation of human fighting with each other by the invention of chess and its subsequent incidents of chess-related deaths). And the reverse direction also holds true.

In relation to the symmetry-asymmetry principle, if there is symmetry (e.g., the co-existence of different kinds of chess players in human history), there is asymmetry (e.g., some chess players accept a gracious loss after a chess game, while others resort to violence after a loss in a chess game). And the reverse direction also holds true.

In relation to the softness-hardness principle, if there is softness (e.g., the peaceful way to compete between two players in a chess game), there is hardness (e.g., the violent way to deal with loss after a chess game for some chess players). And the reverse direction also holds true.

In relation to the regression-progression principle, if there is regression (e.g., the regression made in the history of chess, as shown in the incidents of chess-related deaths in history), there is progress (e.g., the progress made in the history of chess for the mental benefits of chess playing and the fun of playing it). And the reverse direction also holds true.

And in relation to the same-difference principle, if there is similarity in outcome (e.g., the contribution to the shaping of beliefs and behaviors, regardless of whether this be achieved by way of chess playing or by way of physical fighting), there is difference in outcome (e.g., the contribution to the shaping of beliefs and behaviors by way of chess playing for a more peaceful way of competition—but the contribution to the shaping of beliefs and behaviors by way of physical fighting for a more violent way of competition). And the reverse direction also holds true.

## Strategy and Nature

Chess, when examined from the perspective of nature, can reveal something amazing about its possibility and desirability, and a good illustration concerns chess origin and its relationship with cosmology.

The point here is that, as this will be analyzed below, good strategy has much to imitate from the geometrical patterns and number systems in cosmology. And this is also relevant to tactics, for that matter.

## Chess Origin, and Its Relationship with Cosmology

Rangachar Vasantha (2010) precisely argued for this very linkage, that is, between cosmology and chess origin.

To start, Vasantha (2010) pointed out that "[t]he game of chess, which was being played in the ancient India as well as in the present times,...has its origin linked genetically to secret magical and religious rituals in those times," especially in the context of cosmology.

### *The Inner World and the Outer World*

For Vasantha (2010), "[a] cosmos by definition presuppose an ordered universe. Cosmology is the study of the law and intelligence inherent in this ordered universe. There is an inner as well as an outer way of studying cosmology. The outer embraces sensible observation and the inner is expression of cosmological laws within one's own structure. The goal of spiritual disciplines is to unite the inner and outer, the greater and smaller into an inseparable integrity. The language of the archetype laws, which unite the inner and outer cosmos, is that of pattern and in particular number pattern. The game of chess is associated with the number pattern and thus to cosmology."

### *The Inquisitive Mind, and the Invention of Board Games*

For Vasantha (2010), the understanding of the linkage between the inner world and the outer world by the inquisitive human mind later led to the invention of board games.

For instance, "[i]t has always been the inquisitiveness of a human mind to probe into the mysterious world around him and its influence on himself. This led to the discovery of the several fields of knowledge. One such field is the predictive science, which includes astrology, futurology, numerology etc. This probing tendency is reflected on all human activities and day-to-day life style including the recreational activities, which perhaps led to invention of several games and sports. Chess is one such board game, which may be related to the planetary movements and their influence on human being." (Vasantha 2010)

In the context of Hinduism, for Vasantha (2010), chess reflects this linkage between the inner world and the outer world, because the number system in chess is comparable to "the movements of the celestial bodies and their numerical symbolism" in cosmology.

*Number System in Chess, and Numerical Symbolism in Cosmology*

After all, "Hindus strongly believe that the nine planets influence all objects of the material world. Each of these planets is associated with a unique number and the mystic power, and the unique number is represented by a symbol. These geometrical patterns are obtained from a matrix array of the basic nine numbers. This square matrix array known as the Vedic square is the basis of a whole mathematical system, which contained a numerical model of the universe." (Vasantha 2010)

Likewise in chess, there is a number system for the value of each piece and the rules for the move of each piece, to the extent that, as Vasantha (2010) put it, "[c]hess and other board games were derived from, and the moves of the pieces are being closely related to the movements of the celestial bodies and their numerical symbolism. By joining those cells that show the same number, certain geometrical diagrams are obtained which 'represent to ourselves our real life relationships and experiences.' These diagrams are linked to board games in so far as every move of a piece results in similar geometrical patterns."

For this reason, throughout India, as Vasantha (2010) reminded us, there are religious musuems with collections of board games, and good examples include "[a] note on chess and board games [with special reference to the museums of South India Dice] and square board found at Sannthi, a Buddhist site belonging to 1st–2nd cent. AD," "game boards, pieces, and dices mural paintings preserved in Sri Jayachamarajendra art gallery, Mysore, early 18th century," and "chess collection from Salar Jung Museum, Hyderabad (collection ranges from 14th–19th century and belongs to different regions)."

*The Role of Mathematics in Chess and Cosmology*

Thus, just as mathematics plays a major role in cosmology, it likewise does so in chess.

For illustration, a sculpture found in India described by Vasantha (2010) "depicts four men squatting in pairs on opposite sides of a board of 6x6 squares. Besides this, we find a board consisting of 6 squares engraved with dissimilar patterns, a little away from the main board, apart from a single rectangular piece lying close to the board."

What is interesting is that "[t]he topography of the board suggests that...the players distributed on the opposite sides of the board in pairs and are playing with some play materials (pawns) depicted as dots on the board. Since the board consists of even number of square grids, the game ought to have been played with a dice and the players move their individual pawns and try to checkmate the pawns of the opposite player." (Vasantha 2010)

For Vasantha (2010), "the four players sitting in pairs on opposite sides of the board might have started the game from the four corners of the board, each player moving his pawns towards his opponent on the opposite side. The two players on each side possibly played as partners in the game. We may also observe the six pieces board might have served as a 'score board' with each pair of players marking their score on one of the two pieces on a definite side and the total score of the partners marked on the rest of the two squares, adjoining them. The single piece, in rectangular shape, might have served as a dice with four sides, marked with the numbers of points related to each movement of the pawn. The movements might have been horizontal, vertical or diagonal ear-marked for each pawn, and the exit being a checkmate."

Another illustration of mathematics in chess by Vasantha (2010) concerns different "Tricks, Puzzles and Number patterns" which "have tested the intellect of the mathematicians for thousands of years. Mathematics owes many interesting problems to the game of chess."

For instance, "[t]he strange move of the knight in chess makes the operations particularly fascinating. Knight is allowed to occupy any unoccupied space on the board, which is two column and one row or two rows and one column away from the cell he is in, regardless of whether or not the intervening cells are occupied. The attempt to cover all the squares of the Chess board with a knight's move, without going over the same square a second time, is, perhaps, is as old as the invention of game itself." (Vasantha 2010)

After all, "Leonard Euler (1701-1783), a great mathematician, had worked out on the knight's tour and developed several solutions to this problem some of which are closed solutions to the re-entry problem and some are open solutions." (Vasantha 2010)

Another scholar, "Krishna Raja Wodeyar III, (1794-1868) the Maharaja of Mysore, viewed this horse movement on a square board in a different perspective. His main intention was to develop acrostics (chitra kavyas) through horse movement (aswagati). He did not choose always movements, which corresponds to tours covering all the cells. His ingenuity lies in the fact that he chose such movements which depict a schematic designs or geometrical shapes and in all such movements one gets a closed solution to the re-entry provided the uncovered cells are ignored." (Vasantha 2010)

And "the Raja of Mysore has invented many ingenious combinations of the movements of horse. Indeed it would be curious to see a collection of knight's tours in the form of mural paintings at Jayachamaraja Art gallery, Mysore"—and "Krishnaraja Wodeyar is pre-eminent for such discoveries. His attachment to the board games and horse games was so great that he combined both with great ingenuity and skill and developed several interesting games, which apparently depicts animate and inanimate forms of choice." (Vasantha 2010)

*The Criticisms of the Cosmological Argument*

However, for the critics, there are some problems with the cosmological argument in regard to chess origin. Consider some of them below for illustration.

Firstly, one criticism of the cosmological argument in regard to chess origin is that it ignores other dimensions of chess in the context of the mental, social, and cultural arguments (as already introduced in *Sec. 1.4*).

Secondly, another criticism of the cosmological argument in regard to chess origin is that it is not clear which "movements of the celestial bodies and their numerical symbolism" in cosmology (as identified by Vasantha) really correspond exactly to which moves in chess in the first place.

Thirdly, still another criticism of the cosmological argument in regard to chess origin is that Vasantha's view is driven by the Hindu worldview (as explained above).

And fourthly, another criticism of the cosmological argument in regard to chess origin is that it is not clear how much the apparent similarities between chess rules and "the movements of the celestial bodies and their numerical symbolism" are exaggerated (and, for that matter, superficial) and to what extent their differences are much neglected (in order to make a misleading linkage between the two, which may not exist otherwise).

*A More Detached Standpoint*

Surely, these criticisms of the cosmologial argument are not to suggest that it is totally invalid; rather, they show us the two opposing sides of the debate, such that the possibility and desirability of strategy (from the perspective of the nature with the cosmological argument as a case study here) are not to the extent that its respective defenders would like us to believe.

In addition, this introduction to the cosmological argument can cast an illuminating light on the ontological principles in existential dialectics, and good examples include the absoluteness-relativeness principle, the predictability-unpredictability principle, the explicability-inexplicability principle, the finiteness-transfiniteness principle, the preciseness-vagueness principle, the simpleness-complicatedness principle, the openness-hiddenness principle, the denseness-emptiness principle, the change-constancy principle, the slowness-quickness principle, the expansion-contraction principle, the convention-novelty principle, the evolution-transformation principle, the symmetry-asymmetry principle, the regression-progression principle, and the same-difference principle.

For instance, in relation to the absoluteness-relativeness principle, if there is absoluteness (e.g., the absolute view of Vasantha in regard to cosmology in chess origin), there is relativeness (e.g., what is true for Vasantha in regard to the role of cosmology in chess origin is not necessarily so for Cho-Yo in regard to the role of the military institution in chess history, as shown in *Sec. 1.4*). And the reverse direction also holds true.

In relation to the predictability-unpredictability principle, if there is predictability (e.g., the predictable tendency of people like Vasantha to understand chess origin by looking into "the movements of the celestial bodies and their numerical symbolism"), there is unpredictability (e.g., the more difficult task to predict exactly to what extent any particular chess games will necessarily follow the rules as observed in "the movements of the celestial bodies and their numerical symbolism" as claimed by Vasantha). And the reverse direction also holds true.

In relation to the explicability-inexplicability principle, if there is explicability (e.g., the explanation by Vasantha that chess origin can be traced back to cosmology), there is inexplicability (e.g., the lack of sufficient explanation by Vasantha about the reductionistic treatment of chess origin, since, for the critics, other factors like the social, cultural, and mental arguments are ignored in the explanation). And the reverse direction also holds true.

In relation to the finiteness-transfiniteness principle, if there is finiteness (e.g., the finite number of pieces and rules in a chess game), there is transfiniteness (e.g., the transfinite number of possible moves in chess, such that different "Tricks, Puzzles and Number patterns have tested the intellect of the mathematicians for thousands of years"). And the reverse direction also holds true.

In relation to the preciseness-vagueness principle, if there is preciseness (e.g., the precise identificaion, by Vasantha, of "the movements of the celestial bodies and their numerical symbolism" in chess origin), there is vagueness (e.g., the vagueness in Vasantha's identification of the movements, since it is not clear which "movements of the celestial bodies and their numerical symbolism" in cosmology correspond exactly to which moves in chess in the first place). And the reverse direction also holds true.

In relation to the simpleness-complicatedness principle, if there is simpleness (e.g., the relatively simple analysis of chess origin in cosmology by Vasantha), there is complicatedness (e.g., the relatively more complicated analysis of chess origin in cosmology by questioning its assumptions and arguments, for the critics). And the reverse direction also holds true.

In relation to the openness-hiddenness principle, if there is openness (e.g., the open exploration of the relationship between chess origin and cosmology, by Vasantha), there is hiddenness (e.g., the hidden bias in Vasantha's view, because it is driven by the Hindu worldview). And the reverse direction also holds true.

In relation to the denseness-emptiness principle, if there is denseness (e.g., the relatively denser concentration of concerns with cosmology in chess origin for people like Vasantha), there is emptiness (e.g., the relatively less dense, or more empty, concentration of concerns with the mind, culture, and society in chess origin for people like Vasantha). And the reverse direction also holds true.

In relation to the change-constancy principle, if there is change (e.g., the ever-changing nature of "Tricks, Puzzles and Number patterns" which "have tested the intellect of the mathematicians for thousands of years"), there is constancy (e.g., the ever constant existence of "Tricks, Puzzles and Number patterns" in chess). And the reverse direction also holds true.

In relation to the slowness-quickness principle, if there is slowness (e.g., the relatively slower readiness of people like Vasantha to see chess moves in terms of social and cultural patterns), there is quickness (e.g., the relatively quicker readiness of people like Vasantha to see chess moves in terms of cosmological patterns). And the reverse direction also holds true.

In relation to the expansion-contraction principle, if there is expansion (e.g.., the relative expansion of ever new "Tricks, Puzzles and Number patterns" in chess to test "the intellect of the mathematicians for thousands of years"), there is contraction (e.g., the relative contraction of old "Tricks, Puzzles and Number patterns" in chess which have been solved, either wholly or partially, by some solutions like the ones proposed by Leonard Euler on the knight's tour, "some of which are closed solutions to the re-entry problem and some are open solutions"). And the reverse direction also holds true.

In relation to the convention-novelty principle, if there is convention (e.g., the conventional wisdom about chess origin in the military institution, for instance), there is novelty (e.g., the alternative novel challenge to the conventional wisdom about chess origin by Vasantha on the basis of cosmology). And the reverse direction also holds true.

In relation to the evolution-transformation principle, if there is evolution (e.g,. the natural evolution of humans to make observations about the sky and their surroundings in the state of nature), there is transformation (e.g., the technical transformation of human observations about the sky and their surroundings by the invention of cosmology and chess). And the reverse direction also holds true.

In relation to the symmetry-asymmetry principle, if there is symmetry (e.g., the co-existence of different views about chess origins throughout history), there is asymmetry (e.g., the acceptance of the version of chess origin in cosmology by people like Vasantha in India—but the acceptance of the version of the chess linkage with the military institution by people like Cho-Yo in Japan). And the reverse direction also holds true.

In relation to the regression-progression principle, if there is regression (e.g., the regression made in the view about chess origin in cosmology as shown in the criticisms pointed out by the critics), there is progression (e.g., the progress made in the view about chess origin in cosmology to help us understand better the linkage in a different way). And the reverse direction also holds true.

And in relation to the same-difference principle, if there is similarity in outcome (e.g., the contribution to the molding of beliefs and values, regardless of whether this be achieved by way of the appeal to cosmology in chess origin or to the military institution instead), there is difference in outcome (e.g., the contribution to the molding of beliefs and values by way of the appeal to cosmology in chess origin for its emphasis on geometrical patterns and nmber systems—but the contribution to the molding of beliefs and values by way of the appeal to the military institution in chess history for its emphasis on the relevance to warfare). And the reverse direction also holds true.

## Strategy and Society

Strategy, when examined from the perspective of society, can reveal in a different way its possibility and desirability, and this can be shown by way of two case studies, namely, (2.4.1) computer chess, and the impact of technology, and (2.4.2) chess games, and the influence of social structure—to be addressed hereafter, respectively.

## Computer Chess, and the Impact of Technology

A fascinating question to ask here, in the context of understanding chess from the perspective of society, concerns the role of technology. For instance, to what extent can technology contribute to the formation of a sucessful chess strategy?

A good answer, as this will be explained below, is that brilliant strategy has much to learn from computer chess. And this is also relevant to tactics, for that matter.

*Human-Computer Chess Matches*

The brief history of human-computer chess matches in the past decades speaks volumes of the trend for computers to increasingly take a bigger role in shaping the game of chess, alhough with limits.

For instance, in the early days, the conventional wisdom was that no chess program "would ever be able to defeat the expertise of top humans. In 1968, International Master David Levy made a famous bet that no chess computer would be able to beat him within ten years. He won his bet in 1978 by beating Chess 4.7 (the strongest computer at the time), but acknowledged then that it would not be long before he would be surpassed. In 1989, Levy was defeated by the computer Deep Thought in an exhibition match." (WK 2010m)

But "Deep Thought...was still considerably below World Championship Level, as the then reigning world champion Garry Kasparov demonstrated in two sterling wins in 1989. It was not until a 1996 match with IBM's Deep Blue that Kasparov lost his first game to a computer at tournament time controls in Deep Blue–Kasparov, 1996, Game 1. This game was, in fact, the first time a reigning world champion had lost to a computer using regular time controls....In May 1997, an updated version of Deep Blue defeated Kasparov 3½-2½ in a return match." (WK 2010m)

As computers were given faster processing power and more database, they "began to rival top flight players. In 1998, Rebel 10 defeated Viswanathan Anand who at the time was ranked second in the world, by a score of 5-3. However most of those games were not played at normal time controls....At least in fast games, computers played better than humans but at classical time controls—at which a player's rating is determined—the advantage was not so clear." (WK 2010m; RB 2010)

Then, "in October 2002, Vladimir Kramnik and Deep Fritz competed in the eight-game Brains in Bahrain match, which ended in a draw. Kramnik won games 2 and 3 by 'conventional' anti-computer tactics—play conservatively for a long-term advantage the computer is not able to see in its game tree search. Fritz, however, won game 5 after a severe blunder by Kramnik....The final two games were draws. Given the circumstances, most commentators still rate Kramnik the stronger player in the match." (WK 2010m)

A few years later, "in November-December 2006,...Vladimir Kramnik played Deep Fritz [again]. This time the computer won, the match ended 2-4." (WK 2010m)

This brief history of "human-computer chess matches showed the best computer systems overtaking human chess champions in the late 1990s. For the 40 years prior to that, the trend had been that the best machines gained about 40 points per year in the ELO ranking while the best humans only gained roughly 2 points per year." (WK 2010m; S. Coles 2002)

*Computer Strengths*

There are certain noticeable strengths of chess engines (or computers) when they play against humans—as summarized in *Table 2.3*.

These noticeable strengths of chess engines have much to do with the speed of their processing power and the nature of the database in storage, both of which are important for successful strategy (and tactics, for that matter).

For instance, "the increase in processing power...enables calculations to be made to ever greater depths in a given time." (WK 2010n) In one estimation, "doubling the computer speed gains approximately fifty to seventy ELO points in playing strength." (WK 2010m; D. Levy 1991)

Furthermore, "programming techniques have improved enabling the engines to be more selective in the lines that they analyse and to acquire a better positional understanding." (WK 2010n)

As an illustration, nowadays, "chess engines use endgame tablebases to increase their playing strength during the endgame. An endgame tablebase is a database of all possible endgame positions with small groups of material. Each position is conclusively determined as a win, loss, or draw for the player whose turn it is to move, and the number of moves to the end with best play by both sides. Endgame tablebases in all cases identify the absolute best move in all positions included (identifying the move that wins fastest against perfect defense, or the move that loses slowest against optimal opposition). Such tablebases are available for all positions containing three to six pieces (counting the kings) and for some seven-piece combinations." (WK 2010n)

*Computer Weaknesses*

If there are computer strengths, there are likewise computer weaknesses. To understand computer weaknesses, let's first consider some important "implementation issues" in the development of a chess computer programming, namely, (a) "board representation," (b) "search techniques," and (c) "leaf evaluation," to be addressed in what follows (and summarized in *Table 2.3*).

(a) Board Representation

The first implementation issues concerns "board representation," which refers to "how a single position is represented in data structures." (WK 2010m)

For instance, "[t]he data structure used to represent each chess position is key to the performance of move generation and position evaluation. Methods include pieces stored in an array ('mailbox' and '0x88'), piece positions stored in a list ('piece list'), collections of bit-sets for piece locations ('bitboards'), and huffman coded positions for compact long-term storage." (WK 2010m)

In addition, "computer chess programs usually support a number of common de facto standards. Nearly all of today's programs can read and write game moves as Portable Game Notation (PGN), and can read and write individual positions as Forsyth-Edwards Notation (FEN). Older chess programs often only understood long algebraic notation, but today users expect chess programs to understand standard algebraic chess notation." (WK 2010m)

(b) Search Techniques

The second implementation issue concerns "search engines," which refers to "how to identify the possible moves and select the most promising ones for further examination." (WK 2010m)

In general, there are two major types "search strategies," both of which have strengths and weaknesses, and they were called "Type A" and "Type B" by Claude Shannon in 1950. (WK 2010m)

The first type of search strategy, "Type B," uses "a 'brute force' approach, examining every possible position for a fixed number of moves using the minimax algorithm." (WK 2010m)

Type B is good and bad. It is good, because it searches for "every possible position for a fixed number of moves using the minimax algorithm," but it is bad, for two "impractical" reasons. (WK 2010m)

The first "impractical" reason is that it takes a longer time to finish the task of calculation, because, "with approximately thirty moves possible in a typical real-life position,...searching the approximately 109 positions involved in looking three moves ahead for both sides (six plies) would take about sixteen minutes, even in the 'very optimistic' case that the chess computer evaluated a million positions every second." (WK 2010m)

In fact, "the program wastes too much time looking at uninteresting positions….[T]here is a risk cutting out interesting nodes." (WK 2010m) In addition, it is important to remember a rule in chess, in that "[e]ach player receives certain amount of time for thinking. If he spends more time, he loses." (WK 2010o)

The second "impractical" reason is that "it ignored the problem of quiescence, trying to only evaluate a position that is at the end of an exchange of pieces or other important sequence of moves ('lines')," without looking at the larger picture too. (WK 2010m)

By contrast, the second type of search strategy, "Type B," which Shannon favored, avoids the problem of "wasting processing power examining bad or trivial moves" by making two improvements, as shown below. (WK 2010m)

- "Employ a quiescence search."
- "Only look at a few good moves for each position."

This, for Shannon, "would enable them to look further ahead ('deeper') at the most significant lines in a reasonable time. The test of time has borne out…; all modern programs employ a terminal quiescence search before evaluating positions." (WK 2010m)

But the problem with "Type B" is that "it relies on the program being able to decide which moves are good enough to be worthy of consideration ('plausible') in any given position and this proved to be a much harder problem to solve than speeding up type A searches with superior hardware and search extension techniques." (WK 2010m)

For some scholars, Type A can still be better, especially because "technological advances by orders of magnitude in processing power have made the brute force approach [inType A] far more incisive than was the case in the early years. The result is that a very solid, tactical AI player aided by some limited positional knowledge built in by the evaluation function and pruning/extension rules began to match the best players in the world. It turned out to produce excellent results, at least in the field of chess, to let computers do what they do best (calculate) rather than coax them into imitating human thought processes and knowledge. In 1997 Deep Blue defeated World Champion Garry Kasparov, marking the first time a computer has defeated a reigning world chess champion in standard time control." (WK 2010m)

(c) Leaf Evaluation

And the third implementation issue concerns "leaf evaluation," which refers to "how to evaluate the value of a board position, if no further search will be done from that position." (WK 2010m)

This issue is not less challenging to computers, because "[f]or most chess positions, computers cannot look ahead to all final possible positions. Instead, they must look ahead a few plies and then evaluate the final board position. The algorithm that evaluates final board positions is termed the 'evaluation function,' and these algorithms are often vastly different between different chess programs." (WK 2010m)

Evaluation function has to "consider material value along with other factors affecting the strength of each side. When counting up the material for each side, typical values for pieces are 1 point for a pawn, 3 points for a knight or bishop, 5 points for a rook, and 9 points for a queen." (WK 2010m)

And "[t]he king is sometimes given an arbitrary high value such as 200 points (Shannon's paper) or 1,000,000,000 points (1961 USSR program) to ensure that a checkmate outweighs all other factors.... Evaluation functions take many factors into account, such as pawn structure, the fact that a pair of bishops are usually worth more, centralized pieces are worth more, and so on." (WK 2010m; D. Levy 1991)

So, this means that the nature of evaluation function is also debatable.

*An Overall Vantage Point*

Of course, this analysis of the implementation problems of chess engines is not to dismiss their promise in the future of chess but to show us the two opposing sides of the debate, such that the possibility and desirability of strategy (from the perspective of society with computer chess as a case study here) are not to the extent that its respective defenders would like us to believe.

Furthermore, this analysis of computer chess can teach us something valuable about the ontological principles in existential dialectics, and good examples include the formalness-informalness principle, the absoluteness-relativeness principle, the predictability-unpredictability principle, the explicability-inexplicability principle, the finiteness-transfiniteness principle, the preciseness-vagueness principle, the simpleness-complicatedness principle, the openness-hiddenness principle, the denseness-emptiness principle, the order-chaos principle, the slowness-quickness principle, the expansion-contraction principle, the theory-praxis principle, the convention-novelty principle, the evolution-transformation principle, the symmetry-asymmetry principle, the regression-progression principle, and the same-difference principle.

For instance, in relation to the formalness-informalness principle, if there is formalness (e.g., the formal requirement of completeness in a system of ideas or rules), there is informalness (e.g., the non-formal nature of a chess engine to be unable to evaluate or justify the rules and values in chess games, except using them as already explicitly programmed in the computer software). And the reverse direction also holds true.

In relation to the absoluteness-relativeness principle, if there is absoluteness (e.g,. the absolute view by the supporters of Type A research strategy that it is better than Type B), there is relativeness (e.g., what is true for those in Type A in regard to its being better is not necessarily so for the supporters of Type B research strategy, who regard the latter as better). And the reverse direction also holds true.

In relation to the predictability-unpredictability principle, if there is predictability (e.g., the predictable tendency of Type A research strategy to examine "every possible position for a fixed number of moves using the minimax algorithm"), there is unpredictability (e.g., the more difficult task to predict the outcome of a particular chess game, if Type A research strategy is used on that particular occasion). And the reverse direction also holds true.

In relation to the explicability-inexplicability principle, if there is explicability (e.g., the explanation, in the early days of human-computer chess matches, by chess experts, especially after a chess computer in 1998 "defeated Viswanathan Anand who at the time was ranked second in the world," in that "in fast games, computers played better than humans but at classical time controls—at which a player's rating is determined—the advantage was not so clear"), there is inexplicability (e.g., the lack of sufficient explanation by those chess experts who still think that humans are better than chess engines, of why better and better "computer systems" have increasingly beaten more and more "human chess champions" later on, even at classical time controls). And the reverse direction also holds true.

In relation to the finiteness-transfiniteness principle, if there is finiteness (e.g., the finite number of rules that a chess computer needs to memorize), there is transfiniteness (e.g., the transfinite number of possible moves that a chess computer needs to calculate, especially in a Type A search strategy). And the reverse direction also holds true.

In relation to the preciseness-vagueness principle, if there is preciseness (e.g., the relatively precise identification of Type B search strategy by Shannon in 1950 as the better one), there is vagueness (e.g., the relatively vagueness in Shannon's identification of Type B, since it is not clear how a Type B program is "able to decide which moves are good enough to be worthy of consideration...in any given position and this proved to be a much harder problem to solve than speeding up type A searches with superior hardware and search extension techniques"). And the reverse direction also holds true.

In relation to the simpleness-complicatedness principle, if there is simpleness (e.g., the relatively simple analysis of two types of research strategy by Shannon), there is complicatedness (e.g., the relatively more complicated analysis of research strategy by a team from Nortwhestern University in 1973, who invented Chess 4.0, which abandoned Type B while learning something from it, like the use of "quiescence searches" and "extensions and pruning"). (WK 2010m) And the reverse direction also holds true.

In relation to the openness-hiddenness principle, if there is openness (e.g., the open exploration of the relationship between successful strategy and selective search techniques in Type B), there is hiddenness (e.g., the hidden nature of Type B search techniques, because "it was difficult to anticipate which moves the...type B programs would play, and why" in a particular game). And the reverse direction also holds true.

In relation to the denseness-emptiness principle, if there is denseness (e.g., the relatively denser concentration of concerns with "examining every possible position for a fixed number of moves using the minimax algorithm" by the supporters of Type A search strategy), there is emptiness (e.g., the relatively less dense, or more empty, concentration of concerns with "quiescence search" by the supporters of Type A search strategy). And the reverse direction also holds true.

In relation to the order-chaos principle, if there is order (e.g., the relatively orderly nature of chess games in regard to the rules which are explicitly stated), there is chaos (eg., the relatively chaotic nature of chess games in regard to unexpected "mistakes," the difficulty to "anticipate which moves...[computer] programs would play, and why," "the possibilities [of moves] are so numerous that no one can foresee how the game will move," and so on). (WK 2010m; CT 2009) And the reverse direction also holds true.

In relation to the slowness-quickness principle, if there is slowness (e.g., the relatively slower speed of a chess engine to make a move if it has to search for "every possible position for a fixed number of moves using the minimax algorithm" as in Type A search strategy), there is quickness (e.g,. the relatively quicker speed of a chess engine to make a move if it only has to do selective searches as in Type B search strategy). And the reverse direction also holds true.

In relation to the expansion-contraction principle, if there is expansion (e.g, the relatively more developed ability in a computer with Type A search strategy to search for "every possible position for a fixed number of moves using the minimax algorithm"), there is contraction (e.g., the relatively less developed ability in a computer with Type A search strategy to use "quiescence searches" and "extensions and pruning" instead). And the reverse direction also holds true.

In relation to the theory-praxis principle, if there is theory (e.g., the theoretical construction of alternative search strategies for chess engines by the team in Northwestern University in the 1970s), there is praxis (e.g., the practical application of alternative search strategies for chess engines by the team in Northwestern University, such that they won the ACM Computer Chess Championship in that year, with the new program called "Chess 4.0"). (WK 2010m) And the reverse direction also holds true.

In relation to the convention-novelty principle, if there is convention (e.g., the conventional wisdom about Type A and Type B search strategies in the 1950s as proposed by Shannon), there is novelty (e.g,. the alternative novel challenge to the conventional wisdom about search strategies by the team in Northwestern University in the 1973, such that they abandoned Type B while learning something from it and coming up with something new like Chess 4.0). And the reverse direction also holds true.

In relation to the evolution-transformation principle, if there is evolution (e.g., the natural evolution of humans to play with each other in the state of nature), there is transformation (e.g., the technical transformation of human playing with each other by the invention of computers and computer chess). And the reverse direction also holds true.

In relation to the symmetry-asymmetry principle, if there is symmetry (e.g., the co-existence of different search strategies in the history of chess games), there is asymmetry (e.g., the popularity of selective search techniques like Type B search strategy in the older days, when computers were not powerful enough—but the use of more comprehensive search techniques like Type A search strategy nowadays, when "technological advances by orders of magnitude in processing power have made the brute force approach [inType A] far more incisive than was the case in the early years"). And the reverse direction also holds true.

In relation to the regression-progression principle, if there is regression (e.g., the regression made in chess engines as shown by the problems in search strategies pointed out by the critics), there is progression (e.g., the progress made in chess engines, as shown by computers beating human chess champions nowadays). And the reverse direction also holds true.

And in relation to the same-difference principle, if there is similarity in outcome (e.g., the contribution to victory in a chess game, be it done by using Type A or Type B search strategy), there is difference in outcome (e.g., the contribution to victory in a chess game by using Type A search strategy with the search for "every possible position for a fixed number of moves using the minimax algorithm"—but the contribution to victory in a chess game by using Type B search strategy with the use of "quiescence searches" and "extensions and pruning"). And the reverse direction also holds true.

## Chess Games, and the Influence of Social Structure

Another way to evaluate the possibility and desirability of strategy from the perspective of society has to do with the influence of social structure.

A good question to ask here, for instance, is why chess strategy often values the safety of the king and queen, and their protection by the pawns and others.

A good answer, as will be explained below, has to do with the influence of social structure on the development of chess strategy. And this is also relevant to tactics, for that matter.

### *Chess Origin, and the History of Social Structure*

To understand the influence of social structure on the development of chess strategy, it is important to start with a short history of chess origin from the perspective of society.

Although chess (at least in the Western tradition) has a long history, chess historians are still not very sure about "[t]he place of its origin," because "some say it was in Afghanistan and some say it was in India." (WCC 2009)

In any event, "[i]n India it was known by the name of 'Chaturanga' in the medieval times. There are evidences that chess was played in India, Persia along with other Asian countries many centuries ago. The Arabian army known as Moors, learnt chess from the Persians while they invaded Persia. The Moors took it to Spain when they attacked the country.the game was spread to the entire of Europe and rest of the world this way. The Europeans named the chess pieces, the way we pronounce them now might be because they had problems pronouncing the Persian names." (WCC 2009)

Once chess became more formalized, the rules and values in the game reflect the older history of social structure in the society in question. In Western chess, as an illustration, "[i]f one looks into the game intently, it is lot more than just a game. It is a picture of how people lived in the medieval times, their society, beliefs and lifestyle. Just a look at a chessboard and one knows it reflects different classes of the society and people representing them. The pieces on the chessboard are more of characters than mere playthings. It is the narration of ceremonies, wars and battlefields of the medieval time, narrated by the pieces and their moves. It's kind of a cross section of the lives of people of the medieval era." (WCC 2009)

For instance, the lowest-ranked pieces on the chessboard are "the pawns," and "[t]he pawns on the chessboard represent the laborers who used to be at the beck and call of the landowners and were treated like slaves, given up for others. These laborers could be traded or sacrificed to allow the landowners to escape harm, which is why they are the highest number of pieces on the chessboard and are sacrificed for other important pieces." (WCC 2009)

Then, there are the "rooks" and the "knights" near the two corners of the chessboard: "The castle pieces at both corners of the chessboard represent the home or refuge for the soldiers, like it used to be in the medieval times. A chessboard has two castle pieces, which are rightly used for saving the important members of the state. The professional soldiers of the medieval era were called knights who used to protect important members of the state like kings, queens and bishops. There are two knights in a game of chess and they are used to save the important pieces in a game." (WCC 2009)

Then, there are the "bishops" on the chessboard, which "represented the church in medieval times. Church was the most powerful institution in the society and so were the bishops. They represented religion and held one of the most important positions in the society. There are two bishops for each side in the game of chess who are protected by other pieces on the board." (WCC 2009)

Then, there is "the queen" on the chessboard, who "is the only female figure on a chessboard. She is the most powerful piece [after the king] on the board and regulates the game. She is protected by all means. This again gives a clear picture of the medieval society where queen stood the most powerful position [after the king]....There is only one queen on each side of a chessboard." (WCC 2009)

Finally, there is "the king" on the chessboard, which is "the tallest piece on a chessboard," because "the king...holds the strongest position in a chess game. He is defended by all pieces on the board and once taken, ends the game. This is what happened in the medieval times. If the king could be captivated, the war was lost and the kingdom was taken over by the invading army." (WCC 2009)

This summary thus shows "how Chess became more than a game and a clear depiction of life in the medieval time." (WCC 2009)

Even in Eastern chess, it is also no coincidence that scholars like Cho-Yo (1905) studied Chinese and Japanese chess variants and treated them as "the Science and Art of War or Struggle," as Cho-Yo used the examples of Japanese chess (*sho-ngi*) and Chinese chess (*chong-kie* and *i-go*) for the study of the Russo-Japanese War duriing 1904-1905. In so doing, the study depicted how life was like in Eastern societies at the time.

*The Gender Issue, and the Question about Social Structure*

In order to elaborate further the influence of social structure on the development of chess strategy, let's use a particular case study, this time in regard to the gender issue for illustration.

A controversial question which is sometimes asked is why the king holds a higher position than the queen in chess strategy. The answer reveals that, historically, men have been more dominant (and successful) in society than women, both in battlefields and in chess games.

In battlefields, it is widely known that men, since time immemorial, have done almost all of the fights in battlefields. Legendary figures like Joan of Arc are more exceptional cases than the norms; besides, it is not clear how much the story about her is mixed with fiction, as legends often are.

In both *Beyond Nature and Nurture* (*BNN*) and *The Future of Post-Human War and Peace* (*FPHWP*), I already went to great lengths to show the gender differences like this from both biological and non-biological perspectives.

And in chess games, it is also widely known that almost all the best chess champions are men; surely, there are some brilliant female chess players, but they are more exceptional cases than the norms.

This debate continues in our time, as shown in a recent controversial report by a team of researchers from the UK which "has shown that the under-representation of women at the top end in chess is almost exactly what would be expected, given the much greater number of men that participate in the game at all. Researchers Merim Bilalic [2008], et al., have published their research on this statistical sampling explanation in a recent issue of the *Proceedings of the Royal Society B*." (L. Zyga 2009)

More specifically, "[t]he authors analyzed the population of about 120,000 German players as recorded by the German chess federation in April 2007. Based on more than 3,000 tournaments per year, the German chess federation measures the skill level of all competitive and most hobby players in the country (the rating correlates highly with the widely known Elo rating). The sample population included 113,386 men and 7,013 women (a ratio of 16:1)." (L. Ziga 2009)

Then, "the researchers estimated the expected performance of the top 100 male and top 100 female players. Then, they compared the expected differences in points between these high-ranking male and female players with the actual point differences. Theoretically, the size difference between the male and female groups should correspond to the point differences between the top performers in the two groups." (L. Zyga 2009)

Interestingly, "[t]he results showed that the top three women had more points than expected, the next 70 or so pairs showed a small advantage for the men, and the last 20 pairs showed a small advantage for the women. Overall, men performed slightly better than expected, with an average advantage of 353 points, whereas the expected advantage was 341 points. Nevertheless, about 96% of the actual difference between genders could be explained by the statistical fact that the extreme values from a large sample are likely to be larger than those from a small one." (L. Zyga 2009)

However, this study has suffered from some major criticisms. Consider a few of them below, for illustration.

The first criticism against the report is that it does not explain "why so few women participate in chess at all. While it's possible that there exists a self-selection process based on innate biological differences that leads women to drop out of chess early on, this argument rests on a controversial assumption, the researchers say." (L. Zyga 2009)

Unfortunately, the researchers could not provide a better argument to supersede the biological explanation—other than saying that it is controversial.

The second criticism against the report is that it is one thing to say that the argument for the "innate biological differences that leads women to drop out of chess early on" is controversial, it is something else to even suggest that it is therefore false. Although the researchers did not explicitly claim that it is false, they could not prove that it is so. In other words, the researchers were making an unquestioned assumption in their report at the outset that the biological argument is not to be taken seriously.

In *Beyond Nature and Nurture* (BNN), I already went to great lengths to show that the gender issue has both biological and sociocultural dimensions to be considered, and it is a dominant ideology in our time to be intolerant to any talk of biological differences among different groups, be they based on race, gender, ethnicity, age, etc.

And the third criticism against the report is that the researchers also could not answer "[w]hether or not statistical sampling covers all the bases of explaining male superiority in chess." (L. Zyga 2009) They only hoped that their study would be considered by others as a stepping stone for further research.

*The Redemption of the Analysis*

Yet, these criticisms againt the report on the impact of social structure on the development of chess strategy should not be taken to dismiss the social argument; instead, they are used to teach us the two opposing sides of the debate, such that the possibility and desirability of strategy (from the perspective of society with the question about social structure as a case study here) are not to the extent that its respective defenders would like us to believe.

Furthermore, this analysis of the influence of social structure can teach us something valuable about the ontological principles in existential dialectics, and good examples include the formalness-informalness principle, the absoluteness-relativeness principle, the predictability-unpredictability principle, the explicability-inexplicability principle, the preciseness-vagueness principle, the simpleness-complicatedness principle, the openness-hiddenness principle, the denseness-emptiness principle, the slowness-quickness principle, the expansion-contraction principle, the convention-novelty principle, the evolution-transformation principle, the symmetry-asymmetry principle, the softness-hardness principle, the regression-progression principle, and the same-difference principle.

For instance, in relation to the formalness-informalness principle, if there is formalness (e.g., the formal requirement of a system of ideas to be complete), there is informalness (e.g., the nonformal aspects of the report by a team of researchers from the UK led by Merim Bilalic, like its unquestioned assumption that the biological argument is not to be taken seroiusly, without proving that it is so). And the reverse direction also holds true.

In relation to the absoluteness-relativeness principle, if there is absoluteness (e.g., the absolute view by Bilalic and others that the biological argument is "controversial" or untenable), there is relativeness (e.g., what is true for Bilalic and others in regard to the rejection of the biological argument is not so for the critics, who allow the biological argument too). And the reverse direction also holds true.

In relation to the predictability-unpredictability principle, if there is predictability (e.g., the prediction that "men performed slightly better than expected, with an average advantage of 353 points, whereas the expected advantage was 341 points," according to the results by Bilalic et al.), there is unpredictability (e.g., the more difficult task to predict exactly the gender differences in chess performance, if projected outside the database of the year 2007 and outside the sample case of Germany, as originally used by the report). And the reverse direction also holds true.

In relation to the explicability-inexplicability principle, if there is explicability (e.g., the explanation of the gender differences in chess performance by the team of research from the UK on the basis of gender differences in chess participation rate, as shown in the sample population which "included 113,386 men and 7,013 women," that is, "a ratio of 16:1"), there is inexplicability (e.g., the failture of the team of research from the UK to explain "why so few women participate in chess at all" in the first place). And the reverse direction also holds true.

In relation to the preciseness-vagueness principle, if there is preciseness (e.g., the precise identification, by the team of research from the UK, of "the sample population" which "included 113,386 men and 7,013 women," that is, "a ratio of 16:1"), there is vagueness (e.g., the vagueness in the report, in that it does not say "[w]hether or not statistical sampling covers all the bases of explaining male superiority in chess"). And the reverse direction also holds true.

In relation to the simpleness-complicatedness principle, if there is simpleness (e.g., the relatively simple analysis of the gender differences in chess performance by the team of research from the UK on the basis of the gender differences in chess participation rate), there is complicatedness (e.g., the relatively complicated analysis of the results provided by the team of research from the U.S., as shown in the criticisms by the critics who challenge its claims and assumptions). And the reverse direction also holds true.

In relation to the openness-hiddenness principle, if there is openness (e.g., the open exploration of the relationship between gender differences in chess participation rate and gender differences in chess perfromance), there is hiddenness (e.g., the hidden bias in the report by the team of research from the U.S., in that it assumes uncritically that the biological argument is controversial and is not to be taken seriously at the outset of the research). And the reverse direction also holds true.

In relation to the denseness-emptiness principle, if there is denseness (e.g., the relatively denser concentration of chess champions who are males), there is emptiness (e.g., the relatively less dense, or more empty, concentration of chess champions who are females). And the reverse direction also holds true.

In relation to the slowness-quickness principle, if there is slowness (e.g., the relatively slower readiness of a chess player to sacrifice the queen in a chess game), there is quickness (e.g., the relatively quicker readiness of a chess player to sacrifice a pawn in a chess game). And the reverse direction also holds true.

In relation to the expansion-contraction principle, if there is expansion (e.g., the relatively more developed ability of a male to play chess), there is contraction (e.g., the relatively less developed ability of a male to play with dolls). And the reverse direction also holds true.

In relation to the convention-novelty principle, if there is convention (e.g., the conventional wisdom about male success in chess performance), there is novelty (e.g., the alternative novel challenge to the conventional wisdom about chess performance, by the new proposal that gender differences in chess participation rate are related to gender differences in chess perfromance). And the reverse direction also holds true.

In relation to the evolution-transformation principle, if there is evolution (e.g., the natural evolution of humans to fight with each other in the state of nature), there is transformation (e.g., the technical transformation of human fighting with each other by the invention of an elaborated social structure involving soldiers, knights, bishops, queens, kings, etc.). And the reverse direction also holds true.

In relation to the symmetry-asymmetry principle, if there is symmetry (e.g., the co-existence of differrent pieces in a chess game), there is asymmetry (e.g., the value of a knight on the basis of 3 points, but the value of a queen on the basis of 9 points). (WK 2010) And the reverse direction also holds true.

In relation to the softness-hardness principle, if there is softness (e.g., the promotion of a pawn in an endgame), there is hardness (e.g., the sacrifice of a pawn for a more valuable piece). And the reverse direction also holds true.

In relation to the regression-progression principle, if there is regression (e.g., the regression made in the report by a team of research from the UK., as shown in the criticisms about its assumptions and claims), there is progression (e.g., the progress made in the report by a team of research from the UK., in that there is a correlation between gender differences in chess participation rate and gender differences in chess performance). And the reverse direction also holds true.

And in relation to the same-difference principle, if there is similarity in outcome (e.g., the contribution to the molding and regulation of beliefs and values, regardless of whether this is achieved by way of emphasizing gender biological differences or dismissing them), there is difference in outcome (e.g., the contribution to the molding and regulation of beliefs and values by way of emphasizing gender biological differences—but the contribution to the molding and regulation of beliefs and values by way of dismissing gender biological differences). And the reverse direction also holds true.

# Strategy and Culture

Strategy, when examined from the perspective of culture, reveals in a different way its possibility and desirability, and this can be shown by way of two case studies, namely, (2.5.1) chess and morality, and (2.5.2) chess development, and cultural variations—to be addressed hereafter, respectively.

## Chess and Morality

A good question to ask, in the present context of chess and morality, is to what extent chess strategy in chess games reflects some moral values which are culturally contingent.

A good answer, as will be explained below, is that chess strategy reflects some moral values in a culture. And this is also relevant to tactics, for that matter.

An excellent case study of the relationship between chess and morality is Benjamin Franklin's well-known essay titled "The Morals of Chess" published in *Columbian* magazine in December of 1786, which "is the second known writing on chess in America. This essay in praise of chess and prescribing a code of behavior for it has been widely reprinted and translated." (WK 2010p; J. McCrary 2009; D. Hooper 1992)

### *The Introduction of Chess to America*

For Franklin (2010), "chess…is the most ancient and the most universal game known among men; for its original is beyond the memory of history, and it has, for numberless ages, been the amusement of all the civilized nations of Asia, the Persians, the Indians, and the Chinese.

Europe has had it above 1000 years; the Spaniards have spread it over their part of America, and it begins lately to make its appearance in these northern states."

Franklin (2010) appreciated this introduction of chess to America, because chess "is so interesting in itself, as not to need the view of gain to induce engaging in it; and thence it is never played for money. Those, therefore, who have leisure for such diversions, cannot find one that is more innocent; and...it may, in its effects on the mind, be not merely innocent, but advantageous, to the vanquished as well as to the victor."

*The Four Morals of Chess*

For Franklin (2010), "[t]he game of Chess is not merely an idle amusement. Several very valuable qualities of the mind, useful in the course of human life, are to be acquired or strengthened by it, so as to become habits, ready on all occasions. For life is a kind of chess, in which we have often points to gain, and competitors or adversaries to contend with, and in which there is a vast variety of good and ill events, that are, in some degree, the effects of prudence or the want of it."

More specifically, Franklin identified four morals of chess, namely, (a) "foresight," (b) "circumspection," (c) "caution," and (d) "hope," as described below, in that order—and summarized in *Table 2.4*. (B. Franklin 2010)

(a) Foresight

The first moral of chess, for Franklin (2010), is "foresight," which "looks a little into futurity, and considers the consequences that may attend to an action: for it is continually occurring to the player, 'If I move this piece, what will be the advantages of my new situation? What use can my adversary make of it to annoy me? What other moves can I make to support it, and to defend myself from his attacks?'"

(b) Circumspection

The second moral of chess, for Franklin (2010), is "circumspection," which "surveys the whole chess-board, or scene of action, the relations of the several pieces and situations, the dangers they are respectively exposed to, the several possibilities of their aiding each other; the probabilities that the adversary may make this or that move, and attack this or the other piece; and what different means can be used to avoid his stroke, or turn its consequences against him."

(c) Caution

The third moral of chess, for Franklin (2010), is "caution," which "is not to make our moves too hastily. This habit is best acquired by observing strictly the laws of the game, such as, If you touch a piece, you must move it somewhere; if you set it down, you must let it stand. And it is therefore best that these rules should be observed, as the game thereby becomes more the image of human life, and particularly of war; in which, if you have incautiously put yourself into a bad and dangerous position, you cannot obtain your enemy's leave to withdraw your troops, and place them more securely; but you must abide by all the consequences of your rashness."

(d) Hope

And the fourth moral of chess, for Franklin (2010), refers to "hope," in that "we learn by chess the habit of not being discouraged by present bad appearances in the state of our affairs, the habit of hoping for a favorable change, and that of persevering in the search of resources."

After all, "[t]he game is so full of events, there is such a variety of turns in it, the fortune of it is so subject to sudden vicissitudes, and one so frequently, after long contemplation, discovers the means of extricating one's self from a supposed insurmountable difficulty, that one is encouraged to continue the contest to the last, in hopes of victory by our own skill, or, at least, of giving a stale mate, by the negligence of our adversary. And whoever considers, what in chess he often sees instances of, that particular pieces of success are apt to produce presumption, and its consequent, inattention, by which more is afterwards lost than was gained by the preceding advantage; while misfortunes produce more care and attention, by which the loss may be recovered, will learn not to be too much discouraged by the present success of his adversary, nor to despair of final good fortune, upon every little check he receives in the pursuit of it." (B. Franklin 2010)

*The Eight Rules of Chess*

In addition, Franklin (2010) also proposed some rulers for playing chess, such that it is a fair game.

Hereafter are the eight rules, as shown below.

The first rule is: "If it is agreed to play according the strict rules, then those rules are to be exactly observed by both parties; and should not be insisted on for one side, while deviated from by the other; for this is not equitable." (B. Franklin 2010)

The second rule is: "If it is agreed not to observe the rules exactly, but one party demands indulgences, he should be as willing to allow them to the other." (B. Franklin 2010)

The third rule is: "No false move should ever be made to extricate yourself out of a difficulty, or to gain advantage. There can be no pleasure in playing with a person once detected in such unfair practice." (B. Franklin 2010)

The fourth rule is: "If your adversary is long in playing, you ought not to hurry him, or express any uneasiness at his delay. You should not sing, or whistle, nor look at your watch, nor take up a book to read, nor make a tapping with your feet on the floor, or with your fingers on the table, nor do any thing that may disturb his attention. For all these things displease. And they do not show in playing, but your craftiness or your rudeness." (B. Franklin 2010)

The fifth rule is: "You ought not to endeavour to amuse and deceive your adversary, by pretending to have made bad moves, and saying you have now lost the game, in order to make him secure and careless, and inattentive to your schemes; for this is fraud, and deceit, not skill at the game." (B. Franklin 2010)

The sixth rule is: "You must not, when you have gained a victory, use any triumphing or insulting expression, nor show too much pleasure; but endeavour to console your adversary, and make him less dissatisfied with himself by every kind and civil expression, that may be used with truth; such as, You understand the game better than I, but you are a little inattentive; or, You play too fast; or, You had the best of the game, but something happened to divert your thoughts, and that turned it in my favour." (B. Franklin 2010)

The seventh rule is: "If you are a spectator, while others play, observe the most perfect silence. For if you give advice, you offend both parties; him, against whom you may give it, because it may cause the loss of his game; him, in whose favour you give it, because, tho' it may be good, and he follows it, he loses the pleasure he might have had, if you had permitted him to think till it occurred to himself. Even after a move or moves, you must not, by replacing the pieces, show how it might have been played better: for that displeases, and may occasion disputes or doubts about their true situation. All talking to the players, lessens or diverts their attention, and is therefore displeasing; nor should you give the least hint to either party, by any kind of noise or motion.—If you do, you are unworthy to be a spectator.—If you have a mind to exercise or show your judgment, do it in playing your own game when you have an opportunity, not in criticizing or meddling with, or counseling, the play of others." (B. Franklin 2010)

And the eighth rule is: "If the game is not to be played rigorously, according to the rules above mentioned, then moderate your desire of victory over your adversary, and be pleased with one over yourself. Snatch not eagerly at every advantage offered by his unskillfulness or inattention; but point out to him kindly that by such a move he places or leaves a piece in danger and unsupported; that by another he will put his king in a dangerous situation, etc. By this generous civility (so opposite to the unfairness above forbidden) you may indeed happen to lose the game to your opponent, but you will win what is better, his esteem, his respect, and his affection; together with the silent approbation and good will of impartial spectators." (B. Franklin 2010)

*The Other Sides of Chess*

However, there are the other sides of chess which Franklin did not pay attention to, such that his view is not immune to criticisms. Consider some of them below, for illustration—as summarized in *Table 2.4*.

Firstly, one criticism of the "morality" argument is that not everyone agrees with the benefits of chess playing. For instance, in *Sec. 2.2.2*, I already showed that, for some chess players, chess can have some mental benefits which do not have to be moral in nature (e.g., logical thinking, planning, imagination, etc.), while there are other benefits which may have moral implications (e.g., concentration, and self-confidence) but do not have to be the ones as claimed by Franklin.

Secondly, another criticism of the "morality" argument is that it does not consider the "dark" side of chess playing, as shown in chess-related deaths and corrupted lifestyles as already discussed in *Sec. 2.2.2*.

Thirdly, still another criticism of the "morality" argument is that it has not yet been conclusively proved, because "the results of research studies have failed to produce unambiguous support for the intellectual benefits of playing chess." (WK 2010i; T. Redman 2006)

Fourthly, one more criticism of the "morality" argument is that it does not consider other dimensions of chess in the context of the mental, natural, and social arguments (as introduced in *Sec. 1.4*).

Fifthly, still another criticism of the "morality" argument is that it also does not focus on other cultural aspects of chess (like the issue of aesthetics).

And sixthly, still one more criticism of the "morality" argument is that it is not obivous at all that the four morals of chess as claimed by Franklin are really moral at all, as philosophers have debated for thousands of years what moral virtues are (or are not) and, in many cases, do not necessarily include the four as mentioned by Franklin. For instance, is "caution" really a moral virtue, or only a practical convenience?

*Above the Fray*

Once more, these criticisms of the "morality" argument do not mean to dismiss it outright; instead, they are used to show us the two opposing sides of the debate, such that the possibility and desirability of strategy (from the perspective of culture with the "morality" argument as a case study here) are not to the extent that its respective defenders would like us to believe.

In addition, this analysis of the "morality" argument can teach us something important about the ontological principles in existential dialectics, and good examples include the absoluteness-relativeness principle, the predictability-unpredictability principle, the explicability-inexplicability principle, the preciseness-vagueness principle, the simpleness-complicatedness principle, the openness-hiddenness principle, the denseness-emptiness principle, the order-chaos principle, the slowness-quickness principle, the expansion-contraction principle, the theory-praxis principle, the convention-novelty principle, the evolution-transformation principle, the symmetry-asymmetry principle, the softness-hardness principle, the regression-progression principle, and the same-difference principle.

For instance, in relation to the absoluteness-relativeness principle, if there is absoluteness (e.g., the absolute view by Franklin about the morals of chess), there is relativeness (e.g., what is true for Franklin about the morals of chess is not necessarily so for the critics who are more skeptical to his list of the four mental benefits of chess playing). And the reverse direction also holds true.

In relation to the predictability-unpredictability principle, if there is predictability (e.g., the predictable tendency of the followers of Franklin to accept the four mental benefits of chess playing), there is unpredictability (e.g., the more difficult task to predict which particular mental benefits a particular chess player will gain, since "the results of research studies have failed to produce unambiguous support for the intellectual benefits of playing chess"). (WK 2010i; T. Redman 2006) And the reverse direction also holds true.

In relation to the explicability-inexplicability principle, if there is explicability (e.g., the explanation by Franklin that chess playing is beneficial because of the moral consequences), there is inexplicability (e.g., the lack of sufficient explanation, by the defenders of Franklin, of why, if the moral benefits are so obvious, there are chess-related deaths or murders, as pointed out by the critics). And the reverse direction also holds true.

In relation to the preciseness-vagueness principle, if there is preciseness (e.g., the precise identification of the four morals of chess by Franklin), there is vagueness (e.g., the vagueness in the identification, since it is not clear if there are other morals, to what extent every reasonalbe chess player will agree about the list, and whether or not chess can have some mental benefits which do not have to be moral in nature at all). And the reverse direction also holds true.

In relation to the simpleness-complicatedness principle, if there is simpleness (e.g., the relatively simple analysis of the morals of chess by Franklin), there is complicatedness (e.g., the relatively more complicated analysis of the morals of chess, as shown in the criticisms by the critics). And the reverse direction also holds true.

In relation to the openness-hiddenness principle, if there is openness (e.g., the open exploration of the relationships between morals and chess by Franklin), there is hiddenness (e.g., the hidden bias in Franklin's treatement of the morals of chess, because of his strong moral conviction in "civic and personal virtues," especially in the context of his pious Puritan background). (WK 2010p) And the reverse direction also holds true.

In relation to the denseness-emptiness principle, if there is denseness (e.g., the relatively denser concentration of moral manners in pious chess players while playing chess, according to Franklin), there is emptiness (e.g., the relatively less dense, or more empty, concentration of moral manners in deviant chess players who engage in chess-related deaths). And the reverse direction also holds true.

In relation to the order-chaos principle, if there is order (e.g., the existence of certain rules as listed by Franklin for chess games, so as to provide an orderly structure, which can be followed), there is chaos (e.g., the relative chaotic nature of chess games too, such that, as Franklin put it, "the game is so full of events, there is such a variety of turns in it, [and] the fortune of it is so subject to sudden vicissitudes"). And the reverse direction also holds true.

In relation to the slowness-quickness principle, if there is slowness (e.g., the relatively slower chess players, like the "adversary" as identified by Franklin in rule #4), there is quickness (e.g., the relatively quicker chess players, like "you" as identified by Franklin in rule #4). And the reverse direction also holds true.

In relation to the expansion-contraction principle, if there is expansion (e.g., the relatively more developed ability to be "civil" and "moral" by those who follow the morals and rules as laid down by Franklin), there is contraction (e.g., the relatively less developed ability to be "rude" and "unequitable" by those who follow the morals and rules as laid down by Franklin). And the reverse direction also holds true.

In relation to the theory-praxis principle, if there is theory (e.g., the theoretical construction of the morals of chess by Franklin), there is praxis (e.g., the practical application of the morals of chess by encouraging people to play chess for the benefits, as "Franklin was posthumously inducted into the U.S. Chess Hall of Fame in 1999"). (WK 2010p; J. McCrary 2009) And the reverse direction also holds true.

In relation to the convention-novelty principle, if there is convention (e.g., the conventional wisdom about the rules governing chess playing), there is novelty (e.g., the alternative novel challenge to this conventional wisdom by Franklin, in that there is also a close relationship between morality and chess). And the reverse direction also holds true.

In relation to the evolution-transformation principle, if there is evolution (e.g., the natural evolution of humans in the state of nature to play with each other without the hassle of "civic" and "moral" concerns), there is transformation (e.g., the technical transformation of human playing with each other by the invention of "morals" and "rules" as shown in chess playing, in the later stages of the development of human civilization). And the reverse direction also holds true.

In relation to the symmetry-asymmetry principle, if there is symmetry (e.g., the co-existence of different benefits of chess playing), there is asymmetry (e.g., the acceptance of the four morals of chess by the followers of Franklin—but the advocation of other benefits of chess playing as already discussed in the earlier section on "chess games, and the development of mental skills"). And the reverse direction also holds true.

In relation to the softness-hardiness principle, if there is softness (e.g., the praise by rule-abiding chess players to those who follow the rules of chess games), there is hardness (e.g., the reprimand by rule-abiding chess players to those who do not follow the rules of chess games). And the reverse direction also holds true.

In relation to the regression-progression principle, if there is regression (e.g., the regression made in the morals of chess as shown in the criticisms by the critics), there is progression (e.g., the progress made by the morals of chess for more "civil" behaviors in chess playing). And the reverse direction also holds true.

And in relation to the same-difference principle, if there is similarity in outcome (e.g., the contribution to the molding and regulation of beliefs and manners, regardless of whether this be accomplished by way of the morals of chess by Franklin or by way of other values by other scholars), there is difference in outcome (e.g., the contribution to the molding and regulation of beliefs and manners by way of the morals of chess by Franklin—but the contribution to the molding and regulation of beliefs and manners by way of alternative emphasis on different mental benefits as already discussed in the earlier section on "chess games, and the development of mental skills"). And the reverse direction also holds true.

## Chess Development, and Its Cultural Variants

Another way to evaluate the possibility and desirability of strategy from the perspective of culture has to do with the cultural variants of chess development.

A good question to ask here is, To what extent is the development of chess strategy contingent on the impact of a given culture?

A good answer, as will be explained below, is that chess strategy reveals the impact of cultural variants in its history of spreading from its original birthplace in Asia to the rest of the world. And this is also relevant to tactics, for that matter.

For illustration, consider three major cultures around the world which have shaped the development of chess strategy in its history, namely, (a) South Asia, (b) East Asia, and (c) Europe, respectively, in what follows.

*South Asia*

It is often said that "[t]he precursors of chess [like Western chess] originated in India during the Gupta empire, where its early form in the 6th century was known as chaturaṅga, which translates as 'four divisions [of the military]': infantry, cavalry, elephantry, and chariotry, represented by the pieces that would evolve into the modern pawn, knight, bishop, and rook, respectively." (WK 2010r; A. Leibs 2004; D. Forbes 1860; D. Robinson 1996; H. Bird, 1893; H. Murray 1913)

This is so, even though chess historians are still not very sure about "[t]he place of its origin," because "some say it was in Afghanistan and some say it was in India." (WCC 2009)

In any event, the history of chess development in ancient India reveals some cultural variations on chess strategy. Consider a few of them below, for illustration.

(a) Chess Board

In ancient India, "[c]hess was designed for an *ashtāpada* (Sanskrit for 'having eight feet,' i.e., an 8x8 squared board)." (WK 2010r; H. Murray 1913)

But there were also other variants like "the 10×10 *Dasapada* and the 9×9 *Saturankam*." (WK 2010r; S. Wilkins 2002)

In fact, "[o]ther game pieces (speculatively called 'chess pieces') uncovered in archaeological findings are considered as coming from other, distantly related, board games, which may have had boards of 100 squares or more." (WK 2010r; EB 2002)

(b) Dice or No Dice

In earlier times, chess was played with the use of a dice, in that "[s]ome people formerly played chess using a dice to decide which piece to move. There was an unproven theory that chess started as this dice-chess and that the gambling and dice aspects of the game were removed because of Hindu religious objections." (WK 2010r; S. Wilkins 2002)

(c) The Nature of a Win

There are at least three variants on how to decide a win in the Indian history of chess development, that is, (a) by "checkmate," (b) by "stalemate," or (c) by "bare king" ("taking all of an opponent's pieces except the king"). (WK 2010r)

(d) The Moves of the Pieces

Even on the rules governing how to move the pieces, there were different variants over time. For instance, in earlier times, the moves of the pieces were not always exactly what they are now, as shown below: (WK 2010r)

- "King"—as now.
- "Queen"—"one square diagonally, only"
- "Bishop"
  - –"In the version that went into Persia: two squares diagonally (no more or less), but could jump over a piece between"
  - –"In a version sometimes found in India in former times: two squares sideways or front-and-back (no more or less), but could jump over a piece between"
  - –"In versions found in Southeast Asia: one square diagonally, or one square forwards"
- "Knight"—"as now"
- "Rook"—"as now"

- "Pawn"—"one square forwards (not two), capturing one square diagonally forward; promoted to queen only"

Surely, there are other cultural variants, but the examples above suffice for illustration here.

*East Asia*

In East Asia, the history of chess development reveals its cultural variants in regard to chess strategy.

Consider a few of them below, for illustration.

(a) Chess Board

A good example concerns Chinese chess.

Although "[t]he object of the Chinese variation is similar to Chaturanga, i.e., to render helpless the opponent's king, sometimes known as *general*," there is also a big difference, because "Chinese chess also borrows elements from the game of Go, which was played in China since at least the 6th century BC. Owing to the influence of Go, Chinese chess is played on the intersections of the lines on the board, rather than in the squares. Chinese chess pieces are usually flat and resemble those used in checkers." (WK 2010r; EB 2007)

(b) The Moves of the Pieces

Another cultural variation concerns the moves of the pieces in Japanese chess.

For instance, "a prominent variant of chess in East Asia is the game of Shogi, transmitted from India to China and Korea before finally reaching Japan. The two distinguishing features of Shogi are: [a] The captured pieces may be reused by the captor and played as a part of the captor's forces, and [b] Pawns capture as they move, one square straight ahead." (WK 2010r; EB 2002a)

(c) The Names of the Pieces

Even the names of the pieces have cultural variants.

For instance, in Mongolian chess, the following pieces have different names: (WK 2010r)

- "King"—"Noyon–Ноён–lord"
- "Queen"—"Bers/Nohoi–Бэрс/Нохой–dog (to guard the livestock)"
- "Bishop"—"Temē–Тэмээ—camel"
- "Knight"—"Morĭ–Морь–horse"
- "Rook"—"Tereg–Тэрэг–cart"

• "Pawn"—"Hū–Xүү–boy (the piece often showed a puppy)"

Nowadays, however, "[c]hess in Mongolia is...played following the usual international rules." (WK 2010r)

*Europe*

In Europe, the history of chess development also has its cultural variants in regard to chess strategy.

Consider a few of them below, for illustration.

(a) The Names of the Sides of Chess Board

To start, although, by European custom, "the sides are conventionally called White and Black,...in earlier European chess writings, the sides were often called Red and Black because those were the commonly available colors of ink when handwriting drawing a chess game layout. In such layouts, each piece was represented by its name, often abbreviated (e.g. 'ch'r' for French 'chevalier'='knight')." (WK 2010r)

(b) The Time Limits for Chess Games

Another cultural variation has to do with the time limits for chess games (or the speed of playing chess). For instance, "[t]he game, as played during the early Middle Ages, was slow, with many games lasting for days. Some variations in rules began to change the shape of the game...by 1300 AD." (WK 2010r; I. Riddler 1998; EB 2002b)

For illustration, here are some rules which were introduced to make chess games faster: (WK 2010r)

- One rule was introduced to allow the "king jumping once, to make it quicker to put the king safe in a corner. (This eventually led to castling.)" (WK 2010r)
- Another rule was introduced to allow the "queen once moving two squares with jump, diagonally or straight. This right was sometimes extended to a new queen made by promoting a pawn." (WK 2010r)
- Still another rule was introduced to allow "the short assize ['assize'= 'sitting']....Here the pawns started on the third rank; the queens started on d3 and d6 along with the queens' pawns; the players arranged their other pieces as they wished behind their pawns at the start of the game. This idea did not endure." (WK 2010r; H. Murray 1913)

By late 19$^{th}$ century, additional rules were introduced to make chess games faster.

For instance, "the 1851 London Chess tournament raised concerns about the time taken by the players to deliberate their moves. On recording time it was found that players often took hours to analyze moves, and one player took as much as two hours and 20 minutes to think over a single move at the London tournament." (WK 2010r; EN 2002c)

So, additional rules were introduced late, as shown below: (WK 2010r)

- One rule was introduced to allow "speed chess" for "a bank of time to each player in which to play a previously agreed number of moves, e.g. two hours for 30 moves. In the final variant, the player who made the predetermined number of moves in the agreed time received additional time budget for his next moves. Penalties for exceeding a time limit came in form of fines and forfeiture. Since fines were easy to bear for professional players, forfeiture became the only effective penalty; this added 'lost on time' to the traditional means of losing such as checkmate and resigning." (WK 2010r; EB 2002c)
- Another rule was introduced in 1861 to use "sandglasses" for time-checking "in a tournament match at Bristol, England. The sandglasses were later replaced by pendulums. Modern clocks, consisting of two parallel timers with a small button for a player to press after completing a move, were later employed to aid the players. A tiny latch called a flag further helped settle arguments over players exceeding time limit at the turn of the 19th century." (WK 2010r; EB 2002c)
- Still another rule was introduced to resolve "another problem…when adjourning a game for a meal break or overnight. If the players are X and Y, and X moved last before the adjournment, this would make it much easier for Y than for X to analyze the game during the adjournment. Preventing access to a chess set to work out moves during the adjournment in his hotel room or wherever would not stop him from analyzing the position in his head. Various strange ideas were attempted, but the eventual solution was the 'sealed move': X, last thing before the adjournment, does not make his move but writes it on a piece of paper which he hands to a referee, who after the adjournment makes the move, and X and Y then continue the game." (WK 2010r)

Surely, there are other rules too, but the ones aforecited suffice for the present illustration.

(c) The Values of the Pieces

Even the values of the pieces have different cultural variants over time.

For instance, "[t]he queen and bishop remained relatively weak until between 1475 AD and 1500 AD in Spain or Portugal or France or Italy....In some areas (e.g. Russia), the queen could also move like a knight." (WK 2010r)

Later, "[t]his led to much more value being attached to the previously minor tactic of pawn promotion." (WK 2010r; H. Murray 1913)

(d) The Seeding of Players

Still another cultural variation concerns the seeding of players in a tournament.

For example, in the early days of "the previous informal system, the World Champion decided which challenger he would play for the title and the challenger was forced to seek sponsors for the match." (WK 2010r)

Later, "FIDE [the Fédération Internationale des Échecs or World Chess Federation] set up a new system of qualifying tournaments and matches. The world's strongest players were seeded into 'Interzonal tournaments,' where they were joined by players who had qualified from 'Zonal tournaments.' The leading finishers in these Interzonals would go on the 'Candidates' stage, which was initially a tournament, later a series of knock-out matches. The winner of the Candidates would then play the reigning champion for the title. A champion defeated in a match had a right to play a rematch a year later. This system worked on a three-year cycle." (WK 2010r)

By the 1960s, another rule was introduced by FIDE, which "abolished the automatic right of a deposed champion to a rematch, and the next champion, Armenian Tigran Petrosian, a genius of defense and strong positional player, was able to hold the title for two cycles, 1963–1969. His successor, Boris Spassky from Russia (1969–1972), was a player able to win in both positional and sharp tactical style." (WK 2010r; G. Kasparov 2003 & 2004)

*A Broader Horizon*

Needless to say, the historical analysis of the cultural variants of chess (especially in regard to its rules which were changed over time) does not mean to trivialize the game of chess but to show us the ever contentious disagreements among different individuals and authorities on the right way to play chess over time, such that the possibility and desirability of strategy (from the perspective of culture with the cultural variants of chess as a case study here) are not to the extent that its respective defenders would like us to believe. After all, what is right for one cultural variant of chess in regard to certain rules and names is not necessarily so for another cultural variant.

Furthermore, this historical analysis of the cultural variants of chess can teach us something important about the ontological principles in existential dialectics, and good examples include the absoluteness-relativeness principle, the partiality-totality principle, the predictability-unpredictability principle, the explicability-inexplicability principle, the preciseness-vagueness principle, the simpleness-complicatedness principle, the openness-hiddenness principle, the denseness-emptiness principle, the change-constancy principle, the slowness-quickness principle, the expansion-contraction principle, the convention-novelty principle, the evolution-transformation principle, the symmetry-asymmetry principle, the softness-hardness principle, the regression-progression principle, and the same-difference principle.

For instance, in relation to the absoluteness-relativeness principle, if there is absoluteness (e.g., the absolute view by each cultural variant that its rules of chess are the right way to play chess), there is relativeness (e.g,. what is right for Western chess to play "in the squares" is not so for Chinese chess to play "on the intersections of the lines on the board"). And the reverse direction also holds true.

In relation to the partiality-totality principle, if there is partiality (e.g., the partial vantage point of each cultural variant that its rules are the correct way to play chess), there is totality (e.g., the holistic view of the larger picture of all the vantage points of all the cultural variants, such that there is no universal correct way to play chess, so that the whole is not the sum of its parts, that is, the holistic view is not the sum of all the partial viewpoints of the cultural variants). And the reverse direction also holds true.

In relation to the predictability-unpredictability principle, if there is predictability (e.g,. the predictable tendency of each cultural variant to play chess in accordance to its own rules at a given point in time), there is unpredictability (e.g. the more difficult task to predict exactly what the future rules of chess for each cultural variant will be in the long term, because all these rules undergo changes over time). And the reverse direction also holds true.

In relation to the explicability-inexplicability principle, if there is explicability (e.g., the explanation, by chess historians, of the development of modern chess from early India on the basis of its historical origin in *ashtāpada* [Sanskrit for 'having eight feet,' i.e., an 8x8 squared board]"), there is inexplicability (e.g., the lack of sufficient explanation of why modern chess had to begin like that, instead of other alternative origins like "the 10×10 *Dasapada* and the 9×9 *Saturankam*"). And the reverse direction also holds true.

In relation to the preciseness-vagueness principle, if there is preciseness (e.g., the precise identification of the three possibilities of the outcome in a chess game, in the early Indian history of chess development, that is, by "checkmate," by "stalemate," or by "bare king" ["taking all of an opponent's pieces except the king"]), there is vagueness (e.g., the vagueness in this identification, since it is not clear how "forefeiture" can fit in any of them or constitutes something else instead). And the reverse direction also holds true.

In relation to the simpleness-complicatedness principle, if there is simpleness (e.g., the relatively simple understanding of chess rules from the standpoint of each cultural variant), there is complicatedness (e.g., the relatively more complicated understanding of chess rules by considering all cultural variants, many of which are not necessarily compatible with each other). And the reverse direction also holds true.

In relation to the openness-hiddenness principle, if there is openness (e.g., the open exploration of the relationship between chess rules and cultural variants, like the ban of using dice in the later stage of Indian chess development), there is hiddenness (e.g., the hidden bias in the Indian ban of using dice in chess, because of "Hindu religious objections", whereas in some variants of Chinese chess, using a dice, for instance, is allowed). And the reverse direction also holds true.

In relation to the denseness-emptiness principle, if there is denseness (e.g., the relatively denser concentration of chess players in Western chess to play "in the squares"), there is emptiness (e.g., the relatively less dense, or more empty, concentration of chess players in Western chess to play "on the intersections of the lines on the board"). And the reverse direction also holds true.

In relation to the change-constancy principle, if there is change (e.g., the ever new rules for chess over time), there is constancy (e.g., the ever constancy of the existence of certain rules in chess). And the reverse direction also holds true.

In relation to the slowness-quickness principle, if there is slowness (e.g., some players tend to play very slowly, like the example that "one player took as much as two hours and 20 minutes to think over a single move at the London tournament" in 1851), there is quickness (e.g., the relatively quicker chess games for chess players nowadays, like the introduction of "speed chess" for "a bank of time to each player in which to play a previously agreed number of moves, e.g. two hours for 30 moves"). And the reverse direction also holds true.

In relation to the expansion-contraction principle, if there is expansion (e.g, the relative expansion of the chess variant of the 8x8 *ashtāpada* to the rest of the world from its birthplace in India), there is contraction (e.g., the relative decline of the chess variants of "the 10×10 *Dasapada*" or "the 9×9 *Saturankam*" from its birthplace in India). And the reverse direction also holds true.

In relation to the convention-novelty principle, if there is convention (e.g., the conventional wisdom of playing chess "in the squares"), there is novelty (e.g., the alternative novel challenge to this conventional wisdom when chess was introduced to the Chinese, because "Chinese chess also borrows elements from the game of Go" and "is played on the intersections of the lines on the board, rather than in the squares"). And the reverse direction also holds true.

In relation to the evolution-transformation principle, if there is evolution (e.g., the natural evolution of humans to fight or compete with each other in the state of nature), there is transformation (e.g, the technical transformation of human fighting or competing with each other by the invention of chess games like "chaturaṅga, which translates as 'four divisions [of the military]': infantry, cavalry, elephantry, and chariotry, represented by the pieces that would evolve into the modern pawn, knight, bishop, and rook, respectively"). And the reverse direction also holds true.

In relation to the symmetry-asymmetry principle, if there is symmetry (e.g., the co-existence of different cultural variants of chess), there is asymmetry (e.g., the popularity of the Western variant of chess in the West—but the more acceptance of the Chinese variant of chess in mainland China). And the reverse direction also holds true.

In relation to the softness-hardness principle, if there is softness (e.g., the reward of "additional time budget" to a player "who made the predetermined number of moves in the agreed time"), there is hardness (e.g. the penalty of "forfeiture" to a player who exceeds the time limit). And the reverse direction also holds true.

In relation to the regression-progression principle, if there is regression (e.g, the regression made in the history of cultural variants, as shown in the emergence of new problems which later require new rules to fix them, like some players who took too much time to play in earlier times by taking advantage of the older rules), there is progression (e.g., the progress made in chess development, as shown by the new rules to fix old problems). And the reverse direction also holds true.

And in relation to the same-difference principle, if there is similairity in outcome (e.g., the contribution to the molding of thinking and behaving, regardless of whether this be achieved by the time limits on chess games for faster thinking or by no such time limits for slower thinking), there is difference in outcome (e.g., the contribution to the molding of thinking and behaving by way of the imposition of the time limits on chess games so as to encourage faster thinking nowadays—but the contribution to the molding of thinking and behaving by way of no imposition of any time limits on chess games so as to allow slower and more thorough thinking in the older days). And the reverse direction also holds true.

## The Uncertainty of Strategy

This comprehensive analysis of the possibility and desirability of strategy, from the perspectives of the mind, nature, society, and culture, has the scholarly virtue to show that strategy is both possible and desirable, but not to the extent that the spokespersons from each side would like us to believe. And this can apply too to tactics.

Strategy thus has its uncertainty, just as it has its foresight too, as another side of the same coin.

But this is only one part of a larger story, since the other side is tactics. Since this chapter already deals with strategy, the next chapter will therefore address the issue of tactics.

These double analyses of strategy and tactics are important, because "]t]hese two parts of chess thinking cannot be completely separated, because strategic goals are mostly achieved by the means of tactics, while the tactical opportunities are based on the previous strategy of play." (WK 2010)

With this in mind, let's now turn to Chapter Three for the study of tactics and its narrowness—for which we now turn to chapter three.

**Table 2.1. Strategy and Its Inadequacy**

- **Strategy and the Mind**
  —Ex: chess expertise, and psychological explanations
  —Ex: chess games, and the development of mental skills

- **Strategy and Nature**
  —Ex: chess origin, and its relationship with cosmology

- **Strategy and Society**
  —Ex: computer chess, and the impact of technology
  —Ex: chess games, and the influence of social structure

- **Strategy and Culture**
  —Ex: chess and morality
  —Ex: chess development, and cultural variations

*Notes*: The examples in the categories are solely illustrative (not exhaustive), and the comparison is relative (not absolute), nor are they necessarily mutually exclusive. And some can be easily re-classified elsewhere. As generalities, they allow exceptions.
*Source*: A summary of Ch.2 of *FPHCHESS*

**Table 2.2. The Debate on the Mental Benefits of Playing Chess**

- **The Pros**
  —Ex: improving memory
  —Ex: improving concentration
  —Ex: stoking imagination and creativity
  —Ex: developing logical thinking
  —Ex: building self-confidence
  —Ex: improving the ability to plan
  —Ex: cultivating patience (and will power)

- **The Cons**
  —Ex: the benefits remain to be conclusively proved.
  —Ex: the benefits fail to take into account of social and cultural factors.
  —Ex: the benefits ignore chess-related deaths and social problems.

*Notes*: The examples in the categories are solely illustrative (not exhaustive), and the comparison is relative (not absolute), nor are they necessarily mutually exclusive. And some can be easily re-classified elsewhere. As generalities, they allow exceptions.
*Source*: From *Sec. 2.2* of *FPHCHESS*. See book for citations, like CT (2009), WK (2010i), WK (2010k), WK (2010r), and others.

**Table 2.3. Strengths and Weaknesses of Chess Engines**

- **Strengths**
  —Ex: amazing processing speed
  —Ex: huge database in storage
  —Ex: remarkable software programs

- **Weaknesses**
  —Ex: different board representations
  —Ex: no best search strategies across the board
  —Ex: debatable value evaluations

*Notes*: The examples in the categories are solely illustrative (not exhaustive), and the comparison is relative (not absolute), nor are they necessarily mutually exclusive. And some can be easily re-classified elsewhere. As generalities, they allow exceptions.
*Source*: From *Sec. 2.4.1* of *FPHCHESS*. See book for citations, like WK (2010m), WK (2010n), WK (2010o), and others.

**Table 2.4. The Debate on the Moral Benefits of Playing Chess**

- **The Pros**
  - —Ex: foresight
  - —Ex: circumspection
  - —Ex: caution
  - —Ex: hope

- **The Cons**
  - —Ex: not everyone agrees with the benefits of chess playing.
  - —Ex: the claim does not consider the "dark" side of chess playing, as shown in chess-related deaths and corrupted lifestyles.
  - —Ex: the benefits have not yet been conclusively proved,
  - —Ex: the claim does not consider other dimensions of chess in the context of the mental, natural, and social arguments
  - —Ex: the claim also does not focus on other cultural aspects of chess (like the issue of aesthetics).
  - —Ex: it is not obivous at all that the four morals of chess (as claimed by Franklin) are really moral at all, as philosophers have debated for thousands of years what moral virtues are (or are not) and, in many cases, do not necessarily include the four (as mentioned by Franklin).

*Notes*: The examples in the categories are solely illustrative (not exhaustive), and the comparison is relative (not absolute), nor are they necessarily mutually exclusive. And some can be easily re-classified elsewhere. As generalities, they allow exceptions.
*Source*: From *Sec. 2.5.1* of *FPHCHESS*. See book for citations, like B. Franklin (2010), WK (2010i), and others.

**Table 2.5. Three Great Future Transformations of Mind Games**

- **Virtual Games**
  —Ex: online chess
  —Ex: virtual experience

- **Novel Games**
  —Ex: new chess variants
  —Ex: new chess engines

- **Post-Human Mind Games**
  —Ex: the quest for broader/deeper mental benefits of chess playing
  —Ex: games designed for the evolution of the mind into different bodyless forms (e.g., "floating consciousness," "hyper-spatial consciousness," "unfolding unconsciousness")

*Notes*: The examples in the categories are solely illustrative (not exhaustive), and the comparison is relative (not absolute), nor are they necessarily mutually exclusive. And some can be easily re-classified elsewhere. As generalities, they allow exceptions.
*Sources*: From *Sec. 4.16* of *FPHCHESS*. See text for more info. Also, consult *FCD* for strategy and tactics in warfare and *FPHO* for strategy and tactics in organization.

# • Part Three •

## *Tactics*

# CHAPTER 3
# *TACTICS AND ITS NARROWNESS*

When you see a good move, wait—
look for a better one.
—Max Emanuel (WK 2010g)

## The Usefulness of Tactics

In chess, "[t]actics are usually contrasted with strategy, in which advantages take longer to be realized, and the opponent is less constrained in responding." (WK 2010b) The same concept exists in military studies, in that "tactics…are concerned with the conduct of an engagement, while strategy is concerned with how different engagements are linked." (WK 2010)

In this sense, chess playing involves both strategy and tactics, as "[t]hese two parts of chess thinking cannot be completely separated, because strategic goals are mostly achieved by the means of tactics, while the tactical opportunities are based on the previous strategy of play." (WK 2010)

Since the issue of strategy was already addressed in Chapter Two, this chapter here will consider the one of tactics.

However, it should be stressed—once more, as it was already so qualified in Chapter Two—that an examination of tactics cannot be completely separated from that of strategy.

Thus, this book takes the scholarly challenge to analyze tactics in this chapter, with the understanding, however, that one cannot analyze tactics (or strategy, for that matter) separately, so many of the issues in this chapter overlap with those of strategy.

With this reminder in mind—a good way to analyze tactics and its narrowness (in this chapter) is by way of the evaluation of the extent to which tactics are in fact both possible and desirabe.

This chapter organizes a comprehensive analysis of tactics in the context of chess from the four perspectives of the mind, nature, society, and culture (in accordance to my sophisticated methodological holism, as explained in *Sec. 1.8*).

In other words, this chapter is organized in four main sections, in relation to (3.2) tactics and the mind, (3.3) tactics and nature, (3.4) tactics and society, and (3.5) tactics and culture—to be examined hereafter, in that order (and summarized in *Table 3.1*).

# Tactics and the Mind

Tactics, when examined from the perspective of the mind, can show us some amazing features in regard to its possibility and desirability, and this can be shown by way of two case studies, namely, (3.2.1) the game-tree complexity of chess, and the mind, and (3.2.2) chess blindness, and the appeal to reason —to be addressed in what follows, respectively.

## The Game-Tree Complexity of Chess, and the Mind

A good question to ask here is, To what extent can the mind calculate different possible moves and evaluate the resulting positions to find the best move in a chess game?

A realistic answer, as this will be explained below, is that there are limits to the ability of the mind to calculate different possible moves and evaluate the resulting positions to find the best move in a chess game.

### *Mathematics, and the Game Structue of Chess*

The issue about the mind to calculate different possible moves and evaluate the resulting positions to find the best move in a chess game has attracted the interest of mathematicians, especially in relation to the so-called "game-tree complexity of chess." (WK 2010)

The challenge of the "game-tree complexity of chess" is a research topic in "combinatorial game theory," which offers different ways to measure "game complexity." (WK 2010s)

Although there can be many different ways of measuring game complexity, five most often cited ones are shown below, for illustration. (WK 2010s)

- "State-Space Complexity"—refers to "the number of legal game positions reachable from the initial position of the game"; for instance, "an upper bound can often be computed by including illegal positions or positions that can never arise in the course of a game." (WK 2010s)

- "Game Tree Size"—refers to "the total number of possible games that can be played: it's the number of leaf nodes in the game tree rooted at the game's initial position. The game tree is typically vastly larger than the state space because the same positions can occur in many games by making moves in a different order....An upper bound for the size of the game tree can sometimes be computed by simplifying the game in a way that only increases the size of the game tree (for example, by allowing illegal moves) until it becomes tractable. However, for games where the number of moves is not limited (for example by the size of the board, or by a rule about repetition of position) the game tree is infinite." (WK 2010s)
- "Decision Complexity"—refers to one type of "decision tree," which is "a subtree of the game tree, with each position labelled with 'player A wins,' 'player B wins' or 'drawn,' if that position can be proved to have that value (assuming best play by both sides) by examining only other positions in the graph." More specifically, it refers to "the number of leaf nodes in the smallest decision tree that establishes the value of the initial position. Such a tree includes all possible decisions for the player to move second, but only one possibility for each decision for the player who starts the game." (WK 2010s)
- "Game-Tree Complexity"—refers to another type of "decision tree," which is about "the number of leaf nodes in the smallest full-width decision tree that establishes the value of the initial position. A full-width tree includes all nodes at each depth." (WK 2010s)
- "Computational complexity"—refers to "the asymptotic difficulty of a game as it grows arbitrarily large, expressed in big O notation or as membership in a complexity class. This concept doesn't apply to particular games, but rather to games that have been generalized so they can be made arbitrarily large, typically by playing them on an n-by-n board....From the point of view of computational complexity a game on a fixed size of board is a finite problem that can be solved in O(1), for example by a look-up table from positions to the best move in each position." (WK 2010s)

Surely, there can be other ways to measure game complexity, but the five examples aforecited suffice for the present illustration.

In any event, to understand the complexity for the mind to calculate different possible moves and evaluate the resulting positions to find the best move in a chess game, just consider some mind-boggling numbers below.

For instance, in one estimation,, "[t]he number of legal positions in chess is estimated to be between $10^{43}$ and $10^{47}$ (a provable upper bound), with a game-tree complexity of approximately $10^{123}$. The game-tree complexity of chess was first calculated by Claude Shannon as $10^{120}$, a number known as the Shannon number. Typically an average position has thirty to forty possible moves, but there may be as few as zero (in the case of checkmate or stalemate) or as many as 218." (WK 2010; J. Tromp 2010; E. Weisstein 2010; CB 2003)

These numbers are truly mind-boggling, for sure; but the point here is that they pose a formidable challenge to the mind in any attempt to calculate different possible moves and evaluate the resulting positions to find the best move in a chess game

*The Use of Computers for Calculation*

In light of this formidable challenge to the mind—it is thus no wonder, therefore, that computers have been increasingly used to calculate different possible moves and evaluate the resulting positions to find the best move in a chess game.

For instance, it has been quite for a long time that "[t]he most important mathematical challenge of chess is the development of algorithms which can play chess. The idea of creating a chess playing machine dates to the 18th century; around 1769, the chess playing automaton called The Turk became famous before being exposed as a hoax. Serious trials based on automatons, such as El Ajedrecista, were too complex and limited to be useful." (WK 2010; G. Levitt 2000)

This limitation in early automatons had existed until the 1950s, when "the advent of the digital computer" brought new promises, so "chess enthusiasts and computer engineers have built, with increasing degrees of seriousness and success, chess-playing machines and computer programs. The groundbreaking paper on computer chess, 'Programming a Computer for Playing Chess,' was published in 1950 by [Claude] Shannon." (WK 2010)

For Shannon (1950), the use of computers is useful for this task of calculation for four major reasons, as summarized below (and also in *Table 3.2*): (WK 2010)

- The first advantage of computers is that "the problem is sharply defined both in allowed operations (the moves) and in the ultimate goal (checkmate)."

- The second advantage of computers is that "it is neither so simple as to be trivial nor too difficult for satisfactory solution."
- The third advantage of computers is that "chess is generally considered to require 'thinking' for skillful play; a solution of this problem will force us either to admit the possibility of a mechanized thinking or to further restrict our concept of 'thinking'"
- The fourth advantage of computers is that "the discrete structure of chess fits well into the digital nature of modern computers."

The progress which has been made so far in chess machines cannot be dismissed.

For instance, "CHESS 3.0, a chess program from Northwestern University, won the championship. Nowadays chess programs compete in the World Computer Chess Championship, held annually since 1974. At first considered only a curiosity, the best chess playing programs, for example Rybka, have become extremely strong. In 1997 a computer won a match against a reigning World Champion for the first time: IBM's Deep Blue beat Garry Kasparov 3½–2½ (it scored two wins, one loss and three draws). In 2009, a mobile phone won a category 6 tournament with a performance rating 2898: chess engine Hiarcs 13 running on the mobile phone HTC Touch HD won the Copa Mercosur tournament with 9 wins and 1 draw." (WK 2010; F. Hsu 2002; M. Crowther 2009)

*Computational Enhancements for High Performance*

What lies behind the success of using computers to beat human chess champions is more than just the speed and processing power of the chess engines (or the "hardware" of chess engines).

In addition, some computational enhancements for high performance were used, and good examples include "minimax, alpha-beta pruning, killer heuristic, iterative deepening depth-first search, negascout, MTD-f, SSS, null-move heuristic and late move reductions." (WK 2010) These belong to the "software" of chess engines.

(a) Minimax

For instance, in the case of using "minimax"—the focus here concerns "a decision rule used in decision theory, game theory, statistics and philosophy for minimizing the possible loss while maximizing the potential gain. Alternatively, it can be thought of as maximizing the minimum gain (maximin)." (WK 2010t)

Thus, a minimax program, procedure, or "algorithm" is "a recursive algorithm for choosing the next move in an n-player game, usually a two-player game. A value is associated with each position or state of the game. This value is computed by means of a position evaluation function and it indicates how good it would be for a player to reach that position. The player then makes the move that maximizes the minimum value of the position resulting from the opponent's possible following moves. If it is A's turn to move, A gives a value to each of his legal moves." (WK 2010t)

(b) Alpha-Beta Pruning and Killer Heuristic

In the case of using "alpha-beta pruning" and "killer heuristic"—the focus here concerns "a search algorithm which seeks to reduce the number of nodes that are evaluated in the search tree by the minimax algorithm….Further improvement can be achieved without sacrificing accuracy, by using ordering heuristics to search parts of the tree that are likely to force alpha-beta cutoffs early." (WK 2010u)

For instance, "in chess, moves that take pieces may be examined before moves that do not, or moves that have scored highly in earlier passes through the game-tree analysis may be evaluated before others." (WK 2010u)

(c) NegaScout

In the case of using "NegaScout or Principal Variation Search"—the focus here is like (and also unlike) "alpha-beta prunning." (WK 2010v)

On the one hand, NegaScout is like alpha-beta pruning, because it is "a directional search algorithm for computing the minimax value of a node in a tree." (WK 2010v)

On the other hand, NegaScout is unlike alpha-beta pruning, because "it relies on accurate move ordering to capitalize on this advantage, and therefore, it "is a negamax algorithm that can be faster than alpha-beta pruning." (WK 2010v)

(d) Null-Move Heuristic

In the case of using "the null-move heuristic"—the focus here is "a heuristic technique used to enhance the speed of the alpha-beta pruning algorithm." (WK 2010w)

The reason is that "alpha-beta pruning speeds the minimax algorithm by identifying cutoffs, points in the game tree where the current position is so good for the side to move that best play by the other side would have avoided it. Since such positions could not have resulted from best play, they and all branches of the game tree stemming from them can be ignored. The faster the program produces cutoffs, the faster the search runs." (WK 2010w)

So, this is where the null-move heuristic is important, because it is "designed to guess cutoffs with less effort than would otherwise be required, whilst retaining a reasonable level of accuracy. The null-move heuristic is based on the fact that most reasonable chess moves improve the position for the side that played them. So, if the player whose turn it is to move can forfeit the right to move (or make a 'null move'—an illegal action in chess) and still have a position strong enough to produce a cutoff, then the current position would almost certainly produce a cutoff if the current player actually moved." (WK 2010w)

(e) Late Move Reductions

And in the case of using "late move reductions (LMR)"—the focus here concerns "a non-game specific enhancement to the alpha-beta algorithm and its variants which attempts to examine a game search tree more efficiently." (WK 2010x)

It does so by making "the assumption that good game-specific move ordering causes a program to search the most likely moves early. If a cut-off is going to happen in a search, the first few moves are the ones most likely to cause them. In games like chess, most programs search winning captures and 'killers' first. LMR will reduce the search depth for moves searched later at a given node. This allows the program to search deeper along the critical lines, and play better. Most chess programs will search the first several moves at a node to full depth." (WK 2010x)

*The Limits to Computation*

But there are limits to computation in resolving the problem of chess complexity. Consider a few of them below, for illustration—as summarized in *Table 3.2*.

Firstly, one limit to computation is that its four advantages (as cited by Shannon) also constitute its disadvantages, in that computer problems have to be "sharply defined," "not too difficult for satisfactory solution," "mechanized" in thinking, and the like. Not all difficult chess problems meet these criteria.

Secondly, another limit to computation is that it still does not do away the factor of uncertainty in chess, since "the possibilities are so numerous that no one can foresee how the game will move" exactly, which then requires human imagination and creativity (as already addressed in *Sec. 2.2.2*). (CT 2009)

Thirdly, still another limit to computation is that it is subject to "the problem of Bremermann's fundamental limit" as described in my book *The Future of Complexity* (*FC*), in that "H. Bremermann (1977) showed some fundamental constraints on any further improvement of computation,...to the extent that 'there exists a fundamental limit for the speed and efficiency of computing machines which cannot be overcome. Such a fundamental limit stems from the idea that the maximum speed of signal traveling between the inner components of the computing machine is constrained by the speed of light, i.e., 3.108 m/second. The time-lag of signal traveling is determined by the distance between the computing machine inner devices'...(J. Teixeira 1998)."

In addition, "the time-lag is 'in turn constrained by the so-called commutation time. Commutation time is the time-interval involved in processing information (signals) through discrete devices....Even if we supposed the (technological) possibility of building a quite small computer and minimizing/optimizing the trajectory of signal-traveling, such a fundamental limit cannot be overcome [since]...there would remain problems whose complexity can be said to be transcomputable.' (J. Teixeira 1998)"

And fourthly, an additional limit to computation is that it is subject to "the problem of intractability" as already described in my book *The Future of Complexity* (*FC*), in that, "in computational complexity theory,...[p]roblems can be classified by complexity class according to the time it takes for an algorithm to solve them as function of the problem size....Even though a problem may be solvable computationally in principle, but in actual practice it may not be that simple. These problems might require large amounts of time or an inordinate amount of space". (WK 2006 & 2006a)

This then means that "[t]here exist a certain class of problems that although they are solvable in principle they require so much time or space that it is not practical to attempt to solve them. These problems are called Intractable." (WK 2006)

*A More Balanced View*

However, these criticisms of computation should not be misunderstood as a rejection of the usefulness of computers for the better understanding of chess complexity.

Instead, they are shown here so as to reveal the two opposing sides of the debate, such that the possibility and desirability of tactics (from the perspective of the mind with chess complexity as a case study here) are not to the extent that its respective defenders would like us to believe.

In addition, this analysis of chess complexity can reveal something refreshing about the ontological principles in existential dialectics, and good examples include the absoluteness-relativeness principle, the partiality-totality principle, the predictability-unpredictability principle, the explicability-inexplicability principle, the finiteness-transfiniteness principle, the preciseness-vagueness principle, the simpleness-complicatedness principle, the openness-hiddenness principle, the denseness-emptiness principle, the change-constancy principle, the order-chaos principle, the slowness-quickness principle, the expansion-contraction principle, the theory-praxis principle, the convention-novelty principle, the evolution-transformation principle, the symmetry-asymmetry principle, the regression-progression principle, and the same-difference principle.

For instance, in relation to the absoluteness-relativeness principle, if there is absoluteness (e.g., the absolute view of those "chess enthusiasts and computer engineers" like Shanon, who wrote the groundbreaking paper on "programming a computer for playing chess" in 1950), there is relativeness (e.g., what is possible and desirable for Shannon in regard to chess engines is not necessarily so for the critics like Bremermann and others in regard to the fundamental limits to computation). And the reverse direction also holds true.

In relation to the partiality-totality principle, if there is partiality (e.g., the partial view of each player to calculate different possible moves and evaluate the resulting positions to find the best move in a chess game), there is totality (e.g., the holistic view of the entire picture, such that "the possibilities are so numerous that no one can foresee how the game will move" exactly, so that the whole is not the sum of its parts, that is, not the sum of the partial views of the players in a chess game). And the reverse direction also holds true.

In relation to the predictability-unpredictability principle, if there is predictability (e.g., the predictable tendency of those who use the technique of NegaScout to rely "on accurate move ordering to capitalize on this advantage" and therefore to "be faster than alpha-beta pruning" in calculation), there is unpredictability (e.g., the more difficult task to predict exactly how a game will evolve over time, since "the possibilities are so numerous," even with the use of NegaScout). And the reverse direction also holds true.

In relation to the explicability-inexplicability principle, if there is explicability (e.g., the explanation, by Shannon, of the superiority of using chess engines to play, on the basis of some distinctive advantages), there is inexplicability (e.g., the lack of sufficient explanation, by Shannon, of why these advantages will not turn out to be disadvantages too for chess engines, such that there are limits to computation). And the reverse direction also holds true.

In relation to the finiteness-transfiniteness principle, if there is finiteness (e.g., the finite number of squares on a chess board), there is transfiniteness (e.g., the transfinite number of "legal positions in chess…to be between $10^{43}$ and $10^{47}$, with a game-tree complexity of approximately $10^{123}$", for instance). And the reverse direction also holds true.

In relation to the preciseness-vagueness principle, if there is preciseness (e.g., the precise identification of different computational enhancements for high performance, like "minimax," "killer heuristic," etc.), there is vagueness (e.g., the vagueness in the identification of different computational enhancements, since different people have different views on how they can be achieved, like different ways of "minimizing the possible loss while maximizing the potential gain"). And the reverse direction also holds true.

In relation to the simpleness-complicatedness principle, if there is simpleness (e.g., the relatively simple analysis of the advantages of using computers for the understanding of chess complexity), there is complicatedness (e.g., the relatively more complicated analysis of the power of computation for the understanding of chess complexity, as shown in the criticisms on its limits by the critics). And the reverse direction also holds true.

In relation to the openness-hiddenness principle, if there is openness (e.g., the open exploration of the use of computation for the understanding of chess complexity), there is hiddenness (e.g., the hidden bias on the enthusiastic fad of using computation for the understanding of chess complexity, as shown in its neglect of considering some fundamental problems like "Bremermann's fundamental limit" and "the problem of intractability"). And the reverse direction also holds true.

In relation to the denseness-emptiness principle, if there is densess (e.g., the relatively denser concentration of concerns with computational enhancements in those who are enthusiastic supporters of the use of computers for chess games), there is emptiness (e.g., the relatively less dense, or more empty, concentration of concerns with individual subjectivity and creativity in those who are enthusiastic supporters of the use of computers for chess games). And the reverse direction also holds true.

In relation to the change-constancy principle, if there is change (e.g., the ever new innovations of computational enhancements for high performance), there is contancy (e.g., the continued existence of limits to computation, regardless of the innovations). And the reverse direction also holds true.

In relation to the order-chaos principle, if there is order (e.g., the relatively orderly nature of chess games, like the use of computation to calculate different possible moves and evaluate the resulting positions to find the best move in a chess game), there is chaos (e.g,. the relativley chaotic nature of chess games, because "the possibilities are so numerous that no one can foresee how the game will move" exactly). And the reverse direction also holds true.

In relation to the slowness-quickness principle, if there is slowness (e.g., the relatively slower speed of humans to calculate different possible moves and evaluate the resulting positions to find the best move in a chess game), there is quickness (e.g., the relatively faster speed of supercomputers to calculate different possible moves and evaluate the resulting positions to find the best move in a chess game). And the reverse direction also holds true.

In relation to the expansion-contraction principle, if there is expansion (e.g., the relative expansion of the use of computers for the understanding of chess complexity in our time), there is contraction (e.g,.. the relative less reliance on the human mind to calculate chess complexity in our time). And the reverse direction also holds true.

In relation to the theory-praxis principle, if there is theory (e.g,. the theoretical construction of computer chess, as shown by Shannon's groundbreaking paper on computer chess in 1950), there is praxis (e.g., the practical application of Shannon's theoretical construction for the use of chess engines in chess tournaments, like the victory of IBM's Deep Blue over Garry Kasparov). And the reverse direction also holds true.

In relation to the convention-novelty principle, if there is convention (e.g., the conventional wisdom about chess to be played by humans), there is novelty (e.g.., the alternative novel challenge by Shannon to this conventional wisdom, in regard to his groundbreaking article on "Programming a Computer for Playing Chess" published in 1950). And the reverse direction also holds true.

In relation to the evolution-transformation principle, if there is evolution (e.g., the natural evolution of humans to be able to think in the state of nature), there is transformation (e.g., the technical transformation of human ability to think by the invention of supercomputers for computation). And the reverse direction also holds true.

In relation to the symmetry-asymmetry principle, if there is symmetry (e.g., the co-existence of different chess players, be they humans or machines), there is asymmetry (e.g., the popularity of playing chess with humans in the older days, including Kasparov, who was not happy to have lost to Deep Blue—but the popularity of allowing more and more chess engines in chess tournaments nowadays). And the reverse direction also holds true.

In relation to the regression-progression principle, if there is regression (e.g., the regression made in using computation for chess games, as shown in the criticisms by the critics), there is progression (e.g., the progress made in using computation for chess games, as shown by the victory of Deep Blue over Kasparov). And the reverse direction also holds true.

And in relation to the same-difference principle, if there is similarity in outcome (e.g., the contribution to the molding of a certain mode of thinking, regardless of whether this be achieved by computation in chess engines or by natural thinking in humans), there is difference in outcome (e.g., the contribution to the molding of a certain mode of thinking by way of computation in chess engines for a more mechanized thinking—but the contribution to the molding of a certain mode of thinking by way of natural thinking in humans for a more subjective, imaginative thinking). And the reverse direction also holds true.

## Chess Blindness, and the Appeal to Reason

If there are limits to computation for the calculation of different possible moves and the evaluation of the resulting positions to find the best move in a chess game (as discussed in the previous section), there are likewise limits to the mind.

A further illustration of the limits to the mind concerns "chess blindness" by human players in chess games, when they make some stupid mistakes in tactical moves.

### *Chess Blindness and the Kotov Syndrome*

In chess, there is the famous "Kotov syndrome," which refers to "a phenomenon first described in Alexander Kotov's 1971 book *Think Like a Grandmaster*. It occurs when a player thinks very hard for a long time in a complicated position but does not find a clear path. The player then notices they are running low on time, and so quickly make a move, often a blunder, and lose the game." (WK 2010y)

In fact, "many players have agreed that the process is very common. The syndrome has applications in many other deadline-oriented situations as well. For example, a journalist might puzzle over how to explain a complicated political process without finding a simple solution. Realizing that the deadline is approaching, they throw together a story and are disappointed the next morning to realize the story was a failure." (WK 2010y)

In cases like this, there is a temptation to appeal to sober reason, which works better, however, in ideal situations without any distortion due to fatigue, time pressure, overconfidence, carelessness, or else.

*Well-Known Cases of Chess Blindness*

In general, "a blunder [in chess] is a very bad move. It is usually caused by some oversight, whether from time pressure…, overconfidence or carelessness. While a blunder may seem like a stroke of luck for the opposing player, some chess players give their opponent plenty of opportunities to blunder." (WK 2010z)

However, what counts as a "blunder" can be a bit "subjective," since it may vary from case to case. For instance, "a weak move from a novice player might be explained by the player's lack of skill, while the same move from a master might be called a blunder. In chess annotation, blunders are typically marked with a double question mark, '??', after the move." (WK 2010z)

This double question mark "??" can be contrasted with other marks; for instance, "! indicates a good move, !! an excellent move, ? a mistake, ?? a blunder, !? an interesting move that may not be best, or ?! a dubious move, but not easily refuted." (WK 2010)

Nowadays, there is even the triple question mark "???" for the "blunder of the century," after a blunder by Karmnik was awarded "???" by ChessBase on a story covering Kramnik's loss to Deep Fritz in 2006 (as will be explained below).

For illustration, consider the following examples of making chess blunders, which also show that "[s]trong players, even grandmasters, occasionally make elementary blunders"—although lesser players can make the blunders more often. (WK 2010z)

(a) Esteban Canal vs. N.N. in 1934

The first example occurred in 1934, when Esteban Canal (known as "the Peruvian Immortal") played against an unknown amateur in Budapest, Hungary in 1934.

The blunder occurred, when, "after Canal, playing White, played 10. a3, N.N., mistakenly thought that White's 11th possible move, axb4, was out of question. Therefore, N.N. played 10. O-O-O?? [a castling with the queenside rook]. On White's 11th move, Canal did play axb4!!, followed by...11.Qxa1 12.Kd2! Qxh1 13.Qxc6+! bxc6 14.Ba6#. Canal's win finished with a Boden's Mate [a checkmate with the use of bishops]." (WK 2010z)

Some years later, "Iakov Neishtadt wrote that on Black's 10th move, N.N. should've played Nf6, not O-O-O. Yasser Seirawan and Nikolay Minev advised, 'Motto: Think twice before castling on the Queenside!'" (WK 2010z)

(b) Tigran Petrosian vs. David Bronstein in 1956

The second example occurred in 1956, when Tigran Petrosian played against David Bronstein in the Candidates Tournament in Amsterdam.

The blunder by Petrosian was quite unexpected in this game, because at the beginning, "Petrosian, playing White, enjoys a clear advantage with strong knights, active rooks and plenty of mobility while Black's position is congested and hardly able to move. In fact Bronstein, playing Black, has for the last seven moves been making only apparently aimless knight moves, Nc6-d4-c6-d4, and now has played…Nd4-f5, threatening White's queen, while White has been slowly strengthening his position." (WK 2010z)

But then, out of carelessness, Petrosian committed the blunder, because "White can…easily preserve the positional advantage by a move like 36.Qc7, but overlooking that the queen was en prise [undefended], he played 36.Ng5?? and resigned after 36...Nxd6." (WK 2010z)

(c) Viktor Korchnoi vs. Anatoly Karpov in 1978

The third example occurred in 1978, when Viktor Korchnoi played against Anatoly Karpov in 1978 for the World Championship match in Baguio, Philippines.

The blunder by Korchnoi occurred, when "Karpov, playing Black, is hoping for a back rank mate with his rook with the move 39. Rc1#. However, White can prevent this by moving his g- or h-pawn, providing an escape square for his king. Instead, Korchnoi blundered. He played 39. Ra1??, which allowed a nice finish for Karpov, 39. ...Nf3!+, and Korchnoi resigned the game. Otherwise, the game would end as a result of the blunder as 40. gxf3 Rg6+ 41. Kh1 Nf2#. As a result, Karpov kept his chess crown until he was defeated by Garry Kasparov in 1985." (WK 2010z)

(d) Murray Chandler vs. Susan Polgar in 1987

The fourth example occurred in 1987, when Murray Chandler played against Susan Polgar in a tournament in Biel, Switzerland. What is interesting in this case is that Chandler committed the blunder but "did not lose," although he suffered from "an embarrassing draw." (WK 2010z)

The situation began later in the game, when "Susan Polgar has just played the wily trap 53...Ng8-h6!?, looking for the forced draw. Chandler, though, realizes that after 54.gxh6+ Kxh6 he will be left with the considerable material advantage of a rook pawn and bishop against a bare king. However, since the bishop is unable to control the promotion square h8, Black will draw if she is able to get her king to control h8 due to the wrong rook pawn fortress. But Chandler calculates further and realizes that it is he who will win control over the h8 square after 55.Kf6, and thereby win the game." (WK 2010z; A. Soltis 1997)

With this temptation, "Chandler played 54.gxh6+??, but instead of the expected 54...Kxh6 came 54...Kh8! This is in fact almost the same king, bishop, and rook pawn versus bare king situation as Chandler had calculated that he would avoid, and the small difference that White has two rook pawns rather than one has no effect on the result. Black controls the h8 square and cannot be chased or squeezed away from it, and so White cannot promote his pawn. After 55.Bd5 Kh7 56.Kf7 Kh8 the players agreed to a draw." (WK 2010z)

After the game, some chess experts argued that "Chandler had numerous moves which would have maintained his winning position, the fastest according to the Shredder tablebase are 54.h4 and 54.Bf5." (WK 2010z)

(e) Alexander Beliavsky vs. Leif Erlend Johannessen in 2002

The fifth example occurred in 2002, when Alexander Beliavsky played Leif Erlend Johannessens in Linares, Spain.

The blunder by Bliavsky in 2002 is "one of the very rare circumstances where a grandmaster makes the literally worst move on the board, the only one allowing checkmate on the next move. This is a balanced queen endgame, but Beliavsky playing White is careless. After 69.Kf4?? he had overlooked the response 69...Qb8#." (WK 2010z)

(f) Deep Fritz vs. Vladimir Kramnik in 2006

The sixth example occurred in 2006, when the chess engine Deep Fritz played Vladimir Kramnik in Bonn, Germany.

What is interesting about this case is that the blunder by Kramnik is regarded by some as the "blunder of the century." (WK 2010z)

The blunder was also unexpected, because, "[a]fter the first game had ended in a draw, Kramnik, playing Black, was generally considered in a comfortable position in Game 2, and he thought so himself apparently, as he refused a draw by avoiding a potential threefold repetition on 29...Qa7." (WK 2010z)

Thus, "Kramnik's troubles began when he decided to play for a win and pushed his a-pawn, 31...a4. Commentators, including American grandmaster Yasser Seirawan, voiced concerns about Kramnik's intentions and the situation became more uncertain as the game went on with

32.Nxe6 Bxe3+ 33.Kh1 Bxc1 34.Nxf8, turning it into a likely draw. The game could have ended with 34...Kg8 35.Ng6 Bxb2 36.Qd5+ Kh7 37.Nf8+ Kh8 38.Ng6+." (WK 2010z)

Unfortunately, "Kramnik's next move, 34...Qe3?? (a move which was awarded '???' originally, by ChessBase on a story covering Kramnik's blunder), came as a big surprise and was described as the possible 'blunder of the century' by Susan Polgar [2006], as Kramnik—incredibly—overlooked a mate in one. Deep Fritz immediately ended the game with 35.Qh7#, checkmate. Seirawan later called Kramnik's move 'a tragedy.'" (WK 2010z)

After the game, "*ChessBase* described the events as follows, 'Kramnik played the move 34...Qe3 calmly, stood up, picked up his cup and was about to leave the stage to go to his rest room. At least one audio commentator also noticed nothing, while Fritz operator Mathias Feist kept glancing from the board to the screen and back, hardly able to believe that he had input the correct move. Fritz was displaying mate in one, and when Mathias executed it on the board Kramnik briefly grasped his forehead, took a seat to sign the score sheet and left for the press conference,' during which he stated that he had planned the supposedly winning move 34...Qe3 already when playing 29...Qa7, and had rechecked the line after each subsequent move. After an exchange of queens Black would win easily with his distant pawn; after 35.Qxb4 Qe2 or 35.Ng6+ Kh7 36.Nf8+ Kg8 black also wins eventually." (CBN 2006b)

Some chess analysts tried to explain why Kramnik could have made such a blunder; for instance, "Alexander Roshal attempted to explain the blunder by saying that the mating pattern of a queen on h7 protected by a knight on f8 is extremely rare and not contained in a grandmaster's automatic repertoire." (WK 2010z; CBN 2006c)

*A Lesson to Learn*

This analysis of the chess blunders committed by chess players—some of whom are the best and the brightest in their generations or even in the history of the game—should caution us about the limits of the mind to calculate different possible moves and evaluate the resulting positions to find the best move in a chess game.

Of course, the causes of chess blindness can be multiple, be they due to fatigue, time pressure, overconfidence, carelessness, or else—such that the possibility and desirability of tactics (from the perspective of the mind with chess blunders as a case study here) are not to the extent that its respective defenders would like us to believe.

Furthermore, this analysis of chess blunders can reveal, once more, the relevance of ontological principles in existential dialectics, and good examples include the absoluteness-relativeness principle, the partiality-totality principle, the predictability-unpredictability principle, the explicability-inexplicability principle, the preciseness-vagueness principle, the simpleness-complicatedness principle, the openness-hiddenness principle, the denseness-emptiness principle, the order-chaos principle, the slowness-quickness principle, the theory-praxis principle, the convention-novelty principle, the evolution-transformation principle, the symmetry-asymmetry principle, the softness-hardness principle, the regression-progression principle, and the same-difference principle.

For instance, in relation to the absoluteness-relativeness principle, if there is absoluteness (e.g., the absolute view by Kramnik that "he had planned the supposedly winning move 34...Qe3 already when playing 29...Qa7, and had rechecked the line after each subsequent move"), there is relativenes (e.g., what is right for Kramnik in regard to the move 34...Qe3??? is not so for Deep Fritz which countered it with move 35.Qh7#, checkmate). And the reverse direction also holds true.

In relation to the partiality-totality principle, if there is partiality (e.g., the partial view of each player to make the best move, trying to outsmart the other player), there is totality (e.g., the unexpected outcomes of the games as shown in the chess blunders, which surprised many after the matches, such that the whole is not the sum of its parts, that is, not the sum of the partial views of the players at the beginning of the games). And the reverse direction also holds true.

In relation to the predictability-unpredictability principle, if there is predictability (e.g., the predictable bet by chess fans at the beginning of the match in 1934 to favor Esteban Canal, also known as "the Peruvian Immortal," in his play against an unknown amateur), there is unpredictability (e.g., no one predicted that the game would end by a chess blunder when "N.N. played 10. O-O-O??"). And the reverse direction also holds true.

In relation to the explicability-inexplicability principle, if there is explicability (e.g., the explanation, by Alexander Roshal, of the blunder by Kramnik, by saying that Deep Fritz's "mating pattern of a queen on h7 protected by a knight on f8 is extremely rare"), there is inexplicability (e.g., the lack of sufficient explanation by Roshal of why Kramnik could not detect this move in the first place, even if it were rare, since he was such a brilliant grandmaster after all, as pointed out by the critics, who thought there was no excuse for Kramnik to make such a blunder). And the reverse direction also holds true.

In relation to the preciseness-vagueness principle, if there is preciseness (e.g., the precise identification of the tactical moves which are identified as chess blunders, like the move 34...Qe3??? by Kramnik), there is vagueness (e.g., the vagueness in the identification of the chess blunders, since what is regarded as a blunder for one chess analyst may be explained away as quite natural for another chess analyst, as shown by the response by Alexander Roshal, who tried to explain away the blunder by Kramnik by saying that "the mating pattern of a queen on h7 protected by a knight on f8 is extremely rare and not contained in a grandmaster's automatic repertoire"). And the reverse direction also holds true.

In relation to the simpleness-complicatedness principle, if there is simpleness (e.g., the relatively simple analysis of chess blindness on the basis of some common factors like time pressure, fatigue, carelessness, etc.), there is complicatedness (e.g., the relatively complicated analysis of chess blindness by questioning the common causes, as shown in the view by Alexander Roshal, who tried to explain away the blunder by Kramnik by attributing his opponent's move to being "extremely rare and not contained in a grandmaster's automatic repertoire," so Kramnik could not have detected the problem anyway). And the reverse direction also holds true.

In relation to the openness-hiddenness principle, if there is openness (e.g., the move 54.gxh6+ made by Chandler, in broad daylight), there is hiddeness (e.g., the hidden motive by Chandler to make this move 54.gxh6+??, because he calculated, though incorrectly, that, "after 54.gxh6+ Kxh6 he will be left with the considerable material advantage of a rook pawn and bishop against a bare king" and calculated further, also incorrectly, that "he…will win control over the h8 square after 55.Kf6, and thereby win the game," but to his surprise, "instead of the expected 54...Kxh6 came 54...Kh8!"). And the reverse direction also holds true.

In relation to the denseness-emptiness principle, if there is denseness (e.g., the relatively denser concentration of brilliant moves made by grandmasters), there is emptiness (e.g., the relatively less dense, or more empty, concentration of chess blunders made by grandmasters, although they may happen sometimes, as shown in the examples). And the reverse direction also holds true.

In relation to the order-chaos principle, if there is order (e.g., the relatively orderly nature of chess games, because they are governed with different rules), there is chaos (e.g., the relatively chaotic outcomes of chess games, especially when unexpected chess blunders occurred). And the reverse direction also holds true.

In relation to the slowness-quickness principle, if there is slowness (e.g., the relatively slower speed of a chess game to end, if no chess blunder occurs, in general), there is quickness (e.g,. the relatively faster speed of a chess game to end, when chess blunders occur, in general). And the reverse direction also holds true.

In relation to the theory-praxis principle, if there is theory (e.g., the theoretical construction of the "Kotov syndrome" by Alexander Kotov), there is praxis (e.g., the practical application of the "Kotov syndrome…in many other deadline-oriented situations as well," as shown in the example that, when "a journalist might puzzle over how to explain a complicated political process without finding a simple solution" and, as "the deadline is approaching, they throw together a story and are disappointed the next morning to realize the story was a failure"). And the reverse direction also holds true.

In relation to the convention-novelty principle, if there is convention (e.g., the conventional wisdom about playing chess, in accordance to certain rules, with different mating patterns that grandmasters know all too well), there is novelty (e.g., the alternative novel challenge to the conventional wisdom about mating patterns in chess, when Deep Fritz made "the mating pattern of a queen on h7 protected by a knight on f8," which "is extremely rare and not contained in a grandmaster's automatic repertoire"). And the reverse direction also holds true.

In relation to the evolution-transformation principle, if there is evolution (e.g., the natural evolution of humans to have fun with each other in the state of nature), there is transformation (e.g., the technical transformation of having fun by the invention of chess engines like Deep Fritz, which could make "extremely rare…mating patterns" in chess games). And the reverse direction also holds true.

In relation to the symmetry-asymmetry principle, if there is symmetry (e.g., the co-existence of different grandmasters in chess over the centuries), there is asymmetry (e.g., some grandmasters make more chess blunders than some others). And the reverse direction also holds true.

In relation to the softness-hardness principle, there is softness (e.g., the winning in a chess game), there is hardness (e.g., the losing in a chess game). And the reverse direction also holds true.

In relation to the regression-progression principle, if there is regression (e.g., the regression made in the history of chess games, as shown by the continued occurrence of new chess blunders over time), there is progression (e.g., the progress made in the history of chess games for the better understanding of how chess blunders occur, as shown in the examples). And the reverse direction also holds true.

And in relation to the same-difference principle, if there is similarity in outcome (e.g., the contribution to the control and regulation of beliefs and values, regardless of whether this be achieved by way of discouraging chess blunders or by way of forgiving carelessness), there is difference in outcome (e.g., the contribution to the control and regulation of beliefs and values by way of discouraging chess blunders, so that the player in question loses the match or has a draw—but the contribution to the control and regulation of beliefs and values by way of forgiving carelessness, so that the player in question can be given sympathy with the excuse, for example, that his opponent made "extremely rare" tactical moves, which are "not contained in a grandmaster's automatic repertoire"). And the reverse direction also holds true.

# Tactics and Nature

Tactics, when examined from the perspective of nature, can cast a different light on its possibility and desirability, and this can be shown by way of two case studies, namely, (3.3.1) chess, symmetry, and mathematics, and (3.3.2) chess, analogy, and physics—to be addressed hereafter, in that order.

## Chess, Symmetry, and Mathematics

A natural phenomenon in chess which has much appeal to mathematicians concerns symmetry in a series of tactical moves.

It is thus no wonder that, in this sense, "[c]hess and mathematics have been pursued intellectually for centuries by many researchers and scientists, especially mathematicians. Naturally, the logic and symmetry in chess appeal to mathematicians." (WK 2010aa)

For instance, Leonhard Euler, Adrien-Marie Legendre, Abraham de Moivre, and Alexandre-Theophile Vandermonde had been fascinated with the nature of symmetry in the famous problem of "the knight's tour" in chess. (WK 2010aa)

### *The Knight's Tour: Open or Closed*

The chess problem known as "the Knight's Tour" refers to "a mathematical problem involving a knight on a chessboard. The knight is placed on the empty board and, moving according to the rules of chess, must visit each square exactly once." (WK 2010bb)

The knight's tour can be open or closed. On the one hand, "a knight's tour is called a closed tour if the knight ends on a square attacking the square from which it began (so that it may tour the board again immediately with the same path)." (WK 2010bb)

If not, then the knight's tour is "open," but "[t]he exact number of open tours is still unknown. Creating a program to solve the knight's tour is a common problem given to computer science students. Variations of the knight's tour problem involve chessboards of different sizes than the usual 8 × 8, as well as irregular (non-rectangular) boards." (WK 2010bb; H. Deitel 2003)

*A Brief History of the Knight's Tour*

Theoretically speaking, "[t]he knight's tour problem is an instance of the more general Hamiltonian path problem in graph theory. The problem of finding a closed knight's tour is similarly an instance of the Hamiltonian cycle problem," except that, "unlike the general Hamiltonian path problem, the knight's tour problem can be solved in linear time," which is the time required to solve the problem in proportion to the input size. (WK 2010bb; A. Conrad 1994)

The knight's tour problem can be dated "back to the 9th century CE. The pattern of a knight's tour on a half-board has been presented in verse form (as a literary constraint) in the highly stylized Sanskrit poem *Kavyalankara* written by the 9th century Kashmiri poet Rudrata, which discusses the art of poetry, especially with relation to theater (Natyashastra). As was often the practice in ornate Sanskrit poetry, the syllabic patterns of this poem elucidate a completely different motif, in this case an open knight's tour on a half-chessboard." (WK 2010bb)

Some well-known mathematicans had addressed the problem; for instance, "[o]ne of the first mathematicians to investigate the knight's tour was Leonhard Euler. The first algorithm for completing the Knight's Tour was Warnsdorff's algorithm, first described in 1823 by H. C. Warnsdorff." (WK 2010bb)

Later, "in the 20th century the Oulipo group of writers used it among many others. The most notable example is the 10 × 10 Knight's Tour which sets the order of the chapters in Georges Perec's novel *Life: A User's Manual*. The sixth game of the 2010 chess worldchampionship between Anand and Topalov in Sofia, saw Anand making 13 consecutive knight moves—online commentors jested that Anand was trying to solve the Knight's Tour problem during the game." (WK 2010bb)

*A Mathematical Illustration*

For illustration, consider a category of knight's tours, that is, closed knight's tours.

It has been estimated that, "[o]n an 8 × 8 board, there are exactly 26,534,728,821,064 directed closed tours (i.e., two tours along the same path that travel in opposite directions are counted separately). The number of undirected closed tours is half this number, since every tour can be traced in reverse." (WK 2010bb; M. Loebbing 1996; B. McKay 1997; I. Wegener 2000)

However, on a smaller chess board, like "the a 6 × 6 board," the number is smaller too, as "[t]here are 9,862 undirected closed tours." (WK 2010bb; E. Weisstein 2010)

But "[n]o closed tours exist for smaller square boards," and the reason has to do with "Schwenk's Theorem." (WK 2010bb)

In accordance to Schwenk's Theorem, "[f]or any $m \times n$ board with $m$ less than or equal to $n$, a closed knight's tour is always possible unless one or more of these three conditions are true:

- "$m$ and $n$ are both odd; $m$ and $n$ are not both 1"
- "$m$ = 1, 2, or 4; $m$ and $n$ are not both 1"
- "$m$ = 3 and $n$ = 4, 6, or 8"

In other words, these three conditions do not allow closed knight's tours, and they can be explained in what follows.

(a) Condition 1

In condition 1, the prohibition of "closed tours" is "by a simple argument based on parity and coloring. For the standard black-and-white coloring of the chessboard, the knight must move either from a black square to a white square or from a white square to a black square. Thus in a closed tour the knight must visit the same number of white squares as black squares, and the number of squares visited in total must therefore be even. But if $m$ and $n$ are both odd, the total number of squares is odd. (For example, in a 5 × 5 checkerboard there are 13 of one color and 12 of the other.) Therefore closed tours do not exist. Open tours may still exist (with the exception of the trivial case 1 × 1)." (WK 2010bb)

(b) Condition 2

In condition 2, the prohibition of "closed tours" states that "when the shorter side is of length 1, 2, or 4, no closed tour is possible. When $m$ = 1 or 2, no closed tour is possible because the knight would not be able to reach every square (once again, with the exception of the trivial case 1 ×

1). It can be shown that a $4 \times n$ board cannot have a closed tour either, by using a coloring argument." (WK 2010bb)

(c) Condition 3

In condition 3, the prohibition of "closed tours" can be proved just by doing the calculation. For insance, "attempting to construct a closed knight's tour on a 3 by 4, 3 by 6, or 3 by 8 will lead to definite failure. 3 by $n$ boards with $n$ even and greater than 8 can be shown to have closed tours by induction (a repeating pattern)." (WK 2010bb)

*Some Criticisms*

Yet, there are some serious problems with the Schwenk's Theorem. Consider a few of the criticisms below, for illustration.

Firstly, one major criticism of the knight's tour problem is that there is a major weakness in Schwenk's Theorem, in that "[s]mply proving the above three conditions does not prove the theorem; it is still required to prove that all rectangular boards that do not fall in one of the above three categories have closed knight's tours." (WK 2010bb; J. Watkins 2004)

And secondly, another major criticism of the knight's tour problem is that it is not clear if it has any practical use at all for actual players in chess games. It is one thing for mathematicians to be attracted to the natural beauty of symmetry in the knight's tour problem, but it is something else for chess players to apply the knight's tours for best tactical moves.

*The Larger Picture*

But these criticisms do not mean that one should totally ignore the natural beauty of symmetry in the knight's tour problem. Rather, the criticisms give us a larger picture of the possibility and desirability of tactics (from the perspective of nature with the knight's tour as a case study here), such that they are not to the extent that the respective defenders would like us to believe.

In addition, this analysis of the knight's tour can teach us something valuable about the ontological principles in existential dialectics, and good examples include the formalness-informalness principle, the absoluteness-relativeness principle, the predictability-unpredictability principle, the explicability-inexplicability principle, the finiteness-transfiniteness principle, the preciseness-vagueness principle, the simpleness-complicatedness principle, the openness-hiddenness principle, the denseness-emptiness principle, the slowness-quickness principle, the convention-novelty principle, the evolution-transformation principle, the symmetry-asymmetry principle, the regression-progression principle, and the same-difference principle.

For instance, in relation to the formalness-informalness principle, if there is formalness (e.g., the formal requirement of completeness and soundness in a system of ideas), there is informalness (e.g., the informal shortcoming of Schwenk's Theorem, in that it is proved in regard to the three conditions in a way, but in a diferent way, it is not, because "[s]mply proving the…three conditions does not prove the theorem," as "it is still required to prove that all rectangular boards that do not fall in one of the above three categories have closed knight's tours"). And the reverse direction also holds true.

In relation to the absoluteness-relativeness principle, if there is absoluteness (e.g., the absolute view of Schwenk's Theorem, in that "[f]or any $m \times n$ board with $m$ less than or equal to $n$, a closed knight's tour is always possible unless one or more of the…three conditions are true"), there is relativeness (e.g., what is true for Schwenk's Theorem in regard to the proof of the closed knight's tour problem is not necessarily so for the critics, who pointed out its weaknesses). And the reverse direction also holds true.

In relation to the predictability-unpredictability principle, if there is predictability (e.g,. the predictable feature of a knight's tour to be symmetric in a certain way), there is unpredictability (e.g., the more difficult task to predict which particular "rectangular boards that do not fall in one of the above three categories have closed knight's tours," as there are weaknesses in the theorem). And the reverse direction also holds true.

In relation to the explicability-inexplicability principle, if there is explicability (e.g., the explanation, by Schwenk's Theorem, of the impossibility of closed knight's tours if the three conditions are met), there is inexplicability (e.g, the lack of sufficient explanation by Schwenk's Theorem of whether or not "all rectangular boards that do not fall in one of the above three categories have closed knight's tours"). And the reverse direction also holds true.

In relation to the finiteness-transfiniteness principle, if there is finiteness (e.g., the finite number of knights on an $8 \times 8$ chess board), there is transfiniteness (e.g., the transfinite number of 26,534,728,821,064 directed closed tours on an $8 \times 8$ chess board). And the reverse direction also holds true.

In relation to the preciseness-vagueness principle, if there is preciseness (e.g., the precise identification of the three conditions in Schwenk's Theorem), there is vagueness (e.g., the vagueness in the identification of the three conditions, because it is not clear whether or not "all rectangular boards that do not fall in one of the above three categories have closed knight's tours"). And the reverse direction also holds true.

In relation to the simpleness-complicatedness principle, if there is simpleness (e.g., the relatively simple analysis of closed knight's tours on the basis of the three conditions in Schwenk's Theorem), there is complicatedness (e.g., the relatively complicated analysis of closed knight's tours as shown in the criticisms of Schwenk's Theorem by the critics). And the reverse direction also holds true.

In relation to the openness-hiddenness principle, if there is openness (e.g., the open exploration of the natural beauty of symmetry in the phenomenon of the knight's tour), there is hiddenness (e.g., the hidden bias in the mathematical analysis of the knight's tour, because it is not clear if it has any practical use at all for actual players in chess games, since it is one thing for mathematicians to be attracted to the natural beauty of symmetry in the knight's tour problem, but it is something else for chess players to apply the knight's tours for best tactical moves). And the reverse direction also holds true.

In relation to the denseness-emptiness principle, if there is denseness (e.g., the relatively denser concentration of concerns with the beauty of symmetry in the knight's tour by chess mathematicians), there is emptiness (e.g., the relatively less dense, or more empty, concentration of cocnerns with the beauty of symmetry in the knight's tour by chess players in actual chess games). And the reverse direction also holds true.

In relation to the slowness-quickness principle, if there is slowness (e.g., the relatively slower speed of computation to solve Hamiltonian path problems in graph theory, especially in regard to those which are solved in non-linear time), there is quickness (e.g., the relatively faster speed of computation to solve the knight's tour problem, which can be solved "in linear time"). And the reverse direction also holds true.

In relation to the convention-novelty principle, if there is convention (e.g., the conventional wisdom about the knight's tour problem), there is novelty (e.g., the alternative novel challenge to this conventional wisdom about the knight's tour problem, by Schwenk's Theorem, which is original in its specification of three major conditions for them to not be allowed). And the reverse direction also holds true.

In relation to the evolution-transformation principle, if there is evolution (e.g., the natural evolution of humans to be able to calculate in the state of nature), there is transformation (e.g, the technical transformation of human ability to calculate by the invention of computers to calculate complicated problems like the calculation of

26,534,728,821,064 directed closed tours on an 8 × 8 chess board). And the reverse direction also holds true.

In relation to the symmetry-asymmetry principle, if there is symmetry (e.g., the co-existence of different ways to understand the knight's tour problem in history), there is asymmetry (e.g., the understanding of the knight's tour problem "in verse form…in the highly stylized Sanskrit poem *Kavyalankara* written by the 9th century Kashmiri poet Rudrata—but the understanding of the knight's tour problem in computational form like the calculation of 26,534,728,821,064 directed closed tours on an 8 × 8 chess board by computers nowadays). And the reverse direction also holds true.

In relation to the regression-progression principle, if there is regression (e.g,. the regression made in the understanding of the knight's tour problem, as shown in the major weaknesses of Schwenk's Theorem as pointed out by the critics), there is progression (e.g,. the progress made in the understanding of the knight's tour problem, as shown in the Schwenk's Theorem in regard to the specification of the three major conditions). And the reverse direction also holds true.

And in relation to the same-difference principle, if there is similarity in outcome (e.g., the contribution to the molding of beliefs and values, regardless of whether this be done by way of valuing the nature of symmetry by chess mathematicians or by way of valuing instrumental thinking by chess players), there is difference in outcome (e.g., the contribution to the molding of beliefs and values by way of valuing the nature of symmetry by chess mathematicians—but the contribution to the molding of beliefs and values by way of valuing instrumental thinking by chess players in calculating the best tactical move for winning a game). And the reverse direction also holds true.

## Chess, Analogy, and Physics

Another way to understand tactics from the perspective of nature has to do with the relationship between chess and physics.

### *The Nonsense of a Popular Story*

The relationship between chess and physics does not look obvious. If learned correctly, the relationship can enhance our understanding of chess, especially about tactics (and strategy, for that matter). But if learned incorrectly, the relationship can impoverish our understanding of chess.

Let's consider the second case about incorrect learning below (and the first case about correct learning in the next sub-section).

In the case of incorrect learning from the relationship between chess and physics, a good example concerns the nonsense of a popular story often cited in the literature on chess and physics.

The story runs something like this about the Nobel laureate in physics, Richard Feynman, in that he "once noted that discovering the laws of physics is like trying to learn the laws of chess merely by observing chess games. You notice that bishops stay on the same color squares; you write this down as a law of chess. Later, you come up with a better law—bishops move diagonally. And, since diagonal squares are always colored the same, this explains why bishops always stay on the same color. This law is an improvement—it is simpler, and yet explains more. In physics, discovering Einstein's theory of gravity after knowing Newton's theory of gravity is a similar type of discovery." (J. Gott 2002)

Then, Feynman further added that "noticing that pieces don't change their identity in a chess game is similar to discovering the law of mass-and-energy conservation. Eventually, say, you see a chess game in which a pawn reaches the other end of the board and is promoted to become a queen. You say, 'Wait, that violates the laws of chess. Pieces can't just change their identity.' Of course, it does not violate the laws of chess; you just had never seen a game pushed to that extreme before. In time-travel research we are exploring extreme situations in which space and time are warped in unfamiliar ways. That these time-travel solutions may violate 'common sense' makes them intriguing." (J. Gott 2002)

This analogy between physics and chess by Feynman is highly misleading, for two main reasons.

Firstly, the first reason for the misleading nature of the analogy by Feynman concerns his first example about the bishop.

Now, the rule of chess governing the movement of the bishop is that "[t]he bishop can move any number of squares diagonally, but may not leap over other pieces," which also means that it "always stays on the same colored squares." (WK 2010)

So, "moving diagonally" and "staying on the same colored squares" are part of the same definition of the rule governing the movement of the bishop, and neither constitutes a "better" explanation (of the rule of chess governing the movement of the bishop) than the other.

Furthermoe, the rules of chess in general do not try to "explain" anything about reality in nature that the laws of physics do. Thus, it is not appropriate for Feynman to suggest that "diagonal squares…colored the same" can explain better "why bishops always stay on the same color." This has nothing to do with explanation to understand empirical reality, but with the definition of the rule governing the movement of the bishop.

And secondly, the second reason for the misleading nature of the analogy by Feynman has to do with his second example about the pawn.

Now, the rule of chess governing the movement of the pawn is that "[t]he pawn usually moves one step straight forward, but it may move two steps straight forward on its first move unless its path is obstructed….The pawn has two special moves, the en passant capture, and pawn promotion." (WK 2010)

So, in chess, the rule explicitly states, as part of its definition, "pawn promotion," which refers to "the transformation of a pawn that reaches its eighth rank into the player's choice of a queen, knight, rook, or bishop of the same color." (WK 2010cc; T. Just 2003) There is nothing unusual or surprising about "pawn promotion" if the rule governing the pawn is well understood.

But this is not true in physics, since "the law of mass-and-energy conservation" in normal situations belongs to a field of research different from "time-travel research" in "extreme situations"—such that each field has its own laws to explain reality in a different way.

Yet, this analogy fails in the case of chess, since the rule of chess governing the movement of the pawn is clear, nor is there a need for a different rule for pawn promotion in extreme situations. One rule says it all, as part of the definition of pawn movement in chess.

*A Better Understanding of Chess and Physics*

This critique of the misleading analogy between chess and physics by Feynman should caution us against the misuse of analogy between chess and physics.

A better way to understand the relationship between chess and physics is the use of physics for analogy (or conceptual inspiration), with the aim to create new variants of chess.

Joao Neto (1999), for instance, precisely proposed some new chess variants by using physics for analogy (or conceptual inspiration). Consider some examples below, for illustration—as summarized in *Table 3.3*.

(a) Inertia Chess

The first example of a new chess game by using concepts in physics is "inertia chess," which has the following rules (focusing on moving in the same direction): (J. Neto 1999)

- "The FIDE rules apply, except in the following, with the spirit that the last pieces moved continue to move the same way in the next turns."

- "There is no notion of check, checkmate or stalemate. The winner is the player who captures the opponent's king."
- "If you moved a piece without making a capture in your last move, you are forced to play the same piece in the same direction without making a capture. If you can't, you are free to play whatever you want following Rules 1-2."
- "If you made a capture in your last move, you are forced to play the same piece in the same direction and make another capture with it. If you can't, you're free to play whatever you want following rules 1-2."
- "Checking or mating the opponent according to the FIDE rules in the middle of the moves are not valid reasons to stop moving/taking."
- "'Direction' is meant in the oriented sense, so that there are 16 different directions possible in all, including the knight moves."
- "A king is not allowed to castle if he escapes an immediate capture or if the intermediate square he would cross is under immediate attack—like in FIDE rules but with Inertia rules of moves."
- "After castling, player's next move is free: the castling king and rook stay on their new position without having to continue their course."

Surely, there can be other rules to be added, but these were what Neto (1999) could come up with at the time.

(b) Entropy Chess

The second example of a new chess game by using concepts in physics is "entropy chess," which has the following rules (focusing on empty squares), as jointly worked out by Neto (1999) and Benjamin Good:

- "All FIDE rules apply except in the following."
- "After each move, the player can make an optional 'brownian move,' i.e., pick an enemy piece (except the king and the last moved piece) and put it on a empty square not attacked by the player's pieces. Bishops, if moved, must be placed in squares of the same color."
- "A player cannot move the piece that was used in the last brownian move made by the adversary."
- "A player must release the king from a check within the normal move. If that is not possible, the player loses the game."

Surely, there can be other rules to be added, but these were what Neto (1999) and Good could come up with at the time.

(c) Magnetic Chess

The third example of a new chess game by using concepts in physics is "magnetic chess," which has the following rules (focusing on repulsion and attraction): (J. Neto 1999)

- "The FIDE rules apply, except in the following."
- "There is no notion of check, checkmate or stalemate. The player capturing the opponent's king wins."
- "Every piece has a charge generating a magnetic field, except the two kings which have a neutral charge. White pieces are positive, and Black are negative (or vice versa, of course)."
- "When a piece is moved to a specific square, their closest neighbours at the same rank and column (not diagonal) are repelled if they have the same charge (hence the same colour), and attracted if they have an opposite charge (hence have different colours). A king isn't repelled or attracted, doesn't repel or attract and has a blocking effect on the magnetic field lines."
- "There is no en-passant rule."
- "In castling, the changes are made by the rook movement."
- "A white pawn in the 1st or 2nd rank, or a black pawn in the 7th or 8th rank, can move 2 squares ahead, even if it has already been played before."

Surely, there can be other rules to be added, but these were what Neto (1999) could come up with at the time.

(d) Chaos Chess

And the fourth example of a new chess game by using concepts in physics is "chaos chess" (with the focus on chaos or uncertainty). (J. Neto 1999)

What is interesting about "chaos chess" is that it is used "to generalize" or "to iterate the process and make the changes of the moved pieces, again and again, until the board reaches a fixed point or a cycle!" (J. Neto 1999)

So, "[t]his type of game brings a sort of chaos idea, everything determined by fixed rules, but the sensitivity to the initial conditions brings uncertainty to simple moves! Because of this, we suffix the specific game with the 'chaos' word." (J. Neto 1999)

In a less complex way, another variant is "to use the chaos idea, not applying all combinations, but just one chosen by the player, until it reaches a fixed cycle." (J. Neto 1999)

*Looking Beyond*

Surely, there are problems with these new chess games, even if one appreciates their creative nature rooted in concepts from physics for conceptual inspiration. Consider some of them below, for illustration—as summarized in *Table 3.3*.

Firstly, the first problem with the new chess games is that it is not so easily played by humans. For example, in "chaos chess," as Neto (1999) admitted, "this type of game can only be programmed in a computer because the number of iterations can be very large (but certainly finite)."

And secondly, the second problem with the new chess games is that each game has a different flavor, so that it cannot appeal to all kinds of players. For example, those players who do not like "repulsion" and "attraction" are not likely to like "magnetic chess."

However, these criticisms do not mean to dismiss new chess games outright. Rather, the criticisms show us the two opposing sides of the possibility and desirability of tactics (from the perspective of nature with new chess games as a case study here), such that they are not to the extent that the respective defenders would like us to believe.

More importantly, this analysis of new chess games can teach us something refreshing about the ontological principles in existential dialectics, and good examples include the absoluteness-relativeness principle, the partiality-totality principle, the predictability-unpredictability principle, the explicability-inexplicability principle, the finiteness-transfiniteness principle, the preciseness-vagueness principle, the simpleness-complicatedness principle, the openness-hiddenness principle, the denseness-emptiness principle, the change-constancy principle, the order-chaos principle, the slowness-quickness principle, the theory-praxis principle, the convention-novelty principle, the evolution-transformation principle, the symmetry-asymmetry principle, the softness-hardness principle, the regression-progression principle, and the same-difference principle.

For instance, in relation to the absoluteness-relativeness principle, if there is absoluteness (e.g., the absolute view in physics of "extreme situations in which space and time are warped in unfamiliar ways"), there is relativeness (e.g., what is true in physics in regard to extreme situations as opposed to normal situations is not necessarily so in chess, when the rule governing the movement of the pawn also allows pawn promotion). And the reverse direction also holds true.

In relation to the partiality-totality principle, if there is partiality (e.g., the partial view by a player to change some initial conditions in a chaos chess game), there is totality (e.g., the more holistic view of uncertainty for the entire game, when a small change to an initial condition is allowed, because "the sensitivity [in chaos chess] to the initial conditions brings uncertainty to simple moves"). And the reverse direction also holds true.

In relation to the predictability-unpredictability principle, if there is predictability (e.g., the predictable tendency of chaos chess to be fixed by the rules as specified), there is unpredictability (e.g., the more difficult task to predict the outcome of the game if some initial conditions are changed in a chaos chess game). And the reverse direction also holds true.

In relation to the explicability-inexplicability principle, if there is explicability (e.g,. the explanation by Feynman that, "since diagonal squares are always colored the same, this explains why bishops always stay on the same color"), there is inexplicability (e.g., the lack of sufficient explanation by Feynman of what exactly this really "explains" in chess, since the rules of chess in general do not try to "explain" anything about reality in nature that the laws of physics do, and besides, "moving diagonally" and "staying on the same colored squares" are part of the same definition of the rule governing the movement of the bishop, and neither constitutes a "better" explanation than the other"). And the reverse direction also holds true.

In relation to the finiteness-transfiniteness principle, if there is finiteness (e.g., the finite number of rules in chaos chess), there is transfiniteness (e.g., the transfinite number of "iterations" in chaos chess, which "can be very large" and "can only be programmed in a computer"). And the reverse direction also holds true.

In relation to the preciseness-vagueness principle, if there is preciseness (e.g., the precise identification of two examples of "the laws of chess" regarding the bishop and the pawn by Feynman, in his comparison with "the laws of physics"), there is vaguesness (e.g. the vagueness in the identification, since it is not clear of what exactly is meant by the word "law" here in chess, although the word makes sense in regard to the "laws" of physics like a scientific principle to explain reality in nature). And the reverse direction also holds true.

In relation to the simpleness-complicatedness principle, if there is simpleness (e.g., the relatively simple proposal of new chess games by Neto and his colleagues), there is complicatedness (e.g., the relatively more complicated analysis of the new chess games, as shown in the criticisms). And the reverse direction also holds true.

In relation to the openness-hiddenness principle, if there is openness (e.g., the open exploration of new chess games with the use of concepts in physics), there is hiddenness (e.g., the hidden bias in the new chess games, because of their focus on concepts in physics, not concepts in other fields of knowledge). And the reverse direction also holds true.

In relation to the denseness-emptiness principle, if there is denseness (e.g., the relatively denser concentration of moves in the same direction in inertial chess), there is emptiness (e.g., the relatively less dense, or more empty, concentration of moves in the same direction in magnetic chess). And the reverse direction also holds true.

In relation to the change-constancy principle, if there is change (e.g., the ever new variants of chess in the history of chess, with the ones by Neto as latest examples), there is constancy (e.g., the ever constancy of the existence of rules in chess). And the reverse direction also holds true.

In relation to the order-chaos principle, if there is order (e.g., the orderly nature of chaos chess, because of the rules as specified), there is chaos (e.g., the relatively chaotic outcome of chaos chess, when initial conditions are changed). And the reverse direction also holds true.

In relation to the slowness-quickness principle, if there is slowness (e.g., the relatively slower speed of chaos chess to be played by humans without the use of computers), there is quickness (e.g. the relatively quicker speed of normal Western chess to be played by humans). And the reverse direction also holds true.

In relation to the theory-praxis principle, if there is theory (e.g., the theoretical construction in physics, like chaos theory, the law of thermodynamics, etc.), there is praxis (e.g., the practical application of physical theories for the invention of new chess games, as proposed by Neto and his colleagues, for instance). And the reverse direction also holds true.

In relation to the convention-novelty principle, if there is convention (e.g,. the conventional wisdom about Western chess as sanctioned by FIDE), there is novelty (e.g,. the alternative novel challenge to this conventional wisdom about Western chess, by Neto with his invention of new chess games with new ideas from physics). And the reverse direction also holds true.

In relation to the evolution-transformation principle, if there is evolution (e.g., the natural evolution of humans to play with each other in the state of nature), there is transformation (e.g., the technical transformation of human playing with each other by the invention of new chess games with the use of physics). And the reverse direction also holds true.

In relation to the symmetry-asymmetry principle, if there is symmetry (e.g., the co-existence of different forms of chess in history), there is asymmetry (e.g., the popularity of conventional Western chess by those who follow the rules as laid down by FIDE—but the exploration of new kinds of chess which go beyond the rules by FIDE among players like Neto and his followers). And the reverse direction also holds true.

In relation to the softness-hardness principle, if there is softness (e.g., the attraction in magentic chess), there is hardness (e.g,. the repulsion in magnetic chess). And the reverse direction also holds true.

In relation to the regression-progression principle, if there is progression (e.g., the regression in the new forms of chess, as shown in the weaknesses pointed out by the critics), there is progression (e.g,. the progress made in the new forms of chess for playing new chess games with the use of ideas from physics). And the reverse direction also holds true.

And in relation to the same-difference principle, if there is similarity in outcome (e.g., the contribution to the molding of beliefs and values, regardless of whether this be done by way of conforming to the traditional rules of chess as laid down by FIDE or by way of exploring new rules by discontented players), there is difference in outcome (e.g., the contribution to the molding of beliefs and values by way of conforming to the traditional rules of chess as laid down by FIDE—but the contribution to the molding of beliefs and values by way of exploring new rules by discontented players). And the reverse direction also holds true.

# Tactics and Society

Tactics, when examined from the perspective of society, can unconceal in a different way its possibility and desirability, and this can be shown by way of a case study, which concerns the use of computers and online chess for learning chess tactics.

## Chess Tactics, and the Use of Computers and Online Chess

An interesting question to ask here, in the context of understanding chess tactics from the perspective of society, concerns the use of computer games for learning chess tactics.

*Learning Chess Tactics*

It is often said that to play piano well requires that one practices, practices, and practices. The same can be said about playing chess well.

For instance, Muna Wa Wanjiru (2010) argued that, "though many people study chess tactics to improve their games, true true mastery only comes when one has to actually face some crisis cases on the game board."

The reason is that "[t]he efficiency or the futility of a move is more relevant and obvious when analyzed in retrospection; hence, it is highly important that you always keep an eye on the game quality after you conclude it." (M. Wanjiru 2010)

Nowadays, the practice of playing chess is made easier with the advance of technology, especially in regard to chess playing computer games: "From this point of view, computer software that enable the user to see the weak and strong points of his or her chess tactics are an immense help. For instance, poor chess tactics may give rise to awkward situations when some bad moves that you make end up by blocking your own pieces." (M. Wanjiru 2010)

For example, "a frequent chess tactics mistake many players make" is "moving the bishop in front of the pawn during the game opening stage. In the future development of the game you'll have to retreat with the bishop in order to be able to move the pawn freely. Keep in mind that every move needs to have a finality, otherwise it's simply wasted! We cannot speak of chess tactics without a certain strategy in mind, therefore, every move you take needs to fit in the overall context of a general application or approach. To put it in other words it needs to have a purpose." (M. Wanjiru 2010)

This is important, because "[t]here are players who try to built their chess tactics on the mistakes the opponent makes; but waiting as long as that is definitely a bad choice. Of course you may speculate and use any breach in the other's defenses for your advantage, but to actually rely on that completely for building your chess tactics, is definitely wrong. Those who get to learn the secrets of chess, are aware of how important it is to develop all the primary pieces during the game opening. If at the time when you reach the middle of the game, you still have important pieces lying at the back rank, then you're chess tactics are definitely faulty and chances that you win the game are pretty low. If a bishop is kept away from the action, in game economy we could say that you are playing with one piece less than your opponent, and he or she will not hesitate to use that advantage against you." (M. Wanjiru 2010)

### *Using Chess Playing Computer Games*

A good example of using chess playing computer games is "ChessGenius," which is "a chess playing computer program written by Richard Lang who has in the past written programs that have won the Computer World Chess Championship on 10 occasions." (WK 2010dd)

It is amazing to know that "ChessGenius," with its "continuation of a series of programs (which included various incarnations of the Mephisto program," acually "won the World Microcomputer Chess Championship in 1984, 1985, 1986, 1987, 1988, 1989, 1990, 1991 and 1993. ChessGenius was the first computer to beat a world champion (Garry Kasparov) at a non-blitz time limit." (WK 2010dd; ICGA 2009; PC 2009; CP 2009)

ChessGenius not only can be used to play chess (for learning tactics and strategy, for that matter) but also "can read games created in .cbf (Chess Base Format) and .pgn (Portable Game Notation) formats and can analyse games assessing the moves played against its own evaluations. It is also possible to run other chess engines in the ChessGenius interface." (WK 2010dd)

The playing strength of ChessGenius is impressive enough. For instance, "in the early 1990s ChessGenius was 'one of the first master-strength programs.' At the Intel World Chess Grand Prix in London in 1994 ChessGenius achieved a rating performance for the tournament of 2795 ELO." (WK 2010dd; E. Manchester 2009)

However, although "[f]rom 1994 until 1998 ChessGenius remained one of the top chess programs available," the bad news occurred when "[i]n 1999 ChessGenius dropped out of the top ten on the SSDF (Swedish Chess Computer Association) rating list and it continued to slip down the list over the following decade." (WK 2010dd; J. Silman 2009)

One explanation was given by Lang himself, in that "this was because the program does not scale well to faster hardware. Portable versions (for example for Palm and the iPhone) perform well because ChessGenius is particularly strong in rather weak hardware environments. Unlike most other commercial vendors, Richard Lang explicitly forbids including ChessGenius in chess engine rating lists, so it is difficult to gauge its strength compared to modern programs." (WK 2010dd; E. Manchester 2009; C. Kantack 2009)

*Practising Online*

Besides using chess playing computer games, there are other options. One good option, from the standpoint of society, is the use of technology to create a social network for people to learn chess (both tactics and strategy) online.

For instance, "ChessWorld (aka letsplaychess) is a correspondence chess server hosted by the Barnet Chess Club in England and with over 400,000 members in more than two hundred countries. The site appears to be a comprehensive online chess club with games, many forms of tournaments, user homepages, active chess and social forums, and even annual 'meet-ups' where members meet...for chess and socializing. Non-members are permitted a certain number of games and are invited to play in special tournaments." (WK 2010ee)

Just consider some of the services provided by ChessWorld for people to learn chess online, namely, (a) correspondence games, (b) tournaments, (c) teams, (d) fourms, and (e) puzzles and tips—as summarized in *Table 3.4*.

(a) Correspondence Games

The first service is about "correspondence games," in that "ChessWorld offers rated and unrated correspondence games, with time controls of one day per move, two day per move, three day per move, five day per move, 10 moves within 50 days, 10 day per move, and 15 days per move, individual challenges, double round-robin tournaments for any number of players from three to fifteen (in a double round robin, each player in an event plays every other player in the event twice, once with the white pieces and once with the black pieces)," and so on. (WK 2010ee)

In addition, ChessWorld "offered...pyramid events (a three dimensional take on ladder play where players ascend the pyramid by winning) and knockout tournaments, where losing players are eliminated (in the event of a draw, the player with the black pieces advances to the next round). Members may also join teams, captain teams, and participate in team matches." (WK 2010ee)

(b) Tournaments

The second service is about "tournaments," in that "members may create round-robin and knockout tournaments, name them, designate a theme or dedication (e.g., Lasker Memorial Tournament), and fill the tournaments by invitation or allow anyone to join, by putting the tournament on the tournament sign-up page." (WK 2010ee)

In fact, "[m]embers can also join as many tournaments as they wish, but non-members can only be invited to special tournaments. There are two different types of tournaments: knockout tournaments, and all-play-all tournaments. Knockout tournaments allow two players to be paired for a game and whoever wins, goes to the next pairing. The one player who is left, wins. If the game is a draw, black wins. You can either have eight players or sixteen players in a knockout tournament. All-play-all tournaments are when one player plays all the rest of the other players twice at the same time. In other words, everyone plays everyone twice. One white, one black. There can be three to fifteen players in an all-play-all tournament. If you win a tournament you get a tournament norm." (WK 2010ee)

(c) Teams

The third service is about "teams," that is, "a group of chess players that play against other groups of chess players in teams. It sounds confusing but it's just like football, baseball, basketball etc. Every year there is a team tournament of all the teams in that league. Example: League:Pets; Teams: Team Dogs, Team Cats, Team Fish etc. Each team in the league will go against each other to see which team is best. The teams will each play white and black. Example: Team Dogs white vs Team Fish black, Team Dogs black vs Team Fish white." (WK 2010ee)

So, "[i]f you win a team tournament, you and the rest of the team will get a team norm. Of course, this happens only once a year, so in the mean time, the captain of the team can set up friendly team matches to practice for the upcoming season. Players can also make their own team, the player is then called Team Captain. Team captains may recruit players by invitation or accept applications from site members." (WK 2010ee)

(d) Forums

The fourth service is about "fourms," where "a player can post messages, read, and help out by answering questions from other players. A non-member is limited on how many messages he/she can post. Members can post as much as they want but only admins can answer questions. There are four categories which the player can post: Social, Chess, HelpDesk and Suggestions." (WK 2010ee)

(e) Puzzles and Tips

And the fith service is about "puzzles and tips," where "[t]here are thousands of rated chess puzzles which can increase your rating if you get the puzzle right in a limited amount of time. They are open to non-members and members both. There are also hundreds of chess puzzles created by Chessworld members." (WK 2010ee)

In addtion, "[t]here are tips too. There is: Quick Member Tips, Improving Tips, and Winning Tips. Inside each category, there are sub-categories." (WK 2030ee)

*A Sober Reflection*

Surely, a critical question to ask is how effective the use of technology like chess playing computer games and online chess is for chess learning. There is no lack of criticisms; consider some of them below, for illustration—as summarized in *Table 3.4*.

Firstly, a major problem with chess playing computer games is that there is the issue about software and hardware compatiblity.

For instance, in the case of ChessGenius, "the program does not scale well to faster hardware. Portable versions (for example for Palm and the iPhone) perform well because ChessGenius is particularly strong in rather weak hardware environments. Unlike most other commercial vendors, Richard Lang explicitly forbids including ChessGenius in chess engine rating lists, so it is difficult to gauge its strength compared to modern programs." (WK 2010dd; E. Manchester 2009; C. Kantack 2009)

Secondly, anoher major problem with online chess is that online chess fans can spend too much time online, to the point that they become technologically addicted and socially dysfunctional in real life.

This is a serious problem which I already went to great lenghts to explain in my previous books like *The Future of Capitalism and Democracy* (2002) and the 2 volumes *The Future of Human Civilization* (2000)—especially in the context of what I originally called the theory of "e-civic alienation" in Chapter 7 of *FCD*.

Thirdly, still another major problem is that there are limits to computation, such that the use of computer games is not as effective as it may sound, as already discussed in *Sec. 3.2.1*.

And fourthly, another major problem is that using computer games and online chess is not free. The expenses for computer games, online chess, and other related costs are not minimal. For instance, in the U.S., the cost of "family membership" for ChessWorld alone is "$98.00 a year," and "club/team membership" is "$166.00 a year," as of 2010. (WK 2010ee)

Yet, these illustrative problems do not mean to totally reject the usefulness of computer games and online chess for the practice of chess tatics (and strategy too, for that matter). Instead, the indication of the problems is to show us the two opposing sides of the possibility and desirability of tactics (from the perspective of society with computer games and online chess as a case study here), such that they are not to the extent that the respective defenders would like us to believe.

In addition, this analysis of computer games and online chess can teach us something valuable about the ontological principles in existential dialectics, and good examples include the absoluteness-relativeness principle, the predictability-unpredictability principle, the explicability-inexplicability principle, the simpleness-complicatedness principle, the openness-hiddenness principle, the denseness-emptiness principle, the slowness-quickness principle, the theory-praxis principle, the convention-novelty principle, the evolution-transformation principle, the symmetry-asymmetry principle, the softness-hardness principle, the regression-progression principle, and the same-difference principle.

For instance, in relation to the absoluteness-relativeness principle, if there is absoluteness (e.g., the absolute view of addicted fans of online chess in regard to its benefits), there is relativeness (e.g., what is true for the addicted fans of online chess in regard to its benefits is not necessarily so for the critics, as shown in the aforecited problems). And the reverse direction also holds true.

In relation to the predictability-unpredictability principle, if there is predictability (e.g., the predictable tendency of online chess fans to practice their chess tactics online), there is unpredictability (e.g., the more difficult task to predict to what extent and in what eay exactly a particular online chess fan will have his chess tactics improved because of his play online). And the reverse direction also holds true.

In relation to the explicability-inexplicability principle, if there is explicability (e.g., the explanation by Lang of the dropout of ChessGenius of "the top ten on the SSDF…rating list" in 1999 and "over the following decade" on the basis that "the program does not scale well to faster hardware"), there is inexplicability (e.g., the lack of sufficient explanation by Lang of why this must be the reason, not other possible explanations, be they about software weaknesses and something else). And the reverse direction also holds true.

In relation to the simpleness-complicatedness principle, if there is simpleness (e.g., the relatively simple analysis of the promise of online chess for learning chess), there is complicatedness (e.g, the relatively more complicated analysis of online chess by questioning its claims and raising some serious problems). And the reverse direction also holds true.

In relation to the openness-hiddenness principle, if there is openness (e.g, the open exploration of the usefulness of online chess for learnng chess), there is hiddenness (e.g., the hidden bias in online chess because of its impersonality and social isolation, such that the usefuleness has its dark sides not often spoken about). And the reverse direction also holds true.

In relation to the denseness-emptiness principle, if there is denseness (e.g., the relatively denser concentration of online chess players nowadays), there is emptiness (e.g,. the relatively less dense, or more empty, concentration of online chess players in antiquity). And the reverse direction also holds true.

In relation to the slowness-quickness principle, if there is slowness (e.g., the relatively slower speed of a correspondence game which allows "10 days per move"), there is quickness (e.g., the relatively faster speed of a correspondence game which allows "one day per move"). And the reverse direction also holds true.

In relation to the theory-praxis principle, if there is theory (e.g., the theoretical construction of computation and computer programming), there is praxis (e.g., the practical application of computation and computer programming for chess playing computer games). And the reverse direction also holds true.

In relation to the convention-novelty principle, if there is convention (e.g, the conventional wisdom about learning chess by playing it with another player in person), there is novelty (e.g, the alternative novel challenge to the conventional wisdom about face-to-face chess playing, by the use of online chess instead). And the reverse direction also holds true.

In relation to the evolution-transformation principle, if there is evolution (e.g, the natural evolution of humans to play with each other in person in the state of nature), there is transformation (e.g., the natural evolution of humans to play with each other online, by the invention of computers). And the reverse direction also holds true.

In relation to the symmetry-asymmetry principle, if there is symmetry (e.g., the co-existence of different ways to play chess in history), there is asymmetry (e.g,. the popularity of playing chess in person in antiquity—but the popularity of playing chess online nowadays). And the reverse direction also holds true.

In relation to the softness-hardness principle, if there is softness (e.g., to play chess in the traditional way without paying a fee), there is hardness (e.g., to play chess online with ChessWorld with a fee for membership).

In relation to the regression-progression principle, if there is progression (e.g, the regression made in online chess, as shown in the problems pointed out by the critics), there is progression (e.g., the progress made in online chess for different services like tournaments, games, forums, etc.). And the reverse direction also holds true.

And in relation to the same-difference principle, if there is similarity in outcome (e.g., the contribution to the molding of values and behaviors, regardless of how this be achieved by way of playing chess in person or playing chess online), there is difference in outcome (e.g., the contribution to the molding of values and behaviors by way of playing chess in person for more personal interactions—but the contribution to the molding of values and behaviors by way of playing chess online for less personal interactions). And the reverse direction also holds true.

# Tactics and Culture

Tactics, when analyzed from the perspective of culture, can show us in a different way its possibility and desirability, and this can be shown by way of two case studies, namely, (3.5.1) tactical exercises, and the question of beauty, and (3.5.2) chess handicap, and the tradition of odds-giving—to be addressed hereafter, in that order.

## Tactical Exercises, and the Question of Beauty

A fascinating way to illustrate tactics from the perspective of culture has to do with the question of beauty in tactical exercises.

To start, chess tactics can be treated like puzzles for chess composers, who compose any chess problem in the form of "a puzzle set by somebody using chess pieces on a chess board," which then "presents the solver with a particular task to be achieved. For instance, a position might be given with the instruction that white is to move first, and checkmate black in two moves against any possible defense." (WK 2010f)

"Chess problems" so understood reveal a strong connection "between chess compositions on the one hand and puzzles or tactical exercises on the other"; and this is true, even though not all puzzles or tactical exercises in chess have to be related to chess problems in chess composition. (WK 2010f)

*Some Features of Tactical Exercises*

With this last qualification in mind—there are some common features of tacticsl exercises inherent in "chess problems" so composed by chess composers.

Consider some most often cited features of tactical exercises below, for illustration. (WK 2010f)

- "The position is composed"—in that "it has not been taken from an actual game, but has been invented for the specific purpose of providing a problem. Although a constraint on orthodox chess problems is that the original position be reachable via a series of legal moves from the starting position, most problem positions would not arise in over-the-board play."
- "There is a specific stipulation"—in that "a goal to be achieved; for example, to checkmate black within a specified number of moves."

- "There is a theme (or combination of themes)"—in that "the problem has been composed to illustrate: chess problems typically instantiate particular ideas."
- "The problem exhibits economy in its construction"—in that "no greater force is employed than that required to render the problem sound (that is, to guarantee that the problem's intended solution is indeed a solution and that it is the problem's only solution)."
- "The problem has aesthetic value"—in that "[p]roblems are experienced not only as puzzles but as objects of beauty. This is closely related to the fact that problems are organized to exhibit clear ideas in as economical a manner as possible."

Again, it should be stressed that there are two kinds of puzzles or tactical exercises to be distinguished from each other, as explained below.

On the one hand, puzzles or tactical exercises in chess problems are commonly invented by chess composers for the challenge of solving new chess problems.

On the other hand, puzzles or tactical exercies "found in chess columns or magazines" are often taken from "actual" games, where "the task is to find the best move or sequence of moves (usually leading to mate or gain of material) from a given position." (WK 2010f)

*Beauty in Tactical Exercises*

The features of puzzles (or tactical exercises) in chess problems aforecited are fascinating if examined from the vantage point of beauty.

After all, "[t]he role of aesthetic evaluation in the appreciation of chess problems is very significant, and indeed most composers and solvers consider such compositions to be an art form." (WK 2010f)

A good authority on this subject is Vladimir Nabokov, who once "wrote about the 'originality, invention, conciseness, harmony, complexity, and splendid insincerity' of creating chess problems and spent considerable time doing so." (WK 2010f)

Of course, it is often said that "beauty is in the eyes of the beholder." The same can be said about beauty in chess problems, because "[t]here are no official standards by which to distinguish a beautiful problem from a poor one, and such judgments can vary from individual to individual as well as from generation to generation. Such variation is to be expected when it comes to aesthetic appraisal." (WK 2010f)

Even then, this relative view of beauty does not prevent some scholars on chess beauty from identifying some "elements to be important in the aesthetic evaluation of a problem," as shown below (and summarized in *Table 3.5*): (WK 2010f)

- Legality—This means that "[t]he problem position must be legal," in that "the diagram must be reachable by legal moves beginning with the initial game array. It is not considered a defect if the diagram can only be reached via a game containing what over-the-board players would consider gross blunders."
- Uniqueness—This means that "[t]he first move of the problem's solution (the key move or key) must be unique. A problem which has two keys is said to be cooked and is judged to be unsound or defective. (Exceptions are problems which are composed to have more than one solution which are thematically related to one another in some way; this type of problem is particularly common in helpmates.)"
- Ideal Uniqueness—This means that "[i]deally, in directmates, there should be a unique white move after each black move. A choice of white moves (other than the key) is a dual. Duals are often tolerated if the problem is strong in other regards and if the duals occur in lines of play that are subsidiary to the main theme."
- Theme(s)—This means that "[t]he solution should illustrate a theme or themes, rather than emerging from disjointed calculation. Many of the more common themes have been given names by problemists."
- Hiddenness—This means that "[t]he key move of the solution should not be obvious. Obvious moves such as checks, captures, and (in directmates) moves which restrict the movement of the black king make for bad keys. Keys which deprive the black king of some squares to which it could initially move (flight squares), but at the same time make available an equal or greater number of flight squares are acceptable. Key moves which prevent the enemy from playing a checking move are also undesirable, particularly in cases where there is no mate provided after the checking move. In general, the weaker (in terms of ordinary over-the-board play) the key move is the less obvious it will be, and hence the more highly prized it will be."
- No Promotion—This means that "[t]here should be no promoted pawns in the initial position. For example, if White has three knights, one of them must clearly have been promoted; the same is true of two light-square bishops. There are more subtle cases: if f1 is empty, a white bishop stands on b5 and there are white pawns on e2 and g2, then the bishop must be a promoted pawn (there is no way the original bishop could have got past those unmoved pawns). A piece such as

this, which does not leave a player with pieces additional to those at the start of a game, but which nonetheless must have been promoted, is called obtrusive. The presence of obtrusive units constitutes a smaller flaw than the presence of more obviously promoted units."

- Economy—This means that "[t]he problem should be economical. There are several facets to this desideratum. For one thing, every piece on the board should serve a purpose, either to enable the actual solution, or to exclude alternative solutions. Extra units should not be added to create 'red herrings' (this is called dressing the board), except in rare cases where this is part of the theme. If the theme can be shown with fewer total units, it should be. For another, the problem should not employ more moves than is needed to exhibit the particular theme(s) at its heart; if the theme can be shown in fewer moves, it should be."

Surely, there can be other elements for aesthetic evaluation in chess, but these seven as cited above suffice for illustration for the purpose at hand.

*The Problems with the Elements*

Yet, there are some problems with the elements for the aesthetic evaluation of puzzles or tactical exercises in chess problems. Consider a few of them below, for illustration—as summarized in *Table 3.5*.

Firstly, a major problem with the elements for the aesthetic evaluation of tactical exercises in chess problems is that even these seven elements, as chess composers often acknowledge, are not complete. The problem to identify a larger set is difficult enough, because of irreconcilable disagreements among them.

I already went to great lengths in *The Future of Aesthetic Experience* (2007) to discuss the problematic quest for a universal set of criteria for aesthetic experience, especially in the context of my original theory titled "the transformative theory of aesthetic experience."

Secondly, another major problem with the elements for the aesthetic evaluation of tactical exercises in chess problems is that taste is relative, since "beauty is in the eyes of the beholder." As it turns out, the seven elements as cited above represent "modern taste" in chess composition. (WK 2010f)

Thirdly, still another major problem with the elements for the aesthetic evaluation of tactical exercises in chess problems is that the puzzles or tactical exercises to be solved are not taken from actual games but are often "invented for the specific purpose of providing a problem."

(WK 2010f) Therefore, its aesthetic evaluation is often not relevant to actual chess games (and thus limited in practical application).

*Beyond the Opposing Sides*

Yet, these illustrative problems should not be misconstrued as dismissive of the importance of aesthetic evaluaion in chess. On the contrary, they are mentioned above so as to show us the two opposing sides of the possibility and desirability of tactics (from the perspective of culture with the aesthetic evaluation of tactical exercises as a case study here), such that they are not to the extent that the respective defenders would like us to believe.

Furthermore, this analysis of the aesthetic evaluation of tactical exercises in chess problems can teach us something valuable about the ontological principles in existential dialectics, and good examples include the formalness-informalness principle, the absoluteness-relativeness principle, the predictability-unpredictability principle, the explicability-inexplicability principle, the preciseness-vagueness principle, the simpleness-complicatedness principle, the openness-hiddenness principle, the denseness-emptiness principle, the slowness-quickness principle, the expansion-contraction principle, the theory-praxis principle, the convention-novelty principle, the evolution-transformation principle, the symmetry-asymmetry principle, the softness-hardness principle, the regression-progression principle, and the same-difference principle.

For instance, in relation to the formalness-informalness principle, if there is formalness (e.g., the formal logical requirement of completeness and soundness in a system of ideas), there is informalness (e.g., the non-formal inadequacy of the elements for the aesthetic evaluations of tactical exercies in chess problems, in that, although they are recognized by many chess composers and thus not arbitrary, they represent "modern" taste and are not accepted by all as a complete proof). And the reverse direction also holds true.

In relation to the absoluteness-relativeness principle, if there is absoluteness (e.g., the absolute view by those who accept the "modern taste" of the aesthetic evaluations of tactical exercises in chess on the basis of the seven elements), there is relativeness (e.g., what is beautiful for those who accept the 'modern taste" of aesthetical evaluation in regard to the seven elements is not necessarily so for others who do not accept the "modern taste," as "beauty is in the eyes of the beholder"). And the reverse direction also holds true.

In relation to the predictability-unpredictability principle, if there is predictability (e.g., the predictable tendency of chess composers with "modern taste" to judge beauty in tactical exercises on the basis of "uniqueness," "economy," etc.), there is unpredictability (e.g., the more difficult task to predict exactly what future chess composers in a different historical era will judge beauty in tactical exercises, if looked narrowly from the current list of elements only). And the reverse direction also holds true.

In relation to the explicability-inexplicability principle, if there is explicability (e.g., the explanation of beauty in tactical exercises on the basis of the seven elements for aesthetic evaluation), there is inexplicability (e.g., the lack of sufficient explanation of why these seven elements must be both necessary and sufficient, in light of the relativist challenge by the critics, in that "beauty is in the eyes of the beholders"). And the reverse direction also holds true.

In relation to the preciseness-vagueness principle, if there is preciseness (e.g., the precise identification of "economy" as one element for the aesthetic evaluation of tactical exercises), there is vagueness (e.g., the vagueness in the identification of "economy," which has different "facets" and can mean different things to different chess composers). And the reverse direction also holds true.

In relation to the simpleness-complicatedness principle, if there is simpleness (e.g., the relatively simple analysis of the seven elements for the aesthetic evaluation of tactical exercises by "modern" chess composers), there is complicatedness (e.g., the relatively complicated analysis of the seven elements for the aesthetic evaluation of tactical exercises, as shown in the problems pointed out by the critics). And the reverse direction also holds true.

In relation to the openness-hiddenness principle, if there is openness (e.g., the open exploration of the relationship between "beauty" and "tactical exercises" in chess problems), there is hiddenness (e.g., the hidden bias in the seven elements for aesthetic evaluation, because of their "modern taste" as acknowledged by the chess composers). And the reverse direction also holds true.

In relation to the denseness-emptiness principle, if there is denseness (e.g., the relativley denser concentration of the defenders of the seven elements for the aesthetic evaluation of tactical exercises in modern times), there is emptiness (e.g., the relatively less dense, or more empty, concentration of the defenders of the seven elements for the aesthetic evaluation of tactical exercises in ancient times). And the reverse direction also holds true.

In relation to the slowness-quickness principle, if there is slowness (e.g., the relatively slower speed of a tactical exercise to be solved, if "[t]he key move of the solution" is "obvious"), there is quickness (e.g., the relatively quicker speed of a tactical exercise to be solved, if "[t]he key move of the solution" is not "obvious" at all and is hidden deeply inside the puzzle). And the reverse direction also holds true.

In relation to the expansion-contraction principle, if there is expansion (e.g., the relatively more developed ability of a fan of tactical exercises in chess to think logically and aesthetically in relation to the rules as stipulated by chess composers), there is contraction (e.g., the relatively less developed ability of a fan of tactical exercises in chess to think in a way which are not allowed by the rules as stipulated by chess composers in regard to chess logic and aesthetics). And the reverse direction also holds true.

In relation to the theory-praxis principle, if there is theory (e.g., the theoretical construction of aesthetics to understand the nature of beauty), there is praxis (e.g., the practical application of aesthetics by chess composers for the aesthetic evaluation of tactical exercises in chess problems). And the reverse direction also holds true.

In relation to the convention-novelty principle, if there is convention (e.g., the conventional wisdom about beauty, in that "beauty is in the eyes of the beholders"), there is novelty (e.g, the alternative novel challenge to the conventional wisdom about beauty, by way of the "modern taste" of chess composers, who offered some elements to be accepted by those in chess composition, in spite of the relativist problem). And the reverse direction also holds true.

In relation to the evolution-transformation principle, if there is evolution (e.g., the natural evolution of humans to appreciate beautiful objects in the state of nature), there is transformation (e.g., the technical transformation of human appreciation of beauty by the invention of the seven elements for the aesthetic evaluation of tactical exercises in chess problems). And the reverse direction also holds true.

In relation to the symmetry-asymmetry principle, if there is symmetry (e.g., the co-existence of different aesthetic tastes in history), there is asymmetry (e.g., the popularity of aesthetic relativism by those who hold the claim that "beauty is in the eyes of the beholders"—but the popularity of the seven elements for the aesthetic evaluation of tactical exercises in chess problems by chess composers with "modern taste"). And the reverse direction also holds true.

In relation to the softness-hardness principle, if there is softness (e.g., the judgement that a puzzle in chess is "beautiful" on the basis of the seven elements for aesthetic evaluation), there is hardness (e.g., the judgement that a puzzle in chess is not "beautiful," or "ugly" so to speak, because it violates the seven elements for aesthetic evaluation, according to the advocates). And the reverse direction also holds true.

In relation to the regression-progression principle, if there is regression (e.g, the regression made in the aesthetic evaluation of tactical exercies in chess problems, as shown in the criticisms by the critics), there is progression (e.g, the progress made in the aesthetic evaluation of tactical exercies in chess problems, as shown in the proposed seven elements for a better appreciation of chess beauty). And the reverse direction also holds true.

And in relation to the same-difference principle, if there is similarity in outcome (e.g., the contribution to the molding of beliefs and values about beauty, regardless of whether this be achieved by way of aesthetic relativism or by way of aesthetic non-relativism), there is difference in outcome (e.g., the contribution to the molding of beliefs and values about beauty by way of aesthetic relativism, such that "beauty is in the eyers of the beholders"—but the contribution to the molding of beliefs and values about beauty by way of aesthetic non-relativism, such that there are criteria to be used for the appreciation of beauty). And the reverse direction also holds true.

## Chess Handicap, and the Tradition of Odds-Giving

Another way to evaluate the possibility and desirability of tactics from the perspective of culture concerns the tradition of "chess handicap" or "odds-giving" in chess tactics. (WK 2010ff)

### *The Purpose of Odds-Giving in Chess Tactics*

The term "chess handicap" in chess tactics refers to "[a] handicap (or 'odds') in chess" as "a way to enable a weaker player to have a chance of winning against a stronger one." (WK 2010ff)

Of course, "[t]here are many kinds of such handicaps, such as material odds, extra moves (i.e., the weaker player can play the first x number of moves at the beginning of the game), extra time on the chess clock, and special conditions (such as requiring the odds-giver to deliver checkmate with a specified chess piece or pawn). Various permutations of these, such as 'Pawn and two moves,' are also possible." (WK 2010ff)

In any event, "[t]he purpose of a handicap, or odds, is to compensate for the difference in skill between two chess players." (WK 2010ff; D. Hooper 1992; D. Pritchard 1994)

*The Varieties of Odds-Giving in Chess Tactics*

As indicated above, there are many types of odds-giving in chess tactics, and good examples include "material odds; extra moves; time odds; special restrictions (such as pion coiffé); weighting of results (such as 'draw odds'—counting a draw as a loss for the odds-giver); differential stakes; and physical restrictions, such as blindfold chess." (WK 2010ff; D. Hooper 1992; M. Shibut 2004)

In addition, there are "[m]any different permutations of handicaps (for example, a material handicap plus time odds) are also possible, as are countervailing handicaps (for example, a player gives up a piece, but receives one of the opponent's pieces or pawns and/or extra moves, in return)." (WK 2010ff; H. Murray 1913; D. Hooper 1992; H. Golombek 1977; A. Soltis 1978; H. Staunton 1849; I. Chernev 1973)

In spite of the varieties of odds-giving, Harry Golombek took the trouble to rank "the following list of material odds (in increasing handicap level)"—with the clarification that "the odds-giver plays White unless otherwise indicated, and 'pawn odds' normally refers to the f-pawn (i.e. the pawn initially located on the f2-square for White, and on the f7-square for Black)." (WK 2010ff; H. Golombek 1977; H. Staunton 1847; J. Mason 1910)

- "Odds of the Move"—"Weaker player plays White." (WK 2010ff)
- "Two moves"—"Weaker player plays White and starts the game by making two moves." (WK 2010ff)
- "Pawn and Move"—"Weaker player plays White; a black pawn (typically that on f7) is removed from the board." (WK 2010ff)
- "Pawn and Two Moves"—"Weaker player plays the first two moves, and Black's pawn on f7 is removed from the board." (WK 2010ff)
- "Knight Odds"—"One of the stronger player's knights is removed, usually the queen's knight on b1." (WK 2010ff)
- "Rook Odds"—"One of the stronger player's rooks is removed, usually the queen's rook on a1." (WK 2010ff)
- "Rook and Pawn"—"Stronger player's queen rook and f-pawn are removed." (WK 2010ff)
- "Two Minor Pieces"—The odds-giver chooses which two of White's knights and/or bishops to remove." (WK 2010ff)
- "Rook and Knight"—"White's queen rook and queen knight are removed." (WK 2010ff)
- "Queen Odds"—"The stronger player's queen is removed." (WK 2010ff)
- "Two Rooks"—"Both of White's rooks are removed." (WK 2010ff)

- "Strong King"—"The king can move up to 2 squares in any direction." (WK 2010ff)

Surely, there are other kinds of odds-giving too, besides the ones as cited above.

For instance, "Larry Kaufman writes that under the chess tradition of the 18th and 19th centuries, the handicap below knight odds was: 'Pawn and three moves': White plays the first three moves, and Black plays without the f7 pawn. In odds games with extra moves (Pawn and two moves, Pawn and three moves), the odds-receiver cannot move beyond the fourth rank with those moves. Otherwise, White could win immediately with 1.e3 2.Bd3 3.Qh5+ g6 4.Qxg6+ hxg6 5.Bxg6#." (WK 2010ff; H. Staunton 1849; I. Horowitz 1871)

Others add some more types of odds-giving to the list, as shown below: (WK 2010ff; I. Horowitz 1871)

- "Draw Odds"—"The smallest of these handicaps; the stronger player plays White, and draws are counted as wins for Black." (WK 2010ff; I. Horowitz 1871)
- "Queen for a Rook"—"A handicap between Knight odds and Rook odds; the odds-giver's queen, and the odds-receiver's queen rook, are removed." (WK 2010ff; I. Horowitz 1871)
- "Queen for a Knight"—A handicap slightly greater than Rook odds; the odds-giver's queen, and the odds-receiver's queen knight, are removed." (WK 2010ff; I. Horowitz 1871)
- "Blitz Games"—A time handicap, where "[t]he stronger player may be given one or two minutes to play the whole game, while the weaker player receives five minutes or more. Money odds are another way of compensating for a difference in strength; the stronger player puts up some multiple (three, five, ten, etc.) of the amount of money put up by the weaker player." (WK 2010ff; H. Staunton 1849; J. Donaldson 1999)

Of course, even with the additional examples as listed above, the list remains non-exhaustive but illustrative for the purpose at hand.

*A Brief History of the Tradition of Odds-Giving in Chess Tactics*

These examples of odds-giving speak volumes of the tradition of odds-giving in chess tactics. After all, chess handicap is not a new phenomenon in the history of chess.

For Harry Golombek (1977), "odds-giving reached its heyday in the eighteenth century and the early nineteenth century"; for instance, "it was so prevalent in the 18th century that Philidor (1726–95) played the vast majority of his games at odds. About fifteen percent of the known games of Paul Morphy (1837–84) are games in which he gave odds." (WK 2010ff; M. Shibut 2004)

In *The Chess-Player's Handbook* (1847), "Howard Staunton advised inexperienced players to accept odds offered by superior players and, upon improving to the point that they can themselves give odds to some players, to avoid playing such players on even terms, warning that doing so is apt to induce 'an indolent, neglectful habit of play.' In 1849, Staunton published *The Chess-Player's Companion*, a 510-page work 'chiefly directed to the exposition of openings where one party gives odds.' Just over 300 pages were devoted to odds games: Book I (pages 1 to 185) contained games played at various odds, and most of Book V (specifically pages 380-496) discussed various types of odds, including exotic and unusual ones." (WK 2010ff)

By the middle of the 19$^{th}$ century, "Macon Shibut writes that…'chess was a gambling game….Individual matches for stakes were the focus of organized play. Matches between leading players attracted a wide following so masters often succeeded in finding sponsors to back their personal wagers.' However, the available sums were generally relatively meager, and travel was arduous, so the amount of money obtained in this way was not sufficient to enable professional chess players to support themselves financially." (WK 2010ff; M. Shibut 2004)

This remained so, even when, in 1851, "the first major chess tournament was…organized," but "[w]ith tournaments not a reliable means of making a living [at the time], odds-giving became a way for masters to entice amateurs into playing for wagers, since the odds gave the amateur a fighting chance. The odds system even became the earliest rating system: amateurs were graded according to what handicap they needed to compete against a master, and were referred to as a 'Rook player' or 'Pawn and Move player' as we would today speak of players by their Elo ratings, e.g. '1200 player' and '1800 player.'" (WK 2010ff; M. Shibut 2004; R. Byrne 1997; H. Golombek 1976)

However, by late 19$^{th}$ century, "[t]he playing of games at odds gradually grew rarer….Today, odds games, except for those at time odds, have all but disappeared." (WK 2010ff; H. Golombek 1977; M. Shibut 2004)

*The Problems with the Tradition of Odds-Giving in Chess Tactics*

This short history of the tradition of odds-giving in chess tactics can tell us something about the possiblity and desirability of chess handicap, such that they are not to the extent that the respective defenders would like us to believe.

Consider a few problems below, for illustration.

Firstly, a major problem with the practice of odds-giving in chess tactics is that the odds-giver does not always win over the weaker player as expected.

A well-known (but embarassing) example concerns "World Champion Emanuel Lasker," who "had failed at such an endeavor in 1894, losing a match at knight odds to Jackson Showalter's wife; he scored two wins and five losses." (WK 2010ff; G. Diggle 1987) In another example, "in 2001, London businessman Terence Chapman, a master-level player, played a match against former world champion Garry Kasparov with Kasparov giving odds of two pawns in each game (the pawns to be removed being different each time)"; although Kasparov won…two games," he then received "one draw." (WK 2010ff; M. Pein 2001)

Secondly, another major problem with the practice of odds-giving in chess tactics is that it is not always politically correct.

For instance, "the Soviet Union supported chess masters and sponsored chess education, but expected chess masters 'to be cultural icons, not hustlers'"—so that they were not to engage in odds-giving (to prove or promote their superiority). (WK 2010ff ; M. Shibut 2004)

And thirdly, still another major problem with the practice of odds-giving in chess tactics is that it is not considered as "scientific" enough.

In other words, in modern times, "chess began to be treated in a scientific, logical way, 'with an assumption of idealized…best play…to underpin all analysis.' From this perspective, a game beginning from a 'lost' position becomes less interesting, even distasteful. Writings by Wilhelm Steinitz (1836–1900), the first World Champion, and James Mason (1849–1905) are consistent with the last point." (WK 2010ff; M. Shibut 2004; W. Steinitz 1889; J. Mason 1910)

*A Note of Caution*

Yet, these illustrative problems with the practice of odds-giving in chess tactics is not to reject the appeal of chess handicap in chess tactics under some conditions.

Rather, they are described above, so as to show us the two opposing sides of the possibility and desirability of tactics (from the perspective of culture with chess handicap as a case study here), such that they are not to the extent that the respective defenders would like us to believe.

Furthermore, this analysis of chess handicap in chess tactics can reveal to us sometihng useful about the ontological principles in existential dialectics, and good examples include the absoluteness-relativeness principle, the predictability-unpredictability principle, the explicability-inexplicability principle, the preciseness-vagueness principle, the simpleness-complicatedness principle, the openness-hiddenness principle, the denseness-emptiness principle, the change-constancy principle, the slowness-quickness principle, the expansion-contraction principle, the convention-novelty principle, the evolution-transformation principle, the symmetry-asymmetry principle, the softness-hardness principle, the regression-progression principle, and the same-difference principle.

For instance, in relation to the absoluteness-relativeness principle, if there is absoluteness (e.g., the absolute view by Bobby Fisher, who, in an interview "with Ralph Ginzburg published in the January 1962 issue of Harper's Magazine,...was quoted as saying that he could successfully give knight odds to any woman in the world"), there is relativeness (e.g., what is true for Bobby Fisher in regard to his claim that he can play chess handicap against any woman in the world and still beat her is not necessarily so for the critics, like "former World Champion Mikhail Tal," who responded that "Fischer is Fischer, but a knight is a knight!" and doubted that Fisher could easily win as he boastfully claimed). (WK 2010ff; C. Forbes 1992; I. Horowitz 1972) And the reverse direction also holds true.

In relation to the predictability-unpredictability principle, if there is predictability (e.g., the predictable tendency of using chess handicap when "a weaker player" is to play against "a stronger one"), there is unpredictability (e.g., the more difficult task to predict the outcome of a particular chess game with chess handicap, even though world champions often win against lesser players, because the outcomes can be quite unexpected, as shown by the examples of some embarrassing losses or draws). And the reverse direction also holds true.

In relation to the explicability-inexplicability principle, if there is explicability (e.g., the explanation by the authority in the Soviet Union that chess handicap is bad, because "chess masters" should behave like "cultural icons, not hustlers"), there is inexplicability (e.g, the lack of convincing explanation by the authority in the Soviet Union, of why being "cultural icons" in chess should necessarily be incompatible with odds-giving in chess playing in the first place). And the reverse direction also holds true.

In relation to the preciseness-vagueness principle, if there is preciseness (e.g., the precise identification of different kinds of odds-giving in chess), there is vagueness (e.g,. the vagueness in the identification of different kinds of odds-giving in chess, because it is not clear what other possible candidates of odds-giving are excluded in the list, or to what extent chess handicap can also allow "eccentric and peculiar odds"). (WK 2010ff; H. Staunton 1849) And the reverse direction also holds true.

In relation to the simpleness-complicatedness principle, if there is simpleness (e.g., the relatively simple analysis of chess handicap by its proponents), there is complicatedness (e.g, the relatively more complicated analysis of chess handicap, as shown in the problems with the tradition of odds-giving). And the reverse direction also holds true.

In relation to the openness-hiddenness principle, if there is openness (e.g., the open exploration of the relationship between a cultural icon and a chess master who is not to fool around with odds-giving), there is hiddenness (e.g., the hidden bias in the ban of odds-giving in the Soviet Union for chess masters who had to behave like "cultural icons," because of the Soviet ideology). And the reverse direction also holds true.

In relation to the denseness-emptiness principle, if there is denseness (e.g, the relatively denser concentration of odds-giving in its heyday during "the eighteenth century and the early nineteenth century"), there is emptiness (e.g., the relatively less dense, or more empty, concentration of odds-giving among chess masters during the Soviet Union). And the reverse direction also holds true.

In relation to the change-constancy principle, if there is change (e.g., the ever new kinds of odds-giving in chess history), there is constancy (e.g,. the ever constancy of the existence of odds-giving in chess history). And the reverse direction also holds true.

In relation to the slowness-quickness principle, if there is slowness (e.g., the relatively slower speed of a chess master to beat a lesser player if he offers odds-giving), there is quickness (e.g., the relatively faster speed of a chess master to beat a lesser play if he does not offer any odds-giving). And the reverse direction also holds true.

In relation to the expansion-contraction principle, if there is expansion (e.g, the relative expansion of the influence of odds-giving in the $19^{th}$ and $20^{th}$ centuries), there is contraction (e.g,. the relative decline of the influence of odds-giving nowadays). And the reverse direction also holds true.

In relation to the convention-novelty principle, if there is convention (e.g., the conventional wisdom in antiquity about playing chess between two players), there is novelty (e.g,. the alternative novel challenge to the conventional wisdom about chess playing in the 19th and 20th centuries by the institution of odds-giving which reached its peak then). And the reverse direction also holds true.

In relation to the evolution-transformation principle, if there is evolution (e.g., the natural evolution of humans to play with each other in the state of nature), there is transformation (e.g, the technical transformation of humans to play with each other by the invention of odds-giving in chess playing). And the reverse direction also holds true.

In relation to the symmetry-asymmetry principle, if there is symmetry (e.g,. the co-existence of different ways of chess playing, be it with odds-giving or without it), there is asymmetry (e.g., the more popularity of chess playing in antiquity without odds-giving—but the more popularity of chess playing with odds-giving in the 19th and 20th centuries). And the reverse direction also holds true.

In relation to the softness-hardness principle, if there is softness (e.g., to induce someone like a lesser player to play by giving odds in a game), there is hardness (e.g,. to refuse giving odds to someone like a lesser player in a game, because of the political correctness that chess masters are "cultural icons" for society, not "hustlers" for personal promotion). And the reverse direction also holds true.

In relation to the regression-progression principle, if there is regression (e.g, the regression made in the history of odds-giving, as shown in the problems pointed out by the critics), there is progression (e.g., the progress made in the tradition of odds-giving, as a way "to enable a weaker player to have a chance of winning against a stronger one"). And the reverse direction also holds true.

And in relation to the same-difference principle, if there is similarity in outcome (e.g., the contribution to the moldiing of beliefs and values, regardless of whether this be done by way of odds-giving or no-odds-giving), there is difference in outcome (e.g., the contribution to the moldiing of beliefs and values by way of odds-giving for a more level-playing field for fairness—but the contribution to the moldiing of beliefs and values by way of no odds-giving for a more collective sensitivity to cultural iconicity beyond the narrow individual concern with self-promotion). And the reverse direction also holds true.

## The Limitation of Tactics

This comprehensive analysis of the possibility and desirability of tactics, from the perspectives of the mind, nature, society, and culture, is vital to show that tactics is both possible and desirable, but not to the extent that the spokespersons from each side would like us to believe. And this can apply to strategy too, as both are closely related.

So, if tactics has its usefulness, it likewise has its limitation, as another side of the same mirror—just as, if strategy has its foresight, it has its own uncertainty too.

These dual analyses of strategy and tactics are important, because "]t]hese two parts of chess thinking cannot be completely separated, because strategic goals are mostly achieved by the means of tactics, while the tactical opportunities are based on the previous strategy of play." (WK 2010)

Now that this long intellectual journey regarding the nature of strategy and tactics in chess has come to an end, what then can be concluded?

This is the final issue, which the concluding chapter, that is, the next chapter on the future of chess, is to address—for which we now turn to chapter four.

**Table 3.1. Tactics and Its Narrowness**

- **Tactics and the Mind**
  —Ex: the game-tree complexity of chess, and the mind
  —Ex: chess blindness, and the appeal to reason

- **Tactics and Nature**
  —Ex: chess, symmetry, and mathematics
  —Ex: chess, analogy, and physics

- **Tactics and Society**
  —Ex: chess tactics, and the use of computers and online chess

- **Tactics and Culture**
  —Ex: tactical exercises, and the question of beauty
  —Ex: chess handicap, and the tradition of odds-giving

*Notes*: The examples in the categories are solely illustrative (not exhaustive), and the comparison is relative (not absolute), nor are they necessarily mutually exclusive. And some can be easily re-classified elsewhere. As generalities, they allow exceptions.
*Source*: A summary of Ch.3 of *FPHCHESS*

**Table 3.2. The Pros and Cons of Computation for Chess Tactics**

---

• **The Pros**

—Ex: "the problem is sharply defined both in allowed operations (the moves) and in the ultimate goal (checkmate)."

—Ex: "it is neither so simple as to be trivial nor too difficult for satisfactory solution."

—Ex: "chess is generally considered to require 'thinking' for skillful play; a solution of this problem will force us either to admit the possibility of a mechanized thinking or to further restrict our concept of 'thinking'"

—Ex: "the discrete structure of chess fits well into the digital nature of modern computers."

• **The Cons**

—Ex: its four advantages (as cited by Shannon) also constitute its disadvantages. Not all difficult chess problems meet these criteria.

—Ex: it still does not do away the factor of uncertainty in chess. This then requires human imagination and creativity.

—Ex: it is still subject to "the problem of Bremermann's fundamental limit" as described in *FC*.

—Ex: it is subject to "the problem of intractability" as already described in *FC*.

---

*Notes*: The examples in the categories are solely illustrative (not exhaustive), and the comparison is relative (not absolute), nor are they necessarily mutually exclusive. And some can be easily re-classified elsewhere. As generalities, they allow exceptions.

*Source*: From *Sec. 3.2.1* of *FPHCHESS*. See book for citations, like WK (2010), CT (2009), J. Teixeira (1998), and others.

**Table 3.3. Chess Novelties and their Limits**

- **Chess Novelties**
  - —Ex: inertia chess
  - —Ex: entropy chess
  - —Ex: magnetic chess
  - —Ex: chaos chess

- **Their Limits**
  - —Ex: it is not so easily played by humans. For example, in "chaos chess," as Neto (1999) admitted, "this type of game can only be programmed in a computer because the number of iterations can be very large (but certainly finite)."
  - —Ex: each game has a different flavor, so that it cannot appeal to all kinds of players. For example, those players who do not like "repulsion" and "attraction" are not likely to like "magnetic chess."

*Notes*: The examples in the categories are solely illustrative (not exhaustive), and the comparison is relative (not absolute), nor are they necessarily mutually exclusive. And some can be easily re-classified elsewhere. As generalities, they allow exceptions.
*Source*: From *Sec. 3.3.2* of *FPHCHESS*. See book for citations, especially J. Neto (1999).

**Table 3.4. The Promise and Pitfall of Online Chess**

---

- **The Promise**
  —Ex: correspondence games
  —Ex: tournaments
  —Ex: teams
  —Ex: forums
  —Ex: puzzles and tips

- **The Pitfall**
  —Ex: there is the issue about software and hardware compatiblity.
  —Ex: online chess fans can spend too much time online, to the point that they become technologically addicted and socially dysfunctional in real life.
  —Ex: there are limits to computation, such that the use of computer games is not as effective as it may sound, as already discussed in *Sec. 3.2.1*.
  —Ex: using computer games and online chess is not free. The expenses for computer games, online chess, and other related costs are not minimal.

---

*Notes*: The examples in the categories are solely illustrative (not exhaustive), and the comparison is relative (not absolute), nor are they necessarily mutually exclusive. And some can be easily re-classified elsewhere. As generalities, they allow exceptions.
*Source*: From *Sec. 3.4.1* of *FPHCHESS*. See book for citations, like WK (2010ee), WK (2010dd), and others.

**Table 3.5. Beauty in the Eyes of the Beholder in Chess Tactics**

- **The Elements of Beauty**
  —Ex: legality
  —Ex: uniqueness
  —Ex: ideal uniqueness
  —Ex: theme(s)
  —Ex: hiddenness
  —Ex: no promotion
  —Ex: economy

- **The Problems with the Elements of Beauty**
  —Ex: even these elements, as chess composers often acknowledge, are not complete. The problem to identify a larger set is difficult enough, because of irreconcilable disagreements among them.
  —Ex: taste is relative, since "beauty is in the eyes of the beholder." As it turns out, the seven elements as cited above represent "modern taste" in chess composition.
  —Ex: the puzzles or tactical exercises to be solved are not taken from actual games but are often "invented for the specific purpose of providing a problem." (WK 2010f) Therefore, its aesthetic evaluation is often not relevant to actual chess games (and thus limited in practical application).

*Notes*: The examples in the categories are solely illustrative (not exhaustive), and the comparison is relative (not absolute), nor are they necessarily mutually exclusive. And some can be easily re-classified elsewhere. As generalities, they allow exceptions.
*Source*: From *Sec. 3.5.1* of *FPHCHESS*. See book for citations, like WK (2010f), and others.

# • Part Four •

## *Conclusion*

# CHAPTER 4
## *CONCLUSION—THE FUTURE OF CHESS*

---

> Chess is but an imperfect variant of a game that was itself a variant of a germinal game whose origins lie somewhere in the darkness of time.
> —D. B. Pritchard (1994)

## Beyond the Science of Chess and the Art of Chess

The analyses of strategy (in Chapter Two) and tactics (in Chapter Three) are valuable for understanding the extent of their possibility and desirability in chess, in a small sense, and in life, in a broad sense.

These analyses are all the more timely nowadays, in light of the two opposing chess sensibilities (as already described in *Sec. 1.1*).

On the one hand, there is this fashionable sensibility which treats chess as "a set of mathematical rules that obey basic laws of physics"—especially when the chess engine famously known as "IBM Deep Blue" beat the world chess champion Gary Kasparov in a series of chess matches in 1997.

On the other hand, there is a contrasting sensibiltiy in the older days, when many computer-skeptics "found it very, very hard to believe chess engines could ever replace or imitate the best of human chess thinking" or "believed the strongest human grandmasters would always be able to outplay computers when all tactics had been drained out of the position." (A. Moll 2009)

Contrary to these two opposing chess sensibilities (and other views as already discussed in the previous two chapters) on the nature of chess (and its future), the strategy and tactics used in chess are neither so physically mechanical nor artistically subjective to the extent that their respective defenders would like us to believe.

But this unorthodox challenge to the two opposing chess sensibilities (and other views as already discussed in the previous two chapters) does not logically imply that chess is therefore an uninteresting board game, or that chess studies are useless. Of course, neither of these two extreme views is reasonable either.

As already indicated in *Sec. 1.1*, this book provides an alternative (better) way to understand the future of chess, especially in the context of strategy and tactics—while learning from different approaches in the literature but without favoring any one of them (nor integrating them, since they are not necessarily compatible with each other).

Thus, this book offers a new theory (that is, *the mediative-variative theory of chess*) to go beyond the existing approaches in the literature on chess in a new way not conceived before.

This theory was also related to other theories of mine, like the one on organization (namely, *the combinational theory of organization* as already worked out in *FPHO*) and the one on war and peace (namely, *the multifaceted theory of war and peace* as already worked out in *FPHWP*), since the issue concerning strategy and tactics is important not only in the context of chess but also in the context outside chess (e.g., of social organizations and social institutions).

That said—there are two words in the title of my new theory, *mediative* and *variative*.

My theory is "meditative" because it tries to mediate the competing theoretical approaches in the literature (say, on whether chess is a science or an art) by learning from them but without favoring any of them nor trying to integrate them (as they are not necessarily compatible with each other).

And my theory is also "variative," because it offers a visionary prediction of the different variants of chess that the post-humans will use as substitutes for the human ones (hitherto existing in human history) which will no longer be much useful in the post-human future.

With this clarification in mind—the mediative-variative theory of chess has five distinctive features to remember.

Firstly, my theory makes good use of all theoretical approaches in the literature on chess, be they about the mental argument, the natural argument, the social argument, or the cultural argument—especially from the most comprehensive combined perspectives of the mind, nature, society, and culture (as already discussed in Chapter Two and Chapter Three).

Secondly, just like many other theories of mine in my previous books, my theory here does not heavily favor any specific theory over others in the literature, nor trying to integrate them (as they are not necessarily compatible with each other).

Thirdly, my theory treats the issue of chess in the distinctive dialectic context of strategy and tactics—especially in the larger dialectic context of no strategy without tactics, and vice versa, together with the subsequent transcendence of both.

Fourthly, my theory on strategy and tactics is useful not only in the context of chess but also in the context outside chess (e.g., of social organizations and social institutions). In other words, it is important in chess, in a small sense, and in life, in a broad sense (as summarized in *Table 1.1* and *Table 1.2*).

And fifthly, it contains fourteen major theses, namely, (a) the first thesis: the absoluteness-relativeness principle, (b) the second thesis: the predictability-unpredictability principle, (c) the third thesis: the explicability-inexplicability principle, (d) the fourth thesis: the preciseness-vagueness principle, (e) the fifth thesis: the simpleness-complicatedness principle, (f) the sixth thesis: the openness-hiddenness principle, (g) the seventh thesis: the denseness-emptiness principle, (h) the eighth thesis: the slowness-quickness principle, (i) the nineth thesis: the convention-novelty principle, (j) the tenth thesis: the evolution-transformation principle, (k) the eleventh thesis: the symmetry-asymmetry principle, (l) the twelveth thesis: the regression-progression principle, (m) the thirteenth thesis: the sameness-difference principle, and (n) the fourteenth thesis: the post-human variation, in relation to the three great future transformations of mind games—to be elaborated in the rest of the book, with a summary in the concluding chapter

If successful, this seminal project is to fundamentally change the way that we think about chess, from the combined perspectives of the mind, nature, society, and culture, with enormous implications for the human future and what I originally called its "post-human" fate.

## Other Relevant Principles

Lest misunderstanding hastily occurs (as this is something that I regularly did in my previous books, *almost verbatim*), it should be clarified at the outset that, if other ontological principles in existential dialectics (other than the 13 as listed in the 14 theses) are not explicitly included in the listing of the theses in the Table of Contents, it is not because they are not important, but merely because they are not as equally relevant as the others in the current case study (in terms of the number of citations of each principle in the book), although different case studies will, of course, reveal different degrees of relevance of the principles.

Even then, in some cases (like this one), the difference between any two given principles, for instance, in terms of the number of citations in a book, is rather small, so the reason in those cases is more aesthetic (than otherwise), because it looks nicer to list only 13 theses for 13 principles (than 21 theses for all of the 21 principles) in the Table of Contents.

This is true, even if different studies of the same kind can yield different views about the degree of relevance for each ontological principle, depending on the specific nature of a research in question, needless to say. So, if a different author analyzes the subject matter in a different way, the relevance of the ontological principles will be different.

After all, the symmetry-asymmetry principle already implies that not all principles are equally relevant in any topic under examination, as they are not created equal for all cases under all circumstances (which would be an extreme view, of course).

In any event, the relevancy of other principles (besides the ones as already listed in the theory above) for the present study can be summarized below, in relation to the five categories of ontological principles in existential dialectics.

*In the Category of Method*

Firstly, in regard to the formalness-informalness principle (on the formal requirements of logical systems), if there is formalness (e.g., the formal requirement of completeness in a system of ideas or rules, in *Sec. 2.4.1*; the formal requirement of a system of ideas to be complete, in *Sec. 2.4.2*; the formal requirement of completeness and soundness in a system of ideas, in *Sec. 3.3.1*; and the formal logical requirement of completeness and soundness in a system of ideas, in *Sec. 3.5.1*), there is informalenss too (e.g., the non-formal nature of a chess engine to be unable to evaluate or justify the rules and values in chess games, except using them as already explicitly programmed in the computer software, in *Sec. 2.4.1*; the nonformal aspects of the report by a team of researchers from the UK led by Merim Bilalic, like its unquestioned assumption that the biological argument is not to be taken seroiusly, without proving that it is so, in *Sec. 2.4.2*; the informal shortcoming of Schwenk's Theorem, in that it is proved in regard to the three conditions in a way, but in a diferent way, it is not, because "[s]mply proving the...three conditions does not prove the theorem," as "it is still required to prove that all rectangular boards that do not fall in one of the above three categories have closed knight's tours," in *Sec. 3.3.1*; and the non-formal inadequacy of the elements for the aesthetic evaluations of tactical exercies in chess problems, in that, although they are recognized by many chess composers and thus not arbitrary, they represent "modern" taste and are not accepted by all as a complete proof, in *Sec. 3.5.1*). And the reverse direction also holds true.

And secondly, in regard to the partiality-totality principle (on the relationships between whole and parts), if there is partial analysis (e.g., the partial thinking of each player to make moves which would help him beat the other side in a game of chess, in *Sec. 2.2.2*; the partial vantage point of each cultural variant that its rules are the correct way to play chess, in *Sec. 2.5.2*; the partial view of each player to calculate different possible moves and evaluate the resulting positions to find the best move in a chess game, in *Sec. 3.2.1*; the partial view of each player to make the best move, trying to outsmart the other player, in *Sec. 3.2.2*; and the partial view by a player to change some initial conditions in a chaos chess game, in *Sec. 3.3.2*), there is also holistic analysis (e.g., the more totalistic view of the entire game of chess, in that "the possibilities [of moves] are so numerous that no one can foresee how the game will move," such that the whole is not the sum of its parts, that is, not the sum of the individualist views of the players obsessed with winning, in *Sec. 2.2.2*; the holistic view of the larger picture of all the vantage points of all the cultural variants, such that there is no universal correct way to play chess, so that the whole is not the sum of its parts, that is, the holistic view is not the sum of all the partial viewpoints of the cultural variants, in *Sec. 2.5.2*; the holistic view of the entire picture, such that "the possibilities are so numerous that no one can foresee how the game will move" exactly, so that the whole is not the sum of its parts, that is, not the sum of the partial views of the players in a chess game, in *Sec. 3.2.1*; the unexpected outcomes of the games as shown in the chess blunders, which surprised many after the matches, such that the whole is not the sum of its parts, that is, not the sum of the partial views of the players at the beginning of the games, in *Sec. 3.2.2*; and the more holistic view of uncertainty for the entire game, when a small change to an initial condition is allowed, because "the sensitivity [in chaos chess] to the initial conditions brings uncertainty to simple moves," in *Sec. 3.3.2*). And the reverse direction also holds true.

### *In the Category of Structure*

Firstly, in regard to the finiteness-transfiniteness principle (on the nature of numbers), if there is finiteness (e.g., the finite number of the players, that is, 2 in a game, and of the pieces in Western chess, that is, 32 on the chess board, in *Sec. 2.2.2*; the finite number of pieces and rules in a chess game, in *Sec. 2.3.1*; the finite number of rules that a chess computer needs to memorize, in *Sec. 2.4.1*; the finite number of squares on a chess board, in *Sec. 3.2.1*; the finite number of knights on an 8 × 8 chess board, in *Sec. 3.3.1*; and the finite number of rules in chaos chess, in *Sec. 3.3.2*), there is also transfiniteness (e.g., the transfinite number of possible moves in chess, because "the possibilities [of moves] are so numerous that no one can foresee how the game will move," in *Sec. 2.2.2*; the transfinite number of possible moves in chess, such that different "Tricks, Puzzles and Number patterns have tested the intellect of the mathematicians for thousands of years," in *Sec. 2.3.1*; the transfinite number of possible moves that a chess computer needs to calculate, especially in a Type A search strategy, in *Sec. 2.4.1*; the transfinite number of "legal positions in chess...to be between 1043 and 1047, with a game-tree complexity of approximately 10123," for instance, in *Sec. 3.2.1*; the transfinite number of 26,534,728,821,064 directed closed tours on an 8 × 8 chess board, in *Sec. 3.3.1*; and the transfinite number of "iterations" in chaos chess, which "can be very large" and "can only be programmed in a computer," in *Sec. 3.3.2*). And the reverse direction also holds true.

*In the Category of Process*

Firstly, in regard to the change-constancy principle (on the alteration of things), if there is change (e.g., the ever-changing nature of "Tricks, Puzzles and Number patterns" which "have tested the intellect of the mathematicians for thousands of years," in *Sec. 2.3.1*; the ever new rules for chess over time, in *Sec. 2.5.2*; the ever new innovations of computational enhancements for high performance, in *Sec. 3.2.1*; the ever new variants of chess in the history of chess, with the ones by Neto as latest examples, in *Sec .3.3.2*; and the ever new kinds of odds-giving in chess history, in *Sec. 3.5.2*), there is constancy too (e.g., the ever constant existence of "Tricks, Puzzles and Number patterns" in chess, in *Sec. 2.3.1*; the ever constancy of the existence of certain rules in chess, in *Sec. 2.5.2*; the continued existence of limits to computation, regardless of the innovations, in *Sec. 3.2.1*; the ever constancy of the existence of rules in chess, in *Sec. 3.3.2*; and the ever constancy of the existence of odds-giving in chess history, in *Sec. 3.5.2*). And the reverse direction also holds true.

Secondly, in relation to the order-chaos principle (on the pattern of things), if there is order (e.g., the relatively orderly nature of chess games in regard to the rules which are explicitly stated, in *Sec. 2.4.1*; the existence of certain rules as listed by Franklin for chess games, so as to provide an orderly structure, which can be followed, in *Sec. 2.5.1*; the relatively orderly nature of chess games, like the use of computation to calculate different possible moves and evaluate the resulting positions to find the best move in a chess game, in *Sec. 3.2.1*; the relatively orderly nature of chess games, because they are governed with different rules, in *Sec. 3.2.2*; and the orderly nature of chaos chess, because of the rules as specified, in *Sec. 3.3.2*), there is chaos too (e.g., the relatively chaotic nature of chess games in regard to unexpected "mistakes," the difficulty to "anticipate which moves…[computer] programs would play, and why," "the possibilities [of moves] are so numerous that no one can foresee how the game will move," and so on, in *Sec. 2.4.1*; the relative chaotic nature of chess games too, such that, as Franklin put it, "the game is so full of events, there is such a variety of turns in it, [and] the fortune of it is so subject to sudden vicissitudes," in *Sec. 2.5.1*; the relativley chaotic nature of chess games, because "the possibilities are so numerous that no one can foresee how the game will move" exactly, in *Sec. 3.2.1*; the relatively chaotic outcomes of chess games, especially when unexpected chess blunders occurred, in *Sec. 3.2.2*; and the relatively chaotic outcome of chaos chess, when initial conditions are changed, in *Sec. 3.3.2*). And the reverse direction also holds true.

And thirdly, in relation to the expansion-contraction principle (on the growth of things), on the one hand, there is expansion (e.g., the relatively more developed ability of a chess prodigy to rely on his memory, imagination, and perception to play chess, in *Sec. 2.2.1*; the relatively more developed ability of chess players to think logically and creatively, in *Sec. 2.2.2*; the relative expansion of ever new "Tricks, Puzzles and Number patterns" in chess to test "the intellect of the mathematicians for thousands of years," in *Sec. 2.3.1*; the relatively more developed ability in a computer with Type A search strategy to search for "every possible position for a fixed number of moves using the minimax algorithm," in *Sec. 2.4.1*; the relatively more developed ability of a male to play chess, in *Sec .2.4.2*; the relatively more developed ability to be "civil" and "moral" by those who follow the morals and rules as laid down by Franklin, in *Sec. 2.5.1*; the relative expansion of the chess variant of the 8x8 ashtāpada to the rest of the world from its birthplace in India, in *Sec. 2.5.2*; the relative expansion of the use of computers for the understanding of chess complexity in our time, in *Sec. 3.2.1*; the relatively more developed ability of a fan of tactical exercises in chess to think logically and aesthetically in relation to the rules as stipulated by chess composers, in *Sec. 3.5.1*; and the relative expansion of the influence of odds-giving in the 19th and 20th centuries, in *Sec. 3.5.2*).

On the other hand, there is contraction too (e.g., the relatively less developed ability of a chess prodigy to rely on his long history of many years of chess practice to play chess, which he lacks, in *Sec. 2.2.1*; the relatively less developed ability of chess players to think emotionally and mystically, in *Sec. 2.2.2*; the relative contraction of old "Tricks, Puzzles and Number patterns" in chess which have been solved, either wholly or partially, by some solutions like the ones proposed by Leonard Euler on the knight's tour, "some of which are closed solutions to the re-entry problem and some are open solutions," in *Sec. 2.3.1*; the relatively less developed ability in a computer with Type A search strategy to use "quiescence searches" and "extensions and pruning" instead, in *Sec. 2.4.1*; the relatively less developed ability of a male to play with dolls, in *Sec. 2.4.2*; the relatively less developed ability to be "rude" and "unequitable" by those who follow the morals and rules as laid down by Franklin, in *Sec. 2.5.1*; the relative decline of the chess variants of "the 10×10 Dasapada" or "the 9×9 Saturankam" from its birthplace in India, in *Sec. 2.5.2*; the relative less reliance on the human mind to calculate chess complexity in our time, in *Sec. 3.2.1*; and the relative decline of the influence of odds-giving nowadays, in *Sec. 3.5.2*). And the reverse direction also holds true.

*In the Category of Agency*

Firstly, in regard to the theory-praxis principle (on the duality of knowledge), on the one hand, there is theoretical construction (e.g., the theoretical construction of the importance of "deliberate practice" for playing chess well, in *Sec. 2.2.1*; the theoretical construction of the mental benefits of chess playing, in *Sec. 2.2.2*; the theoretical construction of alternative search strategies for chess engines by the team in Northwestern University in the 1970s, in *Sec. 2.4.1*; the theoretical construction of the morals of chess by Franklin, in *Sec. 2.5.1*; the theoretical construction of computer chess, as shown by Shannon's groundbreaking paper on computer chess in 1950, in *Sec. 3.2.1*; the theoretical construction of the "Kotov syndrome" by Alexander Kotov, in *Sec. 3.2.2*; the theoretical construction in physics, like chaos theory, the law of thermodynamics, etc., in *Sec. 3.3.2*; the theoretical construction of computation and computer programming, in *Sec. 3.4.*; and the theoretical construction of aesthetics to understand the nature of beauty, in *Sec. 3.5.1*; ).

On the other hand, there is also its practical application (e.g., the practical application of the theory of "deliberate practice" by such organizations like Chess-In-The-Schools, Inc., the American Foundation for Chess, and others to encourage children to play chess, in *Sec. 2.2.1*; the practical application of the view concerning the mental benefits of chess playing by its advocates to encourage others to play more chess, in *Sec. 2.2.2*; the practical application of alternative search strategies for chess engines by the team in Northwestern University, such that they won the

ACM Computer Chess Championship in that year, with the new program called "Chess 4.0," in *Sec. 2.4.1*; the practical application of the morals of chess by encouraging people to play chess for the benefits, as "Franklin was posthumously inducted into the U.S. Chess Hall of Fame in 1999," in *Sec. 2.5.1*; the practical application of Shannon's theoretical construction for the use of chess engines in chess tournaments, like the victory of IBM's Deep Blue over Garry Kasparov, in *Sec. 3.2.1*; the practical application of the "Kotov syndrome…in many other deadline-oriented situations as well," as shown in the example that, when "a journalist might puzzle over how to explain a complicated political process without finding a simple solution" and, as "the deadline is approaching, they throw together a story and are disappointed the next morning to realize the story was a failure," in *Sec. 3.2.2*; the practical application of physical theories for the invention of new chess games, as proposed by Neto and his colleagues, for instance, in *Sec 3.3.2*; the practical application of computation and computer programming for chess playing computer games, in *Sec. 3.4*; and the practical application of aesthetics by chess composers for the aesthetic evaluation of tactical exercises in chess problems, in *Sec. 3.5.1*). And the reverse direction also holds true.

And secondly, in relation to the softness-hardness principle (on the force of change), if there is softness (e.g., the peaceful way to compete between two players in a chess game, in *Sec. 2.2.2*; the promotion of a pawn in an endgame, in *Sec. 2.4.2*; the praise by rule-abiding chess players to those who follow the rules of chess games, in *Sec. 2.5.1*; the reward of "additional time budget" to a player "who made the predetermined number of moves in the agreed time," in *Sec. 2.5.2*; the winning in a chess game, in Sec. 3.2.2; the attraction in magentic chess, in *Sec. 3.3.2*; to play chess in the traditional way without paying a fee, in *Sec. 3.4*; the judgement that a puzzle in chess is "beautiful" on the basis of the seven elements for aesthetic evaluation, in *Sec. 3.5.1*; and to induce someone like a lesser player to play by giving odds in a game, in *Sec. 3.5.2*), there is hardness too (e.g., the violent way to deal with loss after a chess game for some chess players, in *Sec. 2.2.2*; the sacrifice of a pawn for a more valuable piece, in *Sec. 2.4.2*; the reprimand by rule-abiding chess players to those who do not follow the rules of chess games, in *Sec. 2.5.1*; the penalty of "forfeiture" to a player who exceeds the time limit, in *Sec. 2.5.2*; the losing in a chess game, in *Sec. 3.2.2*; the repulsion in magnetic chess, in *Sec. 3.3.2*; to play chess online with ChessWorld with a fee for membership, in *Sec. 3.4*; the judgement that a puzzle in chess is not "beautiful," or "ugly" so to speak, because it violates the seven elements for aesthetic evaluation, according to the advocates, in *Sec. 3.5.1*; and to refuse giving odds to someone like a lesser player in a game, because of the political correctness that chess masters are "cultural icons" for society, not "hustlers" for personal promotion, in *Sec. 3.5.2*). And the reverse direction also holds true.

## 1st Thesis: The Absoluteness-Relativeness Principle

The first thesis refers to the absoluteness-relativeness principle (on the multiplicity of things) in the category of "method" in existential dialectics, which was first proposed in *FPHK*, in that there is the multiplicity of things in reality, be they about entities, qualities (or properties), and relationships. If there is something absolute, there is likewise something relative. And there is no absoluteness without relativeness—and vice versa.

Both absoluteness and relativeness here are also relevant to different modalities often cited in the literature on ontology, such as possibility (e.g., something "can" happen) and its opposite (e.g., impossibility), probability (e.g., something "will" happen) and its opposite (e.g., improbability), and necessity (e.g., something "should" happen) and its opposite (e.g., contingency).

For instance, on the one hand, there are absolute viewpoints (e.g., the absolute view of de Groot that perception and memory are important for brilliant chess playing, in *Sec. 2.2.1*; the absolute view of the advocates of chess playing in regard to the mental benefits of playing chess, in *Sec. 2.2.2*; the absolute view of Vasantha in regard to cosmology in chess origin, in *Sec. 2.3.1*; the absolute view by the supporters of Type A research strategy that it is better than Type B, in *Sec. 2.4.1*; the absolute view by Bilalic and others that the biological argument is "controversial" or untenable, in *Sec. 2.4.2*; the absolute view by Franklin about the morals of chess, in *Sec. 2.5.1*; the absolute view by each cultural variant that its rules of chess are the right way to play chess, in *Sec. 2.5.2*; the absolute view of those "chess enthusiasts and computer engineers" like Shanon, who wrote the groundbreaking paper on "programming a computer for playing chess" in 1950, in *Sec. 3.2.1*; the absolute view by Kramnik that "he had planned the supposedly winning move 34...Qe3 already when playing 29...Qa7, and had rechecked the line after each subsequent move," in *Sec. 3.2.2*; the absolute view of Schwenk's Theorem, in that "[f]or any m × n board with m less than or equal to n, a closed knight's tour is always possible unless one or more of the…three conditions are true," in *Sec.*

*3.3.1*; the absolute view in physics of "extreme situations in which space and time are warped in unfamiliar ways," in *Sec. 3.3.2*; the absolute view of addicted fans of online chess in regard to its benefits, in *Sec. 3.4*; the absolute view by those who accept the "modern taste" of the aesthetic evaluations of tactical exercises in chess on the basis of the seven elements, in *Sec. 3.5.1*; and the absolute view by Bobby Fisher, who, in an interview "with Ralph Ginzburg published in the January 1962 issue of Harper's Magazine,...was quoted as saying that he could successfully give knight odds to any woman in the world," in *Sec. 3.5.2*).

On the other hand, there are likewise relative counterparts (e.g., what is important for de Groot in regard to perception and memory in chess playing is not necessary so for Ericsson and his colleagues, who emphasize more "deliberate practice," in *Sec. 2.2.1*; what is true for the advocates of chess playing in regard to the mental benefits is not necessarily sure for the critics, who pointed out that "the results of research studies have failed to produce unambiguous support for the intellectual benefits of playing chess," in *Sec. 2.2.2*; what is true for Vasantha in regard to the role of cosmology in chess origin is not necessarily so for Cho-Yo in regard to the role of the military institution in chess history, in *Sec. 2.3.1*; what is true for those in Type A in regard to its being better is not necessarily so for the supporters of Type B research strategy, who regard the latter as better, in *Sec. 2.4.1*; what is true for Bilalic and others in regard to the rejection of the biological argument is not so for the critics, who allow the biological argument too, in *Sec. 2.4.2*; what is true for Franklin about the morals of chess is not necessarily so for the critics who are more skeptical to his list of the four mental benefits of chess playing, in *Sec. 2.5.1*; what is right for Western chess to play "in the squares" is not so for Chinese chess to play "on the intersections of the lines on the board," in *Sec. 2.5.2*; what is possible and desirable for Shannon in regard to chess engines is not necessarily so for the critics like Bremermann and others in regard to the fundamental limits to computation, in *Sec. 3.2.1*; what is right for Kramnik in regard to the move 34...Qe3??? is not so for Deep Fritz which countered it with move 35.Qh7#, checkmate, in *Sec. 3.2.2*; what is true for Schwenk's Theorem in regard to the proof of the closed knight's tour problem is not necessarily so for the critics, who pointed out its weaknesses, in *Sec. 3.3.1*; what is true in physics in regard to extreme situations as opposed to normal situations is not necessarily so in chess, when the rule governing the movement of the pawn also allows pawn promotion, in *Sec. 3.3.2*; what is true for the addicted fans of online chess in regard to its benefits is not necessarily so for the critics, as shown in the aforecited problems, in *Sec.*

*3.4*; what is beautiful for those who accept the 'modern taste" of aesthetical evaluation in regard to the seven elements is not necessarily so for others who do not accept the "modern taste," as "beauty is in the eyes of the beholder," in *Sec. 3.5.1*; and what is true for Bobby Fisher in regard to his claim that he can play chess handicap against any woman in the world and still beat her is not necessarily so for the critics, like "former World Champion Mikhail Tal," who responded that "Fischer is Fischer, but a knight is a knight!" and doubted that Fisher could easily win as he boastfully claimed, in *Sec. 3.5.2*).

And the reverse direction also holds true.

However, it should be stressed (as this is something that I regularly did in my previous books, *almost verbatim*) that there are different shades of gray (or different degrees of truth) in the two opposites, and the classification is not necessarily mutually exclusive either. In fact, even when some combinations of the two occur, they only end up sharing the same dialectic relationship, but in a different degree.

Also, there are some other relationships (as a kind of family resemblance) which have something in common with the principle, but they are not exactly the same but only more or less comparable, which varies from case to case.

Good examples of family resemblance in relation to the principle are uniformity-diversity, internalness-externalness, immanence-transcendence and so forth.

# 2nd Thesis: The Predictability-Unpredictability Principle

The second thesis is called the predictability-unpredictability principle (on the occurrence of events) in the category of "method" in existential dialectics (which was first proposed in *FC* and, later, other books of mine), in that both predictability and unpredictability have a major role to play in the occurrence of things, so that neither determinism nor indeterminism wins the centuries-old fight. There is no predictability without unpredictability—and vice versa.

There are events which are predictable, just as there are those which are not. Or what is regarded as unpredictable at one point in time may turn out to be predictable later, and, conversely, what is deemed as predictable may turn out to not be so predictable. Even in predictability, outcomes are subject to uncertainty, the degree of which varies from case to case.

For instance, on the one hand, there is predictability (e.g., the predictable tendency of a chess prodigy to "play chess so well" that he can "beat Masters and even Grandmasters, often at a very young age," in *Sec. 2.2.1*; the predictable tendency of a chess player to try his best to make moves which, in his judgement, would help him beat his opponent in a chess game, in *Sec. 2.2.2*; the predictable tendency of people like Vasantha to understand chess origin by looking into "the movements of the celestial bodies and their numerical symbolism," in *Sec. 2.3.1*; the predictable tendency of Type A research strategy to examine "every possible position for a fixed number of moves using the minimax algorithm," in *Sec. 2.4.1*; the prediction that "men performed slightly better than expected, with an average advantage of 353 points, whereas the expected advantage was 341 points," according to the results by Bilalic et al., in *Sec. 2.4.2*; the predictable tendency of the followers of Franklin to accept the four mental benefits of chess playing, in *Sec. 2.5.1*; the predictable tendency of each cultural variant to play chess in accordance to its own rules at a given point in time, in *Sec. 2.5.2*; the predictable tendency of those who use the technique of NegaScout to rely "on accurate move ordering to capitalize on this advantage" and therefore to "be faster than alpha-beta pruning" in calculation, in *Sec. 3.2.1*; the predictable bet by chess fans at the beginning of the match in 1934 to favor Esteban Canal, also known as "the Peruvian Immortal," in his play against an unknown amateur, in *Sec. 3.2.2*; the predictable feature of a knight's tour to be symmetric in a certain way, in *Sec. 3.3.1*; the predictable tendency of chaos chess to be fixed by the rules as specified, in *Sec. 3.3.2*; the predictable tendency of online chess fans to practice their chess tactics online, in *Sec. 3.4*; the predictable tendency of chess composers with "modern taste" to judge beauty in tactical exercises on the basis of "uniqueness," "economy," etc., in *Sec. 3.5.1*; and the predictable tendency of using chess handicap when "a weaker player" is to play against "a stronger one," in *Sec. 3.5.2*).

On the other hand, there is unpredictability (e.g., the more difficult task to predict which particular chess prodigy will end up making a major contribution to the world of chess, because "while some become World Champions, others fail to make progress in adulthood," in *Sec. 2.2.1*; the more difficult task to predict which particular chess player would behave to his opponent in a particular way after the end of a particular game, in light of the history of chess-related deaths or violence, in *Sec. 2.2.2*; the more difficult task to predict exactly to what extent any particular chess games will necessarily follow the rules as observed in "the movements of the celestial bodies and their numerical symbolism" as claimed by Vasantha, in *Sec. 2.3.1*; the more difficult task to predict the outcome of a particular chess game, if Type A research strategy is used on that particular occasion, in *Sec. 2.4.1*; the more difficult task to predict exactly the gender differences in chess performance, if projected outside the database of the year 2007 and outside the sample case of Germany, as originally used by the report, in *Sec. 2.4.2*; the more difficult task to predict which particular mental benefits a particular chess player will gain, since "the results of research studies have failed to produce unambiguous support for the intellectual benefits of playing chess," in *Sec. 2.5.1*; the more difficult task to predict exactly what the future rules of chess for each cultural variant will be in the long term, because all these rules undergo changes over time, in *Sec. 2.5.2*; the more difficult task to predict exactly how a game will evolve over time, since "the possibilities are so numerous," even with the use of NegaScout, in *Sec. 3.2.1*; no one predicted that the game would end by a chess blunder when "N.N. played 10. O-O-O??," in *Sec. 3.2.2*; the more difficult task to predict which particular "rectangular boards that do not fall in one of the above three categories have closed knight's tours," as there are weaknesses in the theorem, in *Sec. 3.3.1*; the more difficult task to predict the outcome of the game if some initial conditions are changed in a chaos chess game, in *Sec. 3.3.2*; the more difficult task to predict to what extent and in what eay exactly a particular online chess fan will have his chess tactics improved because of his play online, in *Sec. 3.4*; the more difficult task to predict exactly what future chess composers in a different historical era will judge beauty in tactical exercises, if looked narrowly from the current list of elements only, in *Sec. 3.5.1*; and the more difficult task to predict the outcome of a particular chess game with chess handicap, even though world champions often win against lesser players, because the outcomes can be quite unexpected, as shown by the examples of some embarrassing losses or draws, in *Sec. 3.5.2*).

And the reverse direction also holds true.

Again, it should be stressed, however, that there are different shades of gray (or different degrees of truth) in the two opposites, and the classification is not necessarily mutually exclusive either. In fact, even when some combinations of the two occur, they only end up sharing the same dialectic relationship, but in a different degree.

Also, there are some other relationships (as a kind of family resemblance) which have something in common with the principle, but they are not exactly the same but only more or less comparable, which varies from case to case.

Good examples of family resemblance in relation to the principle are sureness-arbitrariness, and so forth.

# 3rd Thesis: The Explicability-Inexplicability Principle

The third thesis is called the explicability-inexplicability principle (on the underlying mechanisms of things) in the category of "method" in existential dialectics (which was proposed in *FPHU* and other books of mine), in that both explicability and inexplicability are part of the understanding of things. There is no explicability without inexplicability—and vice versa.

This principle tells us the dual nature of the research dilemma, in that, if reality can be explained in some ways, it also has its other ways which are not quite explainable, at a given point in time.

For instance, on the one hand, there is explicability (e.g., the explanation by scholars like Ericsson and his colleagues that "deliberate practice is sufficient for reaching high levels of expertise, like master in chess," in *Sec. 2.2.1*; the explanation of the importance of chess playing on the basis of the mental benefits, as claimed by the advocates, in *Sec. 2.2.2*; the explanation by Vasantha that chess origin can be traced back to cosmology, in *Sec. 2.3.1*; the explanation, in the early days of human-computer chess matches, by chess experts, especially after a chess computer in 1998 "defeated Viswanathan Anand who at the time was ranked second in the world," in that "in fast games, computers played better than humans but at classical time controls—at which a player's rating is determined—the advantage was not so clear," in *Sec. 2.4.1*; the explanation of the gender differences in chess performance by the team of research from the UK on the basis of gender differences in chess participation rate, as shown in the sample population which "included 113,386 men and 7,013 women," that is, "a ratio of 16:1," in *Sec .2.4.2*; the explanation by Franklin that chess playing is beneficial because of the moral consequences, in *Sec. 2.5.1*; the explanation, by chess historians, of the development of modern chess from early India on the basis of its historical origin in ashtāpada [Sanskrit for 'having eight feet,' i.e., an 8x8 squared board]," in *Sec. 2.5.2*; the explanation, by Shannon, of the superiority of using chess engines to play, on the basis of some distinctive advantages, in *Sec. 3.2.1*; the explanation, by Alexander Roshal, of the blunder by Kramnik, by saying that Deep Fritz's "mating pattern of a queen on h7 protected by a knight on f8 is extremely rare," in *Sec. 3.2.2*; the explanation, by Schwenk's Theorem, of the impossibility of closed knight's tours if the three conditions are met, in *Sec. 3.3.1*; the explanation by Feynman that, "since diagonal squares are always colored the same, this explains why bishops always stay on the same color," in *Sec. 3.3.2*; the explanation by Lang of the dropout of ChessGenius of "the top ten on the SSDF...rating list" in 1999 and "over the following decade" on the basis that "the program does not scale well to faster hardware," in *Sec. 3.4*; the explanation of beauty in tactical exercises on the basis of the seven elements for aesthetic evaluation, in *Sec. 3.5.1*; and the explanation by the authority in the Soviet Union that chess handicap is bad, because "chess masters" should behave like "cultural icons, not hustlers," in *Sec. 3.5.2*).

On the other hand, there is inexplicability (e.g., the lack of sufficient explanation, if looked within the worldview of Ericsson and his colleagues, of why chess prodigies succeed in playing chess brilliantly "at a very young age," even without the long history of many years of practice that other players may have, by comparison, in *Sec. 2.2.1*; the lack of sufficient explanation of the intriguing discrepancy that chess playing is not very popular in black Africa but is highly popular in the former Soviet republics and East Asian countries, if the mental benefits of playing chess are so obvious as its advocates often claim, in *Sec .2.2.2*; the lack of sufficient explanation by Vasantha about the reductionistic treatment of chess origin, since, for the critics, other factors like the social, cultural, and mental arguments are ignored in the explanation, in *Sec. 2.3.1*; the lack of sufficient explanation by those chess experts who still think that humans are better than chess engines, of why better and better "computer systems" have increasingly beaten more and more "human chess champions" later on, even at classical time controls, in *Sec. 2.4.1*; the failture of the team of research from the UK to explain "why so few women participate in chess at all" in the first place, in *Sec. 2.4.2*; the lack of sufficient explanation, by the defenders of Franklin, of why, if the moral benefits are so obvious, there are chess-related deaths or murders, as pointed out by the critics, in *Sec. 2.5.1*; the lack of sufficient explanation of why modern chess had to begin like that, instead of other alternative origins like "the 10×10 Dasapada and the 9×9 Saturankam," in *Sec. 2.5.2*; the lack of sufficient explanation, by Shannon, of why these advantages will not turn out to be disadvantages too for chess engines, such that there are limits to computation, in *Sec. 3.2.1*; the lack of sufficient explanation by Roshal of why Kramnik could not detect this move in the first place, even if it were rare, since he was such a brilliant grandmaster after all, as pointed out by the critics, who thought there was no excuse for Kramnik to make such a blunder, in *Sec. 3.2.2*; the lack of sufficient explanation by Schwenk's Theorem of whether or not "all rectangular boards that do not fall in one of the above three categories have closed knight's tours," in *Sec. 3.3.1*; the lack of sufficient explanation by Feynman of what exactly this really "explains" in chess, since the rules of chess in general do not try to "explain" anything about reality in nature that the laws of physics do, and besides, "moving diagonally" and "staying on the same colored squares" are part of the same definition of the rule governing the movement of the bishop, and neither constitutes a "better" explanation than the other," in *Sec. 3.3.2*; the lack of sufficient explanation by Lang of why this must be the reason, not other possible explanations, be they about software weaknesses and something else, in *Sec. 3.4*; the lack of sufficient explanation of why these seven elements must be both necessary and sufficient, in light of the relativist challenge by the critics, in that "beauty is in the eyes of the beholders," in *Sec. 3.5.1*; and the lack of convincing explanation by the authority in the Soviet Union, of why being "cultural icons" in chess should necessarily be incompatible with odds-giving in chess playing in the first place, in *Sec. 3.5.2*).

And the reverse direction also holds true.

Once more, it should be stressed, however, that there are different shades of gray (or different degrees of truth) in the two opposites, and the classification is not necessarily mutually exclusive either. In fact, even when some combinations of the two occur, they only end up sharing the same dialectic relationship, but in a different degree.

Also, there are some other relationships (as a kind of family resemblance) which have something in common with the principle, but they are not exactly the same but only more or less comparable, which varies from case to case.

Good examples of family resemblance in relation to the principle are underlyingness-regularness, causation-regularity, causation-correlation and so forth.

## 4th Thesis: The Preciseness-Vagueness Principle

The fourth thesis is called the preciseness-vagueness principle (on the refinement of things) in the category of "structure" in existential dialectics (which was first out worked out in *FIA*), in that both preciseness and vagueness are important, not that one is better than the other, but that both are used, in different degrees of preference, in accordance to the contextual application from the perspectives of nature, the mind, culture, and society. There is no preciseness without vagueness—and vice versa.

For instance, on the one hand, there is preciseness (e.g., the precise identification of "a very young age" as one major measurement of chess prodigies, in *Sec. 2.2.1*; the precise identification of the seven ways that playing chess can benefit the mind, by the advocates, in *Sec. 2.2.2*; the precise identificaion, by Vasantha, of "the movements of the celestial bodies and their numerical symbolism" in chess origin, in *Sec. 2.3.1*; the relatively precise identification of Type B search strategy by Shannon in 1950 as the better one, in *Sec. 2.4.1*; the precise identification, by the team of research from the UK, of "the sample population" which "included 113,386 men and 7,013 women," that is, "a ratio of 16:1," in *Sec. 2.4.2*; the precise identification of the four morals of chess by Franklin, in *Sec. 2.5.1*; the precise identification of the three possibilities of the outcome in a chess game, in the early Indian history of chess development, that is, by "checkmate," by "stalemate," or by "bare king" ["taking all of an opponent's pieces except the king"], in *Sec. 2.5.2*; the precise identification of different computational enhancements for high performance, like "minimax," "killer heuristic," etc., in *Sec. 3.2.1*; the precise identification of the tactical moves which are identified as chess blunders, like the move 34...Qe3??? by Kramnik, in *Sec. 3.2.2*; the precise identification of the three conditions in Schwenk's Theorem, in *Sec. 3.3.1*; the precise identification of two examples of "the laws of chess" regarding the bishop and the pawn by Feynman, in his comparison with "the laws of physics," in *Sec. 3.3.2*; the precise identification of "economy" as one element for the aesthetic evaluation of tactical exercises, in *Sec. 3.5.1*; and the precise identification of different kinds of odds-giving in chess, in *Sec. 3.5.2*).

On the other hand, there is vagueness (e.g., the vagueness of this criterion itself, since "a very young age" can be 12 years, or 13 years, or 14 years, etc., as shown in the list of child prodigies, in *Sec. 2.2.1*; the vagueness in the seven beneficial ways of chess playing to the mind, because they can mean different things to different scholars and players, like how exactly imagination or creativity can help a chess player beat his opponent in a particular game, in *Sec. 2.2.2*; the vagueness in Vasantha's identification of the movements, since it is not clear which "movements of the celestial bodies and their numerical symbolism" in cosmology correspond exactly to which moves in chess in the first place, in *Sec. 2.3.1*; the relatively vagueness in Shannon's identification of Type B, since it is not clear how a Type B program is "able to decide which moves are good enough to be worthy of consideration...in any given position and this proved to be a much harder problem to solve than speeding up type A searches with superior hardware and search extension techniques," in *Sec. 2.4.1*; the vagueness in the report, in that it does not say "[w]hether or not statistical sampling covers all the bases of explaining male superiority in chess," in *Sec. 2.4.2*; the vagueness in the identification, since it is not clear if there are other morals, to what extent every reasonalbe chess player will agree about the list, and whether or not chess can have some mental benefits which do not have to be moral in nature at all, in *Sec. 2.5.1*; the vagueness in this identification, since it is not clear how "forefeiture" can fit in any of them or constitutes something else instead, in *Sec. 2.5.2*; the vagueness in the identification of different computational enhancements, since different people have different views on how they can be achieved, like different ways of "minimizing the possible loss while maximizing the potential gain," in *Sec. 3.2.1*; the vagueness in the identification of the chess blunders, since what is regarded as a blunder for one chess analyst may be explained away as quite natural for another chess analyst, as shown by the response by Alexander Roshal, who tried to explain away the blunder by Kramnik by saying that "the mating pattern of a queen on h7 protected by a knight on f8 is extremely rare and not contained in a grandmaster's automatic repertoire," in *Sec. 3.2.2*; the vagueness in the identification of the three conditions, because it is not clear whether or not "all rectangular boards that do not fall in one of the above three categories have closed knight's tours," in *Sec. 3.3.1*; the vagueness in the identification, since it is not clear of what exactly is meant by the word "law" here in chess, although the word makes sense in regard to the "laws" of physics like a scientific principle to explain reality in nature, in *Sec. 3.3.2*; the vagueness in the identification of "economy," which has different "facets" and can mean different things to different chess composers, in *Sec. 3.5.1*; and the vagueness in the identification of different kinds of odds-giving in chess, because it is not clear what other possible candidates of odds-giving are excluded in the list, or to what extent chess handicap can also allow "eccentric and peculiar odds," in *Sec. 3.5.2*).

And the reverse direction also holds true.

Once more, it should be stressed, however, that there are different shades of gray (or different degrees of truth) in the two opposites, and the classification is not necessarily mutually exclusive either. In fact, even when some combinations of the two occur, they only end up sharing the same dialectic relationship, but in a different degree.

Also, there are some other relationships (as a kind of family resemblance) which have something in common with the principle, but they are not exactly the same but only more or less comparable, which varies from case to case.

Good examples of family resemblance in relation to the principle are clarity-ambiguity, directness-indirectness, quantitativeness-qualitativeness, specificity-obscurity, describability-nondescribability, specificity-generality, thickness-thinness, concretness-abstractness, and the like.

# 5th Thesis: The Simpleness-Complicatedness Principle

The fifth thesis refers to the simpleness-complicatedness principle (on the interconnection among things) in the category of "structure" in existential dialectics (which was first out worked out in *FIA*), in that both simpleness and complicatedness are vital, without favoring one over the other, and each is utilized, depending on the basis of the perspectives of nature, the mind, culture, and society. There is no simpleness without complicatedness—and vice versa.

For instance, on the one hand, there is simple analysis (e.g., the relatively simple view of Ericsson and his colleagues that "deliberate practice is sufficient for reaching high levels of expertise, like master in chess," in *Sec. 2.2.1*; the relatively simple analysis of the benefits of playing chess in seven ways, as claimed by the advocates, in *Sec. 2.2.2*; the relatively simple analysis of chess origin in cosmology by Vasantha, in *Sec. 2.3.1*; the relatively simple analysis of two types of research strategy by Shannon, in *Sec. 2.4.1*; the relatively simple analysis of the gender differences in chess performance by the team of research from the UK on the basis of the gender differences in chess participation rate, in *Sec. 2.4.2*; the relatively simple analysis of the morals of chess by Franklin, in *Sec .2.5.1*; the relatively simple understanding of chess rules from the standpoint of each cultural variant, in *Sec. 2.5.2*; the relatively simple analysis of the advantages of using computers for the understanding of chess complexity, in *Sec. 3.2.1*; the relatively simple analysis of chess blindness on the basis of some common factors like time pressure, fatigue, carelessness, etc., in *Sec. 3.2.2*; the relatively simple analysis of closed knight's tours on the basis of the three conditions in Schwenk's Theorem, in *Sec. 3.3.1*; the relatively simple proposal of new chess games by Neto and his colleagues, in *Sec. 3.3.2*; the relatively simple analysis of the promise of online chess for learning chess, in *Sec. 3.4.*; the relatively simple analysis of the seven elements for the aesthetic evaluation of tactical exercises by "modern" chess composers, in *Sec. 3.5.1*; and the relatively simple analysis of chess handicap by its proponents, in *Sec. 3.5.2*).

On the other hand, there is also complicated counterpart (e.g., the relatively complicated view of Alfred Binet and his colleagues that brilliant chess performance requires not only memory, but also imagination and experience, for instance, in *Sec. 2.2.1*; the relatively more complicated analysis of the relationship between chess playing and its mental impact, as shown in the criticisms by the critics, in *Sec. 2.2.2*; the relatively more complicated analysis of chess origin in cosmology by questioning its assumptions and arguments, for the critics, in *Sec. 2.3.1*; the relatively more complicated analysis of research strategy by a team from Nortwhestern University in 1973, who invented Chess 4.0, which abandoned Type B while learning something from it, like the use of "quiescence searches" and "extensions and pruning," in *Sec. 2.4.1*; the relatively complicated analysis of the results provided by the team of research from the U.S., as shown in the criticisms by the critics who challenge its claims and assumptions, in *Sec. 2.4.2*; the relatively more complicated analysis of the morals of chess, as shown in the criticisms by the critics, in *Sec. 2.5.1*; the relatively more complicated understanding of chess rules by considering all cultural variants, many of which are not necessarily compatible with each other, in *Sec. 2.5.2*; the relatively more complicated analysis of the power of computation for the understanding of chess complexity, as shown in the criticisms on its limits by the critics, in *Sec. 3.2.1*; the relatively complicated analysis of chess blindness by questioning the common causes, as shown in the view by Alexander Roshal, who tried to explain away the blunder by Kramnik by attributing his opponent's move to being "extremely rare and not contained in a grandmaster's automatic repertoire," so Kramnik could not have detected the problem anyway, in *Sec. 3.2.2*; the relatively complicated analysis of closed knight's tours as shown in the criticisms of Schwenk's Theorem by the critics, in *Sec. 3.3.1*; the relatively more complicated analysis of the new chess games, as shown in the criticisms, in *Sec. 3.3.2*; the relatively more complicated analysis of online chess by questioning its claims and raising some serious problems, in *Sec. 3.4*; the relatively complicated analysis of the seven elements for the aesthetic evaluation of tactical exercises, as shown in the problems pointed out by the critics, in *Sec. 3.5.1*; and the relatively more complicated analysis of chess handicap, as shown in the problems with the tradition of odds-giving, in *Sec. 3.5.2*).

And the reverse direction also holds true.

Again, it should be stressed, however, that there are different shades of gray (or different degrees of truth) in the two opposites, and the classification is not necessarily mutually exclusive either. In fact, even when some combinations of the two occur, they only end up sharing the same dialectic relationship, but in a different degree.

Also, there are some other relationships (as a kind of family resemblance) which have something in common with the principle, but they are not exactly the same but only more or less comparable, which varies from case to case.

Good examples of family resemblance in relation to the principle are inflexibility-flexibility, standardization-specialization, imperfectness-perfectness, superficiality-depth, shallowness-deepness, economicalness-elaboratedness, plainness-circumspection, onesidedness-multisidedness, and the like.

## 6th Thesis: The Openness-Hiddenness Principle

The sixth thesis refers to the openness-hiddenenss principle (on the detection of things) in the category of "structure" in existential dialectics (which was already worked out in my previous books, especially in *FPHU*), in that reality has its hidden face, just as it is open to outside view in some other ways. There is no openness without hiddenness—and vice versa.

For instance, on the one hand, there is openness (e.g., the open exploration of the relationship between practice and good performance in the research on chess psychology by Ericsson and his colleagues, in *Sec. 2.2.1*; the open exploration of the relationship between chess playing and its mental benefits, in *Sec. 2.2.2*; the open exploration of the relationship between chess origin and cosmology, by Vasantha, in *Sec. 2.3.1*; the open exploration of the relationship between successful strategy and selective search techniques in Type B, in *Sec. 2.4.1*; the open exploration of the relationship between gender differences in chess participation rate and gender differences in chess perfromance, in *Sec. 2.4.2*; the open exploration of the relationships between morals and chess by Franklin, in *Sec. 2.5.1*; the open exploration of the relationship between chess rules and cultural variants, like the ban of using dice in the later stage of Indian chess development, in *Sec. 2.5.2*; the open exploration of the use of computation for the understanding of chess complexity, in *Sec. 3.2.1*; the move 54.gxh6+ made by Chandler, in broad daylight, in *Sec. 3.2.2*; the open exploration of the natural beauty of symmetry in the phenomenon of the knight's tour, in *Sec. 3.3.1*; the open exploration of new chess games with the use of concepts in physics, in *Sec. 3.3.2*; the open exploration of the usefulness of online chess for learnng chess, in *Sec. 3.4*; the open exploration of the relationship between "beauty" and "tactical exercises" in chess problems, in *Sec. 3.5.1*; and the open exploration of the relationship between a cultural icon and a chess master who is not to fool around with odds-giving, in *Sec. 3.5.2*).

On the other hand, there is also hiddenness (e.g., the hidden bias in the research on chess psychology by Ericsson and his colleagues, in that for the critics, "the results of research studies have failed to produce unambiguous support for the intellectual benefits of playing chess," in *Sec. 2.2.1*; the hidden bias in the claim by the advocates of chess playing in regard to the mental benefits, because they do not focus on other factors of chess development and the dark side of chess playing, like the incidents of chess-related deaths, in *Sec. 2.2.2*; the hidden bias in Vasantha's view, because it is driven by the Hindu worldview, in *Sec. 2.3.1*; the hidden nature of Type B search techniques, because "it was difficult to anticipate which moves the...type B programs would play, and why" in a particular game, in *Sec. 2.4.1*; the hidden bias in the report by the team of research from the U.S., in that it assumes uncritically that the biological argument is controversial and is not to be taken seriously at the outset of the research, in *Sec. 2.4.2*; the hidden bias in Franklin's treatement of the morals of chess, because of his strong moral conviction in "civic and personal virtues," especially in the context of his pious Puritan background, in *Sec. 2.5.1*; the hidden bias in the Indian ban of using dice in chess, because of "Hindu religious objections," whereas in some variants of Chinese chess, using a dice, for instance, is allowed, in *Sec. 2.5.2*; the hidden bias on the enthusiastic fad of using computation for the understanding of chess complexity, as shown in its neglect of considering some fundamental problems like "Bremermann's fundamental limit" and "the problem of intractability," in *Sec. 3.2.1*; the hidden motive by Chandler to make this move 54.gxh6+??, because he calculated, though incorrectly, that, "after 54.gxh6+ Kxh6 he will be left with the considerable material advantage of a rook pawn and bishop against a bare king" and calculated further, also incorrectly, that "he...will win control over the h8 square after 55.Kf6, and thereby win the game," but to his surprise, "instead of the expected 54...Kxh6 came 54...Kh8!," in *Sec. 3.2.2*; the hidden bias in the mathematical analysis of the knight's tour, because it is not clear if it has any practical use at all for actual players in chess games, since it is one thing for mathematicians to be attracted to the natural beauty of symmetry in the knight's tour problem, but it is something else for chess players to apply the knight's tours for best tactical moves, in *Sec .3.3.1*; the hidden bias in the new chess games, because of their focus on concepts in physics, not concepts in other fields of knowledge, in *Sec. 3.3.2*; the hidden bias in online chess because of its impersonality and social isolation, such that the usefuleness has its dark sides not often spoken about, in *Sec. 3.4*; the hidden bias in the seven elements for aesthetic evaluation, because of their "modern taste" as acknowledged by the chess composers, in *Sec. 3.5.1*; and the hidden bias in the ban of odds-giving in the Soviet Union for chess masters who had to behave like "cultural icons," because of the Soviet ideology, in *Sec. 3.5.2*).

And the reverse direction also holds true.

Again, it should be stressed, however, that there are different shades of gray (or different degrees of truth) in the two opposites, and the classification is not necessarily mutually exclusive either. In fact, even when some combinations of the two occur, they only end up sharing the same dialectic relationship, but in a different degree.

Also, there are some other relationships (as a kind of family resemblance) which have something in common with the principle, but they are not exactly the same but only more or less comparable, which varies from case to case.

Good examples of family resemblance in relation to the principle are overtness-covertness, publicness-privateness, openness-closedness, transparency-secrecy, openness-biasedness, and so on.

# 7th Thesis: The Denseness-Emptiness Principle

The seventh thesis concerns the denseness-emptiness principle (on the distribution of entities in space) in the category of "structure" in existential dialectics, which is first proposed in *FPHUP*, in that both density and void are needed, in relation to the mind, nature, culture, and society, albeit in different ways. There is no denseness without emptiness—and vice versa.

Lest any misunderstanding occurs, the term "void" is used here only as an approximation of emptiness (depending on the degree of the lack of density), since, in physics, it is well known that "empty" space is not really empty all the way, because it can be full of energy (e.g., random quantum fluctuations at the sub-atomic level, and, for that matter, dark energy in the universe) and matter (e.g., different versions of sub-atomic particles, and, for that matter, dark matter in the universe). (F. Wilczek 2008)

For instance, on the one hand, there is denseness (e.g., the relatively denser concentration of chess prodigies in Eastern countries like the former Soviet republics and China, as shown in the list of modern International Grandmasters, in *Sec. 2.2.1*; the relatively denser concentration of individuals who play chess and also believe in the mental benefits of chess playing, in *Sec. 2.2.2*; the relatively denser concentration of concerns with cosmology in chess origin for people like Vasantha, in *Sec. 2.3.1*; the relatively denser concentration of concerns with "examining every possible position for a fixed number of moves using the minimax algorithm" by the supporters of Type A search strategy, in *Sec. 2.4.1*; the relatively denser concentration of chess champions who are males, in *Sec. 2.4.2*; the relatively denser concentration of moral manners in pious chess players while playing chess, according to Franklin, in *Sec. 2.5.1*; the relatively denser concentration of chess players in Western chess to play "in the squares," in *Sec. 2.5.2*; the relatively denser concentration of concerns with computational enhancements in those who are enthusiastic supporters of the use of computers for chess games, in *Sec. 3.2.1*; the relatively denser concentration of brilliant moves made by grandmasters, in *Sec. 3.2.2*; the relatively denser concentration of concerns with the beauty of symmetry in the knight's tour by chess mathematicians, in *Sec. 3.3.1*; the relatively denser concentration of moves in the same direction in inertial chess, in *Sec. 3.3.2*; the relatively denser concentration of online chess players nowadays, in *Sec. 3.4*; the relativley denser concentration of the defenders of the seven elements for the aesthetic evaluation of tactical exercises in modern times, in *Sec. 3.5.1*; and the relatively denser concentration of odds-giving in its heyday during "the eighteenth century and the early nineteenth century," in *Sec. 3.5.2*).

On the other hand, there is emptiness (e.g., the relatively less dense, or more empty, concentration of chess prodigies in Africa, as shown in the list of modern International Grandmasters, in *Sec. 2.2.1*; the relatively less dense, or more empty, concentration of individuals who play chess but seriously question the claim about the mental benefits of chess playing, in *Sec. 2.2.2*; the relatively less dense, or more empty, concentration of concerns with the mind, culture, and society in chess origin for people like Vasantha, in *Sec. 2.3.1*; the relatively less dense, or more empty, concentration of concerns with "quiescence search" by the supporters of Type A search strategy, in *Sec. 2.4.1*; the relatively less dense, or more empty, concentration of chess champions who are females, in *Sec. 2.4.2*; the relatively less dense, or more empty, concentration of moral manners in deviant chess players who engage in chess-related deaths, in *Sec. 2.5.1*; the relatively less dense, or more empty, concentration of chess players in Western chess to play "on the intersections of the lines on the board," in *Sec. 2.5.2*; the relatively less dense, or more empty, concentration of concerns with individual subjectivity and creativity in those who are enthusiastic supporters of the use of computers for chess games, in *Sec. 3.2.1*; the relatively less dense, or more empty, concentration of chess blunders made by grandmasters, although they may happen sometimes, as shown in the examples, in *Sec. 3.2.2*; the relatively less dense, or more empty, concentration of cocnerns with the beauty of symmetry in the knight's tour by chess players in actual chess games, in *Sec. 3.3.1*; the relatively less dense, or more empty, concentration of moves in the same direction in magnetic chess, in *Sec. 3.3.2*; the relatively less dense, or more empty, concentration of online chess players in antiquity, in *Sec. 3.4*; the relatively less dense, or more empty, concentration of the defenders of the seven elements for the aesthetic evaluation of tactical exercises in ancient times, in *Sec. 3.5.1*; and the relatively less dense, or more empty, concentration of odds-giving among chess masters during the Soviet Union, in *Sec. 3.5.2*).

And the reverse direction also holds true.

Again, it should be stressed, however, that there are different shades of gray (or different degrees of truth) in the two opposites, and the classification is not necessarily mutually exclusive either. In fact, even when some combinations of the two occur, they only end up sharing the same dialectic relationship, but in a different degree.

Also, there are some other relationships (as a kind of family resemblance) which have something in common with the principle, but they are not exactly the same but only more or less comparable, which varies from case to case.

Good examples of family resemblance in relation to the principle are fullness-voidness, nearness-farness, concentration-dispersion, and the like.

## 8$^{th}$ Thesis: The Slowness-Quickness Principle

The eighth thesis concerns the slowness-quickness principle (on the speed of change) in the category of "process") in existential dialectics (which was first worked out in *FIA*), in that both slowness and quickness co-exist, with their own internal tension, to the extent that each fights for its own relevance with the other, in accordance to the perspectives of nature, the mind, culture, and society, without one being the victor and the other being the vanquished in the long haul. There is no slowness without quickness—and vice versa.

For instance, on the one hand, there is slowness (e.g., the relatively slower speed of an ordinary chess player to win in a competitive game of chess, in *Sec. 2.2.1*; the relatively slower readiness of poets to plan logically, think calculatively, and wait strategically both in poetry and other life activities, in *Sec. 2.2.2*; the relatively slower readiness of people like Vasantha to see chess moves in terms of social and cultural patterns, in *Sec. 2.3.1*; the relatively slower speed of a chess engine to make a move if it has to search for "every possible position for a fixed number of moves using the minimax algorithm" as in Type A search strategy, in *Sec. 2.4.1*; the relatively slower readiness of a chess player to sacrifice the queen in a chess game, in *Sec. 2.4.2*; the relatively slower chess players, like the "adversary" as identified by Franklin in rule #4, in *Sec. 2.5.1*; some players tend to play very slowly, like the example that "one player took as much as two hours and 20 minutes to think over a single move at the London tournament" in 1851, in *Sec. 2.5.2*; the relatively slower speed of humans to calculate different possible moves and evaluate the resulting positions to find the best move in a chess game, in *Sec. 3.2.1*; the relatively slower speed of a chess game to end, if no chess blunder occurs, in general, in *Sec. 3.2.2*; the relatively slower speed of computation to solve Hamiltonian path problems in graph theory, especially in regard to those which are solved in non-linear time, in *Sec. 3.3.1*; the relatively slower speed of chaos chess to be played by humans without the use of computers, in *Sec. 3.3.2*; the relatively slower speed of a correspondence game which allows "10 days per move," in *Sec. 3.4.*; the relatively slower speed of a tactical exercise to be solved, if "[t]he key move of the solution" is "obvious," in *Sec. 3.5.1*; and the relatively slower speed of a chess master to beat a lesser player if he offers odds-giving, in *Sec. 3.5.2*).

On the other hand, there is quickness (e.g., the relativley faster speed of a chess prodigy to win in a competitive game of chess, in *Sec. 2.2.1*; the relatively quicker readiness of chess players to plan logically, think calculatively, and wait strategically both in chess playing and other life activities, in *Sec. 2.2.2*; the relatively quicker readiness of people like Vasantha to see chess moves in terms of cosmological patterns, in *Sec. 2.3.1*; the relatively quicker speed of a chess engine to make a move if it only has to do selective searches as in Type B search strategy, in *Sec. 2.4.1*; the relatively quicker readiness of a chess player to sacrifice a pawn in a chess game, in *Sec. 2.4.2*; the relatively quicker chess players, like "you" as identified by Franklin in rule #4, in *Sec. 2.5.1*; the relatively quicker chess games for chess players nowadays, like the introduction of "speed chess" for "a bank of time to each player in which to play a previously agreed number of moves, e.g. two hours for 30 moves," in *Sec. 2.5.2*; the relatively faster speed of supercomputers to calculate different possible moves and evaluate the resulting positions to find the best move in a chess game, in *Sec. 3.2.1*; the relatively faster speed of a chess game to end, when chess blunders occur, in general, in *Sec. 3.2.2*; the relatively faster speed of computation to solve the knight's tour problem, which can be solved "in linear time," in *Sec. 3.3.1*; the relatively quicker speed of normal Western chess to be played by humans, in *Sec. 3.3.2*; the relatively faster speed of a correspondence game which allows "one day per move," in *Sec. 3.4*; the relatively quicker speed of a tactical exercise to be solved, if "[t]he key move of the solution" is not "obvious" at all and is hidden deeply inside the puzzle, in *Sec. 3.5.1*; and the relatively faster speed of a chess master to beat a lesser play if he does not offer any odds-giving, in *Sec. 3.5.2*).

And the reverse direction also holds true.

Again, it should be stressed, however, that there are different shades of gray (or different degrees of truth) in the two opposites, and the classification is not necessarily mutually exclusive either. In fact, even when some combinations of the two occur, they only end up sharing the same dialectic relationship, but in a different degree.

Also, there are some other relationships (as a kind of family resemblance) which have something in common with the principle, but they are not exactly the same but only more or less comparable, which varies from case to case.

Good examples of family resemblance in relation to the principle are inconvenience-convenience, passiveness-activenessness, gradualness-abruptness, deceleration-acceleration, and the like.

# 9th Thesis: The Convention-Novelty Principle

The nineth thesis is the convention-novelty principle (on the nature of creative thinking) in the category of "agency" in existential dialectics (which was first worked out in *FPHCT*), in that, if there is conventional wisdom, there is likewise novel challenge, to the extent that both convergent and divergent thinking are part of life. There is no convention without novelty—and vice versa.

As srummarized from *FPHCT* (almost verbatim here), there are (a) "creative techniques" and (b) "creative traits," which, when satisfied—in relation to the larger context of the mind, nature, society, and culture—can be used to enhance creative works.

In addition, creative thinking has its own possibilities and limits (in relation to invention), just as it has its own promises and pitfalls (in relation to innovation)— as already analyzed in *FPHCT*).

In the end, creative thinking has its own desirability and dark sides (as also already analyzed in *FPHCT*).

With these clarifications in mind—there are good empirical examples for the convention-novelty princple.

For instance, on the one hand, there is convention (e.g., the conventional wisdom about chess prodigies, in regard to their unusual memory, perception, and imagination in playing chess, in *Sec. 2.2.1*; the conventional wisdom about the mental benefits of chess playing, in *Sec. 2.2.2*; the conventional wisdom about chess origin in the military institution, for instance, in *Sec. 2.3.1*; the conventional wisdom about Type

A and Type B search strategies in the 1950s as proposed by Shannon, in *Sec. 2.4.1*; the conventional wisdom about male success in chess performance, in *Sec. 2.4.2*; the conventional wisdom about the rules governing chess playing, in *Sec. 2.5.1*; the conventional wisdom of playing chess "in the squares," in *Sec. 2.5.2*; the conventional wisdom about chess to be played by humans, in *Sec. 3.2.1*; the conventional wisdom about playing chess, in accordance to certain rules, with different mating patterns that grandmasters know all too well, in *Sec. 3.2.2*; the conventional wisdom about the knight's tour problem, in *Sec. 3.3.1*; the conventional wisdom about Western chess as sanctioned by FIDE, in *Sec. 3.3.2*; the conventional wisdom about learning chess by playing it with another player in person, in *Sec. 3.4*; the conventional wisdom about beauty, in that "beauty is in the eyes of the beholders," in *Sec. 3.5.1*; and the conventional wisdom in antiquity about playing chess between two players, in *Sec. 3.5.2*).

On the other hand, there is novelty (e.g., the alternative novel challenge to the conventional wisdom by Ericsson and his colleagues, in that "deliberate practice" is very important for brilliant chess performance instead, in *Sec. 2.2.1*; the alternative novel challenge to the conventional wisdom about chess playing by the critics, who pointed out the other side of chess playing like the incidents of chess-related deaths, in *Sec. 2.2.2*; the alternative novel challenge to the conventional wisdom about chess origin by Vasantha on the basis of cosmology, in *Sec. 2.3.1*; the alternative novel challenge to the conventional wisdom about search strategies by the team in Northwestern University in the 1973, such that they abandoned Type B while learning something from it and coming up with something new like Chess 4.0, in *Sec. 2.4.1*; the alternative novel challenge to the conventional wisdom about chess performance, by the new proposal that gender differences in chess participation rate are related to gender differences in chess perfromance, in *Sec. 2.4.2*; the alternative novel challenge to this conventional wisdom by Franklin, in that there is also a close relationship between morality and chess, in *Sec. 2.5.1*; the alternative novel challenge to this conventional wisdom when chess was introduced to the Chinese, because "Chinese chess also borrows elements from the game of Go" and "is played on the intersections of the lines on the board, rather than in the squares," in *Sec. 2.5.2*; the alternative novel challenge by Shannon to this conventional wisdom, in regard to his groundbreaking article on "Programming a Computer for Playing Chess" published in 1950, in *Sec. 3.2.1*; the alternative novel challenge to the conventional wisdom about mating patterns in chess, when Deep Fritz made "the mating pattern of a queen on h7 protected by a knight on f8," which "is extremely rare and not contained in a grandmaster's automatic repertoire," in *Sec. 3.2.2*; the alternative novel challenge to this conventional wisdom about the knight's tour problem, by Schwenk's Theorem, which is original in its specification of three major conditions for them to not be allowed, in *Sec. 3.3.1*; the alternative novel challenge to this conventional wisdom about Western chess, by Neto with his invention of new chess games with new ideas from physics, in *Sec. 3.3.2*; the alternative novel challenge to the conventional wisdom about face-to-face chess playing, by the use of online chess instead, in *Sec. 3.4*; the alternative novel challenge to the conventional wisdom about beauty, by way of the "modern taste" of chess composers, who offered some elements to be accepted by those in chess composition, in spite of the relativist problem, in *Sec. 3.5.1*; and the alternative novel challenge to the conventional wisdom about chess playing in the 19th and 20th centuries by the institution of odds-giving which reached its peak then, in *Sec. 3.5.2*).

And the reverse direction also holds true.

Again, it should be stressed, however, that there are different shades of gray (or different degrees of truth) in the two opposites, and the classification is not necessarily mutually exclusive either. In fact, even when some combinations of the two occur, they only end up sharing the same dialectic relationship, but in a different degree.

Also, there are some other relationships (as a kind of family resemblance) which have something in common with the principle, but they are not exactly the same but only more or less comparable, which varies from case to case.

Good examples of family resemblance in relation to the principle are convergence-divergence (or convergent thinking vs. divergent thinking), normalness-nonnormalness, conformity-nonconformity, and so on.

## 10$^{th}$ Thesis: The Evolution-Transformation Principle

The tenth thesis refers to the evolution-transformation principle (on the multiple kinds of agency). in the category of "agency" in existential dialectics (which was first worked out in *FAE* and then in other books of mine).

This principle—and the symmetry-asymmetry principle, for instance—are both about agency. More precisely, the word "agency," in a formal definition, refers to "a person or thing through which power is exerted or an end is achieved." (MWD 2007) It therefore does not have to necessarily involve an intelligent lifeform.

Because of this dual meaning in agency, the evolution-transformation principle is more concerned with *the kind of agency*, that is, both about the *evolution* in the state of nature (e.g., an object of natural beauty) and the *transformation* in the world of intelligent lifeforms (e.g., a work of art). There is no evolution without transformation—and vice versa.

In classical Darwinian evolutionary theory (as more detailedly analyzed in *BNN*), evolution is "blind." But in the human world, change often takes place because of the conscious intervention of humans in transforming society and culture, just to cite two instances.

And the transformative part of the principle precisely refers to the other dimension in the dual meaning of agency, in giving technology (as invented by intelligent lifeforms like humans and, soon, post-humans) a major role to play in the change of the world, which is something that I extensively analyzed in *FHC* in the context of the technophilic lifeworld, espeically though not exclusively since modern times.

For instance, on the one hand, there is evolution (e.g., the natural evolution of humans to play with each other for fun in the state of nature, in *Sec. 2.2.1*; the natural evolution of humans in the state of nature to fight with each other from time to time, in *Sec. 2.2.2*; the natural evolution of humans to make observations about the sky and their surroundings in the state of nature, in *Sec. 2.3.1*; the natural evolution of humans to play with each other in the state of nature, in *Sec. 2.4.1*; the natural evolution of humans to fight with each other in the state of nature, in *Sec. 2.4.2*; the natural evolution of humans in the state of nature to play with each other without the hassle of "civic" and "moral" concerns, in *Sec. 2.5.1*; the natural evolution of humans to fight or compete with each other in the state of nature, in *Sec. 2.5.2*; the natural evolution of humans to be able to think in the state of nature, in *Sec. 3.2.1*; the natural evolution of humans to have fun with each other in the state of nature, in *Sec. 3.2.2*; the natural evolution of humans to be able to calculate in the state of nature, in *Sec. 3.3.1*; the natural evolution of humans to play with each other in the state of nature, in *Sec. 3.3.2*; the natural evolution of humans to play with each other in person in the state of nature, in *Sec. 3.4*; the natural evolution of humans to appreciate beautiful objects in the state of nature, in *Sec. 3.5.1*; and the natural evolution of humans to play with each other in the state of nature, in *Sec. 3.5.2*).

On the other hand, there is transformation (e.g., the technical transformation of human playing with each other by the invention of chess, in *Sec. 2.2.1*; the technical transformation of human fighting with each other by the invention of chess and its subsequent incidents of chess-related deaths, in *Sec. 2.2.2*; the technical transformation of human observations about the sky and their surroundings by the invention of cosmology and chess, in *Sec. 2.3.1*; the technical transformation of human playing with each other by the invention of computers and computer chess, in *Sec. 2.4.1*; the technical transformation of human fighting with each other by the invention of an elaborated social structure involving soldiers, knights, bishops, queens, kings, etc., in *Sec. 2.4.2*; the technical transformation of human playing with each other by the invention of "morals" and "rules" as shown in chess playing, in the later stages of the development of human civilization, in *Sec. 2.5.1*; the technical transformation of human fighting or competing with each other by the invention of chess games like "chaturaṅga, which translates as 'four divisions [of the military]': infantry, cavalry, elephantry, and chariotry, represented by the pieces that would evolve into the modern pawn, knight, bishop, and rook, respectively," in *Sec. 2.5.2*; the technical transformation of human ability to think by the invention of supercomputers for computation, in *Sec 3.2.1*; the technical transformation of having fun by the invention of chess engines like Deep Fritz, which could make "extremely rare…mating patterns" in chess games, in *Sec. 3.2.2*; the technical transformation of human ability to calculate by the invention of computers to calculate complicated problems like the calculation of 26,534,728,821,064 directed closed tours on an 8 × 8 chess board, in *Sec. 3.3.1*; the technical transformation of human playing with each other by the invention of new chess games with the use of physics, in *Sec. 3.3.2*; the natural evolution of humans to play with each other online, by the invention of computers, in *Sec. 3.4*; the technical transformation of human appreciation of beauty by the invention of the seven elements for the aesthetic evaluation of tactical exercises in chess problems, in *Sec. 3.5.1*; and the technical transformation of humans to play with each other by the invention of odds-giving in chess playing, in *Sec. 3.5.2*).

And the reverse direction also holds true.

Again, it should be stressed, however, that there are different shades of gray (or different degrees of truth) in the two opposites, and the classification is not necessarily mutually exclusive either. In fact, even when some combinations of the two occur, they only end up sharing the same dialectic relationship, but in a different degree.

Also, there are some other relationships (as a kind of family resemblance) which have something in common with the principle, but they are not exactly the same but only more or less comparable, which varies from case to case.

Good examples of family resemblance in relation to the principle are inorganicness-volition, nonwillingness-willingness, naturalness-technologicalness, naturalness-nonnaturalness, and so on.

## 11th Thesis: The Symmetry-Asymmetry Principle

The eleventh thesis is labeled as the symmetry-asymmetry principle (on the relationships among existents) in the category of "agency" in existential dialectics (which was also already worked out in my previous books), in that there is no asymmetry without symmetry—and vice versa.

For instance, the Same can be symmetric and asymmetric towards the Others. But in case of asymmetry (as analyzed in *BDPD*), oppression and self-oppression can occur. So, when the Same is asymmetric towards the Others, the Same can also be relatively asymmetric towards itself in self-oppression, just as the Others can be likewise towards themselves. The subsequent oppressiveness is dualistic, as much by the Same against the Others and itself, as by the Others against the Same and themselves. Both oppression and self-oppression can be achieved by way of downgrading differences between the Same and the Others and of accentuating them.

This is true, even though not all forms of asymmetry have to be about oppression and self-oppression.

In addition, from Chapter Three of *FPHG*, symmetry is not perfect, to be understood in an approximate sense under many life circumstances. With this clarification in mind, hereafter are some empirical examples.

For instance, on the one hand, there is symmetry (e.g., the co-existence of different chess players in society, in *Sec. 2.2.1*; the co-existence of different kinds of chess players in human history, in *Sec. 2.2.2*; the co-existence of different views about chess origins throughout history, in *Sec. 2.3.1*; the co-existence of different search strategies in the history of chess games, in *Sec. 2.4.1*; the co-existence of differrent pieces in a chess game, in *Sec. 2.4.2*; the co-existence of different benefits of chess playing, in *Sec. 2.5.1*; the co-existence of different cultural variants of chess, in *Sec. 2.5.2*; the co-existence of different chess players, be they humans or machines, in *Sec. 3.2.1*; the co-existence of different grandmasters in chess over the centuries, in *Sec. 3.2.2*; the co-existence of different ways to understand the knight's tour problem in history, in *Sec. 3.3.1*; the co-existence of different forms of chess in history, in *Sec. 3.3.2*; the co-existence of different ways to play chess in history, in *Sec. 3.4*; the co-existence of different aesthetic tastes in history, in *Sec. 3.5.1*; and the co-existence of different ways of chess playing, be it with odds-giving or without it, in *Sec. 3.5.2*).

On the other hand, there is also asymmetry (e.g., the high regard towards chess prodigies—but the low regard towards mediocre chess players, in *Sec. 2.2.1*; some chess players accept a gracious loss after a chess game—while others resort to violence after a loss in a chess game, in *Sec. 2.2.2*; the acceptance of the version of chess origin in cosmology by people like Vasantha in India—but the acceptance of the version of the chess linkage with the military institution by people like Cho-Yo in Japan, in *Sec. 2.3.1*; the popularity of selective search techniques like Type B search strategy in the older days, when computers were not powerful enough—but the use of more comprehensive search techniques like Type A search strategy nowadays, when "technological advances by orders of magnitude in processing power have made the brute force approach [inType A] far more incisive than was the case in the early years," in *Sec. 2.4.1*; the value of a knight on the basis of 3 points—but the value of a queen on the basis of 9 points, in *Sec. 2.4.2*; the acceptance of the four morals of chess by the followers of Franklin—but the advocation of other benefits of chess playing as already discussed in the earlier section on "chess games, and the development of mental skills," in *Sec. 2.5.1*; the popularity of the Western variant of chess in the West—but the more acceptance of the Chinese variant of chess in mainland China, in *Sec. 2.5.2*; the popularity of playing chess with humans in the older days, including Kasparov, who was not happy to have lost to Deep Blue—but the popularity of allowing more and more chess engines in chess tournaments nowadays, in *Sec. 3.2.1*; some grandmasters make more chess blunders than some others, in *Sec. 3.2.2*; the highly stylized Sanskrit poem Kavyalankara written by the 9th century Kashmiri poet Rudrata—but the understanding of the knight's tour problem in computational form like the calculation of 26,534,728,821,064 directed closed tours on an 8 × 8 chess board by computers nowadays, in *Sec. 3.3.1*; the popularity of conventional Western chess by those who follow the rules as laid down by FIDE—but the exploration of new kinds of chess which go beyond the rules by FIDE among players like Neto and his followers, in *Sec. 3.3.2*; the popularity of playing chess in person in antiquity, but the popularity of playing chess online nowadays, in *Sec. 3.4*; the popularity of aesthetic relativism by those who hold the claim that "beauty is in the eyes of the beholders"—but the popularity of the seven elements for the aesthetic evaluation of tactical exercises in chess problems by chess composers with "modern taste," in *Sec. 3.5.1*; and the more popularity of chess playing in antiquity without odds-giving—but the more popularity of chess playing with odds-giving in the 19th and 20th centuries, in *Sec. 3.5.2*).

And the reverse direction also holds true.

Again, for the last time, it should be stressed, however, that there are different shades of gray (or different degrees of truth) in the two opposites, and the classification is not necessarily mutually exclusive either. In fact, even when some combinations of the two occur, they only end up sharing the same dialectic relationship, but in a different degree.

Also, there are some other relationships (as a kind of family resemblance) which have something in common with the principle, but they are not exactly the same but only more or less comparable, which varies from case to case.

Good examples of family resemblance in relation to the principle are strength-weakness, balance-extremity, harmoniousness-unharrmoniousness, potence-impotence, reflexiveness-unreflexiveness, equality-inequality, and something like that.

# 12th Thesis: The Regression-Progression Principle

The twelveth thesis is called the regression-progression princple (on the direction of history) in the category of "outcome" in existential dialectics (which was also already worked out in my previous books), in that neither the cyclical nor the linear views are adequate for explaining many phenomena at all levels. There is no regression without progression—and vice versa.

History progresses to more advanced forms, but with a regressive touch. Examples include no freedom without unfreedom, no equality without inequality, and no civilization without barbarity. This is not an inevitable law, but merely a highly likely empirical trend.

For instance, on the one hand, there is regression (e.g., the regression made in the practice of chess, in that "the results of research studies have failed to produce unambiguous support for the intellectual benefits of playing chess," in *Sec. 2.2.1*; the regression made in the history of chess, as shown in the incidents of chess-related deaths in history, in *Sec. 2.2.2*; the regression made in the view about chess origin in cosmology as shown in the criticisms pointed out by the critics, in *Sec. 2.3.1*; the regression made in chess engines as shown by the problems in search strategies pointed out by the critics, in *Sec. 2.4.1*; the regression made in the report by a team of research from the UK., as shown in the criticisms about its assumptions and claims, in *Sec. 2.4.2*; the regression made in the morals of chess as shown in the criticisms by the critics, in *Sec. 2.5.1*; the regression made in the history of cultural variants, as shown in the emergence of new problems which later require new rules to fix them, like some players who took too much time to play in earlier times by taking advantage of the older rules, in *Sec. 2.5.2*; the regression made in using computation for chess games, as shown in the criticisms by the critics, in *Sec. 3.2.1*; the regression made in the history of chess games, as shown by the continued occurrence of new chess blunders over time, in *Sec. 3.2.2*; the regression made in the understanding of the knight's tour problem, as shown in the major weaknesses of Schwenk's Theorem as pointed out by the critics, in *Sec. 3.3.1*; the regression in the new forms of chess, as shown in the weaknesses pointed out by the critics, in *Sec. 3.3.2*; the regression made in online chess, as shown in the problems pointed out by the critics, in *Sec. 3.4*; the regression made in the aesthetic evaluation of tactical exercies in chess problems, as shown in the criticisms by the critics, in *Sec. 3.5.1*; and the regression made in the history of odds-giving, as shown in the problems pointed out by the critics, in *Sec. 3.5.2*).

On the other hand, there is progress (e.g., the progress made in the practice of chess for better performance, as shown by the work of Ericsson and his colleagues, in *Sec. 2.2.1*; the progress made in the history of chess for the mental benefits of chess playing and the fun of playing it, in *Sec. 2.2.2*; the progress made in the view about chess origin in cosmology to help us understand better the linkage in a different way, in *Sec. 2.3.1*; the progress made in chess engines, as shown by computers beating human chess champions nowadays, in *Sec. 2.4.1*; the progress made in the report by a team of research from the UK., in that there is a correlation between gender differences in chess participation rate and gender differences in chess performance, in *Sec. 2.4.2*; the progress made by the morals of chess for more "civil" behaviors in chess playing, in *Sec. 2.5.1*; the progress made in chess development, as shown by the new rules to fix old problems, in *Sec. 2.5.2*; the progress made in using computation for chess games, as shown by the victory of Deep Blue over Kasparov, in *Sec. 3.2.1*; the progress made in the history of chess games for the better understanding of how chess blunders occur, as shown in the examples, in *Sec. 3.2.2*; the progress made in the understanding of the knight's tour problem, as shown in the Schwenk's Theorem in regard to the specification of the three major conditions, in *Sec. 3.3.1*; the progress made in the new forms of chess for playing new chess games with the use of ideas from physics, in *Sec. 3.3.2*; the progress made in online chess for different services like tournaments, games, forums, etc., in *Sec. 3.4*; the progress made in the aesthetic evaluation of tactical exercies in chess problems, as shown in the proposed seven elements for a better appreciation of chess beauty, in *Sec. 3.5.1*; and the progress made in the tradition of odds-giving, as a way "to enable a weaker player to have a chance of winning against a stronger one," in *Sec. 3.5.2*).

And the reverse direction also holds true.

Again, it should be stressed, however, that there are different shades of gray (or different degrees of truth) in the two opposites, and the classification is not necessarily mutually exclusive either. In fact, even when some combinations of the two occur, they only end up sharing the same dialectic relationship, but in a different degree.

Also, there are some other relationships (as a kind of family resemblance) which have something in common with the principle, but they are not exactly the same but only more or less comparable, which varies from case to case.

Good examples of family resemblance in relation to the principle are cost-benefit, undesirability-desirability, badness-goodness, risk-opportunity, and so forth.

## 13th Thesis: The Same-Difference Principle

The thirteen thesis refers to the same-difference principle (on the metamorphosis of change) in the category of "outcome" in existential dialectics (which was worked out in *ALD* and other books of mine), in that an entity, as it evolves over time, can be both different from and similar to its opposing alternatives and does not have to be solely more different from them over time. There is no similarity without difference—and vice versa.

Opposites are not absolute in a black-or-white fashion; so, an entity can become relatively more similar to (or more different from) its opposite over time.

This is further constrained by another principle, that is, the symmetry-asymmetry principle about the relationships among existents under the category about agency in existential dialectics, in that if there is symmetry (equality) between two entities, there is likewise asymmetry (inequality) emerging in a different way.

For instance, on the one hand, there is similarity in outcome (e.g., the contribution to the molding and regulation of beliefs and behaviors, regardless of whether this be achieved by way of playing chess or by way of making war, in *Sec. 2.2.1*; the contribution to the shaping of beliefs and behaviors, regardless of whether this be achieved by way of chess playing or by way of physical fighting, in *Sec. 2.2.2*; the contribution to the molding of beliefs and values, regardless of whether this be achieved by way of the appeal to cosmology in chess origin or to the military institution instead, in *Sec. 2.3.1*; the contribution to victory in a chess game, be it done by using Type A or Type B search strategy, in *Sec. 2.4.1*;

the contribution to the molding and regulation of beliefs and values, regardless of whether this is achieved by way of emphasizing gender biological differences or dismissing them, in *Sec. 2.4.2*; the contribution to the molding and regulation of beliefs and manners, regardless of whether this be accomplished by way of the morals of chess by Franklin or by way of other values by other scholars, in *Sec. 2.5.1*; the contribution to the molding of thinking and behaving, regardless of whether this be achieved by the time limits on chess games for faster thinking or by no such time limits for slower thinking, in *Sec. 2.5.2*; the contribution to the molding of a certain mode of thinking, regardless of whether this be achieved by computation in chess engines or by natural thinking in humans, in *Sec. 3.2.1*; the contribution to the control and regulation of beliefs and values, regardless of whether this be achieved by way of discouraging chess blunders or by way of forgiving carelessness, in *Sec. 3.2.2*; the contribution to the molding of beliefs and values, regardless of whether this be done by way of valuing the nature of symmetry by chess mathematicians or by way of valuing instrumental thinking by chess players, in *Sec. 3.3.1*; the contribution to the molding of beliefs and values, regardless of whether this be done by way of conforming to the traditional rules of chess as laid down by FIDE or by way of exploring new rules by discontented players, in *Sec. 3.3.2*; the contribution to the molding of values and behaviors, regardless of how this be achieved by way of playing chess in person or playing chess online, in *Sec. 3.4*; the contribution to the molding of beliefs and values about beauty, regardless of whether this be achieved by way of aesthetic relativism or by way of aesthetic non-relativism, in *Sec. 3.5.1*; and the contribution to the moldiing of beliefs and values, regardless of whether this be done by way of odds-giving or no-odds-giving, in *Sec. 3.5.2*).

On the other hand, there is difference in outcome (e.g., the contribution to the molding and regulation of beliefs and behaviors by way of playing chess, when one side fights with the other side only on the chess board—but the contribution to the molding and regulation of beliefs and behaviors by way of making war, when one side fights with the other side in the battlefield, in *Sec. 2.2.1*; the contribution to the shaping of beliefs and behaviors by way of chess playing for a more peaceful way of competition—but the contribution to the shaping of beliefs and behaviors by way of physical fighting for a more violent way of competition, in *Sec. 2.2.2*; the contribution to the molding of beliefs and values by way of the appeal to cosmology in chess origin for its emphasis on geometrical patterns and nmber systems—but the contribution to the molding of beliefs and values by way of the appeal to the military institution in chess history for its emphasis on the relevance to warfare, in *Sec. 2.3.1*; the contribution to victory in a chess game by using Type A search strategy with the search for "every possible position for a fixed number of moves using the minimax algorithm"—but the contribution to victory in a chess game by using Type B search strategy with the use of "quiescence searches" and "extensions and pruning," in *Sec. 2.4.1*; the contribution to the molding and regulation of beliefs and values by way of emphasizing gender biological differences—but the contribution to the molding and regulation of beliefs and values by way of dismissing gender biological differences, in *Sec. 2.4.2*; the contribution to the molding and regulation of beliefs and manners by way of the morals of chess by Franklin—but the contribution to the molding and regulation of beliefs and manners by way of alternative emphasis on different mental benefits as already discussed in the earlier section on "chess games, and the development of mental skills," in *Sec. 2.5.1*; the contribution to the molding of thinking and behaving by way of the imposition of the time limits on chess games so as to encourage faster thinking nowadays—but the contribution to the molding of thinking and behaving by way of no imposition of any time limits on chess games so as to allow slower and more thorough thinking in the older days, in *Sec. 2.5.2*; the contribution to the molding of a certain mode of thinking by way of computation in chess engines for a more mechanized thinking—but the contribution to the molding of a certain mode of thinking by way of natural thinking in humans for a more subjective, imaginative thinking, in *Sec. 3.2.1*; the contribution to the control and regulation of beliefs and values by way of discouraging chess blunders, so that the player in question loses the match or has a draw—but the contribution to the control and regulation of beliefs and values by way of forgiving carelessness, so that the player in question can be given sympathy with the excuse, for example, that his opponent made "extremely rare" tactical moves, which are "not contained in a grandmaster's automatic repertoire," in *Sec. 3.2.2*; the contribution to the molding of beliefs and values by way of valuing the nature of symmetry by chess mathematicians—but the contribution to the molding of beliefs and values by way of valuing instrumental thinking by chess players in calculating the best tactical move for winning a game, in *Sec. 3.3.1*; the contribution to the molding of beliefs and values by way of conforming to the traditional rules of chess as laid down by FIDE—but the contribution to the molding of beliefs and values by way of exploring new rules by discontented players, in *Sec. 3.3.2*; the contribution to the molding of values and behaviors by way of playing chess in person for more personal interactions—but the contribution to the molding of values and behaviors by way of playing chess online for less personal interactions, in *Sec. 3.4*; the contribution to the molding of beliefs and values about beauty by way of aesthetic relativism, such that "beauty is in the eyers of the beholders"—but the contribution to the molding of beliefs and values about beauty by way of aesthetic non-relativism, such that there are criteria to be used for the appreciation of beauty, in *Sec. 3.5.1*; and the contribution to the moldiing of beliefs and values by way of odds-giving for a more level-playing field for fairness—but the contribution to the moldiing of beliefs and values by way of no odds-giving for a more collective sensitivity to cultural iconicity beyond the narrow individual concern with self-promotion, in *Sec. 3.5.2*).

And the reverse direction also holds true.

Again, it should be stressed, however, that there are different shades of gray (or different degrees of truth) in the two opposites, and the classification is not necessarily mutually exclusive either. In fact, even when some combinations of the two occur, they only end up sharing the same dialectic relationship, but in a different degree.

Also, there are some other relationships (as a kind of family resemblance) which have something in common with the principle, but they are not exactly the same but only more or less comparable, which varies from case to case.

Good examples of family resemblance in relation to the principle are homogeneity-heterogeneity, we-they, and so forth.

# 14th Thesis: The Post-Human Variation

And the fourteenth thesis is about the role of "post-humans," which I originally proposed in my previous books, starting with the first book titled *The Future of Human Civilization* in 2000 and all others afterwards.

As already pointed out in *Sec. 1.7*, I need to emphasize, as this is something that I used to repeat (*almost verbatim*) from my previous books, two clarifications here about the term "post-human" as a neologism in my works.

Firstly, the word "post-human" here should *not* be confused with another term which looks similar but has a totally different meaning in the literature of postmodernism, namely, "post-humanism"—which constitutes a critique of "humanism" as tradionally understood (especially, though not exclusively, in relation to the idea of progress in science and reason in the Enlightenment project). (WK 2008)

My works reject the project of "postmodernism" and propose the future world of what I orignally called "after-postmodernity" in *FHC* and *FCD*, for instance.

And secondly, the word "post-human" here should also *not* be confused with a similar term which is used to champion the ideology of technology for the future co-existence between humans and cyborgs in "trans-humanism." (WK 2008a)

Instead, my term "post-human" in relation to "posthuman-ism" also rejects "transhumanism" (especially, though not exclusively, in relation to the promises of technology) and refers to something else altogether, that is, the future extinction of humans and its post-human successors in deep space and beyond unto multiverses.

My critique of "transhumanism" was more extensively elabroated in *Sec. 2.4.1* of *BEPE*.

With these two clarifications in mind (as summarized in *Table 1.31*)—the post-human variation in chess can be addressed in relation to six main directions, based on the mediative-variative theory of chess as analysed in this book (together with my visions as already worked out in my previous books).

Firstly, the "post-human" variation in chess will learn from the mediative-variative theory of chess in relation to the dialectic context of strategy and tactics—especially in the larger dialectic context of no strategy without tactics (and vice versa).

Secondly, the "post-human" variation in chess will also learn from the mediative-variative theory of chess in relation to the four perspectives of the mind, nature, society, and culture.

Thirdly, the "post-human" variation in chess will also learn from the mediative-variative theory of chess in relation to the non-privilege of any specific theory over others in the literature and the non-integration of them all (as they are not necessarily compatible with each other).

Fourthly, the "post-human" variation in chess will also learn from the mediative-variative theory of chess in relation to its usefulness not only in the context of chess but also in the context outside chess (e.g., of social organizations and social institutions). In other words, it is important in chess, in a small sense, and in life, in a broad sense (as summarized in *Table 1.1* and *Table 1.2*).

Fifthly, the "post-human" variation in chess will have to confront the new challenges as posed by the long-term civilizational development of intelligent life in the distant future, both here on this planet Earth and elsewhere in deep space until multiverses.

As this is something that I regularly pointed out in my previous books for background information (and summarize here, *almost verbatim*), I already worked out, in my numerous books, what these new challenges will be and provide different original visions to meet them—especially, in regard to the future evolutions of the mind (e.g., "the contrarian personality," "the hyper-martial body," "the hyper-sexual body," "floating consciousness," "hyper-spatial consciousness," "unfolding unconsciousness," "thinking machines," "thinking robots," "genetically altered superior beings," "cyborgs," and others), of nature (e.g., "post-Earth geology," "post-cosmology," "the alteration of space-time," "the creation of new matter-energy," "selective geometry," and the like), of society (e.g., "multifaceted war and peace," virtual organizations, different versions of "post-capitalism" and "post-democracy," the movement of "cyclical progression" at both structural and systemic levels, "ambivalent technology," and so forth), and of culture (e.g., the rise of "post-civilization," "transformative aesthetic experience," "contrastive mathematical-logic," and the whatnot) in history (e.g., the age of "after-postmodernity"), in the context of my approaches in relation to methodology (e.g., "sophisticated methodologial holism," "critical-dialectic formal science," etc,) and ontology (e.g., "existential dialectics," "contrastive rationality").

Of course, the examples (as listed above) are not exhaustive but illustrative, since my numerous books have worked out many other visions (as already roughly summarized in *Sec.1.7* and *Sec. 1.8*). Many, though not all, of my visions on the mind, nature, society, culture, history, methodology, and ontology (as cited above) are summarized in the tables as shown in Chapter One and Chapter Four (especially in *Table 4.43* about my original theories on numerous topics).

And finally, or sixthly, the "post-human" variation in chess will also learn from the mediative-variative theory of chess in regard to the need to move towards the new mind games (as further elaborated in the next section).

## Towards the New Mind Games

This "post-human" variation in chess from the mediative-variative theory of chess to transcend both strategy and tactics towards the new mind games can be understood in terms of three great future transformation of mind games (as summarized in *Table 2.5*).

## Three Great Future Transformations of Mind Games

The three great future transformations of mind games can be explained in what follows, in that order.

*The First Great Future Transformation*

The first great future transformation of mind games concerns what I originally call *virtual games*, which have two distinctive features to remember.

Firstly, one distinctive feature of "virtual games" has to do with the further development of online chess (as already discussed in *Sec. 3.4*).

And secondly, another distinctive feature of "virtual games" has to do with the future development of what I call *virtual experience* of chess, which is comparable to "the technological use of *virtual memory* (or *virtual recall*) in terms of uploading info into the brain for sexual fantasy" (which I already discussed in *Sec. 3.4.2* of *The Future of Post-Human Sexuality*).

In addiition, in *The Future of Human Civilization* (2000), I already analyzed the different technological breakthroughs in the post-human future for uploading info into the brain for different purposes (not just for sexuality).

In the current context of new mind games, these technological breakthroughs will further transform chess at the virtual level never experienced before in human history.

*The Second Great Future Transformation*

The second great future transformation of mind games concerns what I originally call *novel games*, which have two distinctive features to remember.

Firstly, one distinctive feature of "novel games" has to do with the ever new chess variants (as already discussed in *Sec. 3.3.2*). Future new chess variants will be radically different from the ones as discussed in *Sec. 3.3.2* (which were used in Chapter Three solely for illustration only, so as to exercise our current imagination).

And secondly, another distinctive feature of "novel games" has to do with the future improvement of "chess engines" (as already discussed in *Sec. 2.4.1*; *Sec. 3.2.1*, and *Sec. 3.4*). Of course, future new chess engines will be more powerful from the ones as discussed in *Sec. 2.4.1*; *Sec. 3.2.1*, and *Sec. 3.4* (which were used solely for illustration only, so as to exercise our current imagination).

*The Third Great Future Transformation*

The third great future transformation of mind games concerns what I originally call *the post-human mind games*, which have two distinctive features to remember.

Firstly, one distinctive feature of "the post-human mind games" has to do with the quest for broader and deeper mental benefits of playing chess (as already discussed in *Sec. 2.2.2* and *Sec. 2.5.1*). Future exploration of broader and deeper mental benefits of playing chess will be radically different from the ones as discussed in *Sec. 2.2.2* and *Sec. 2.5.1* (which were described in Chapter Two solely for illustration only, so as to exercise our current imagination). The words "broader" and "deeper" are used on purpose here, so as to have a multifaceted view of mental benefits, which can be intellectual, aesthetic, moral, political, etc.

And secondly, another distinctive feature of "the post-human mind games" has to do with the future evolution of the mind into different forms, ultimately without the body—as already analyzed in my prevoius books in regard to my different visions of the future mind like "floating consciousness" (as in *FPHC* and *FCD*), "hyper-spatial consciousness" (as in *FPHC* and *FPHG*), "unfolding unconsciousness" (as in *FPHU*), and the like.

## Path-Breaking Implications

These future great transformations of mind games have path-breaking implications for some soul-searching questions which have bothered many of the best and the brightest minds in chess history—Is chess an art or a science? Are chess engines better than human players, in general? What use of chess playing is there for society and culture, let alone the mind? And....?

Questions like this remain relevant today, as shown in the contentious tale of two chess sensibilities and other views (as already introduced in *Sec. 1.1* and in the previous two chapters), which, unfortunately, have much impoverished the mind from exploring the other (different) sides of both strategy and tactics not only in chess but also outside chess.

Contrary to the two opposing chess sensibilities and other views (as already further analyzed in the context of the theoretical debate in *Sec. 1.4* and the rest of the book), the foresight of strategy (as already discussed in Chapter Two) and the usefulness of tactics (as already analyzed in Chapter Three) are neither possible nor desirable to the extent that their respective ideologues would like us to believe.

If strategy has its foresight, it has its uncertainty too, as the other side of the same coin. Conversely, if tactics has its usefulness, it likewise has its limitation, as the other side of the same mirror.

This conclusion can be carried out to its logical endpoint, in two major ways.

Firstly, strategy and tactics cannot exist without each other. There is no strategy without tactics—just as there is no tactics without strategy. This is important not only for chess but also outside chess—for life.

And secondly, in the long run of civilizational development for intelligent life, both on earth and, later, even more importantly, in deep space unto multiverses—both strategy and tactics are to be transcended into different spiritual concerns with the further development of the mind that our world in this relativley primitive historical era has never known.

Chess will still be played in the post-human future, just as it has existed for much of human history. But each age has its own versions of chess, to the extent that what counts as a chess variant varies from one age to another.

In due time, chess will likewise be superseded by the post-human forms of mind games for future spirituality in deep space until multivrses.

Although some variants of post-human mind games are proposed here, they constitute only a small fraction of other future transformations of the mind, nature, society, and culture (as already addressed in my other books on my other different visions of the future).

Once more, lest any human fantasy takes its hold, the post-human future of chess (or mind games in general) will be neither utopian nor dystopian, as they will instead be contextualized within the constraints of existential dialectics.

Welcome to this future post-human world of chess (or mind games in general), beyond strategy and tactics, away from the current narrow dichotomy of their being an art or a science, towards the higher spirituality of the mind, for better and for worse—not to be accepted on the basis of the wishes and beliefs of present humans but to be further developed on the basis of the values and beliefs of future post-humans in a way that the entire history of the human world hitherto existing has never known.

**Table 4.1. Sophisticated Methodological Holism (Part I)**

• "My methodological holism implies the partiality-totality principle in the ontology of existential dialectics (see the table on the partiality-totality principle for summary), which is against the varieties of (a) reductionism and (b) reverse-reductionism, in relation to (i) concept, (ii) theory, (iii) methodology, and (iv) ontology." (*FC*)

• "[M]y methodological holism here is not opposed to methodological individualism but includes it (and, for that matter, other methodologies too)…." (*FPHC*) For this reason (and others too, as summarized hereafter), my version of methodological holism is sophisticated—not vulgar as sometimes used by inapt scholars using the same term. (*FC*)

• "[M]y methodological holism does not democratically presume that all levels are equally valid, as all levels are not created equal. In other words, in relation to issue X, level A may be more relevant than level B, but in relation to Y, level B can be more relevant than level A instead." (*FPHC*) One excellent example of this vulgar democratic presumption is what I called in *BNN* "the compromise fallacy." (*FC*)

• My methodological holism does not presume that a lower level of analysis is more important than a higher level, solely because the former serves as the foundation for the latter—and vice versa, for that matter. One excellent example of this reductionistic presumption is what I called in *FPHST* "the foundation fallacy." (*FPHST*, *FC*)

*(continued on next page)*

**Table 4.1. Sophisticated Methodological Holism (Part II)**

• "[M]y methodological holism does not make any a-priori postulation that there must be a definite (and, for that matter, indefinite) number of levels" in any analysis. (*FPHC*) Nor does it dogmatically require that there must be a certain combination of levels of analysis in a given inquiry. (*FC*)

• "[M]ethodological holism, in my usage, does not assume that all levels...can necessarily be integrated, since methodological holism is not aimed to search for the holy grail of 'an integral theory...' (as is the case for Wilber). In other words, it allows that sometimes some levels may experience irreducible gaps between them, to be understood, at best, as empirical correlations, not as causal relations...." (*FPHC*)

• "[D]ifferent levels may overlap and even interact with each other in a given context (but sometimes may not), and the fact that I even proposed different ways of re-classifying the levels (whenever needed) in *FDC* reinforces this point....The dual danger here is either forcefully making different levels interact when they are just different (or, metaphorically speaking, apples and oranges) or inappropriately ignoring their interactions when some situations instead require them." (*FPHC*)

• "[T]o understand different levels from their own (unique) perspectives (as required by my methodological holism) is not the same as trying to reduce them to a preferred level in the process of learning from other levels. This second kind of multidisciplinary work is not genuine and does no justice to the unique complexities and merits inherent at each level." (*FPHC*)

*(continued on next page)*

**Table 4.1. Sophisticated Methodological Holism (Part III)**

• "[My] methodological holism walks a fine line between the artificial classification (separation) of levels and the simultaneous incorporation of them, if only for the sake of human scholarly endeavor. It should be reminded that nature does not impose upon itself the academic classification of the levels of analysis as humans have. The enterprise of classification is therefore anthropocentric." (*FPHC*)

• "[M]y methodological holism advocates neither epistemic subjectivism nor epistemic non-subjectivism (e.g., realism, idealism, and historicism), neither epistemic relativism (e.g., subjectivism, historicism) nor epistemic absolutism (e.g., realism, positivism), neither epistemic reductionism nor epistemic emergencism, and neither epistemic objectivism (e.g., realism, idealism) nor epistemic historicism....Neither does methodological holism, in my usage, accept the false meta-conceptual dichotomy between nominalism and realism....These false dichotomies...are to be transcended. In other words, methodological holism does not fully accept epistemic realism, positivism (a form of epistemic idealism), historicism, subjectivism, and reductionism in epistemology and philosophy of science but learns from the strengths and weaknesses of all of the opposing approaches without siding with any of them...." (*FPHC*)

• "Sophisticated methodological holism is subject to the constraints as imposed by the syntax of existential dialectics (e.g., the partiality-totality principle and the predictability-unpredictability principle). Even in predictability, outcomes are subject to uncertainty, the degree of which varies from case to case." (*FC*)

*(continued on next page)*

**Table 4.1. Sophisticated Methodological Holism (Part IV)**

• "Sophisticated methodological holism—when applied, especially though not exclusively, as illustrated in my numerous works—can enrich the understanding of reality in some distinctive ways. Here are three examples (as revealed in each of my books)." (*FPHCT*; *FPHL*)

—"Firstly, it provides a comprehensive analysis of a subject matter, from which much can be learned about reality," "both in relation to the perspectives of the mind, nature, society, and culture and also in relation to a new classification of the subject matter."

—"Secondly, it suggests some visions of the future in relation to the subject matter in question."

—"And thirdly, it proposes some insights on meta-theory (e.g., methodology and ontology) in general—with the clear understanding, however, of the dilemma of specific vs. general ontology (as shown in the table on the syntax of existential dialectics in the context of the dilemma of ontology). For this very reason, all of these ways are important, without reducing one into the analysis of another."

• "In the end, my meta-theory (both sophisticated methodological holism and existential dialectics) serves as a foundation to unify all domains of knowledge into *an unified theory of everything* (by way of some ontological principles and the comprehensive perspectives of the mind, nature, society, and culture). This is so, without commiting the sins of reductionism and reverse-reductionism (as often seen in many holistic approaches, with the fad of sysems approach—be it about systems theory, chaos theory, complexity theory, or else—as a most recent notorious example, which I debunked in *The Future of Complexity* and also in *The Future of Post-Human Formal Science*). My distinctive approach makes good use of different schools of thought without favoring any of them nor trying to integrate them (as they are not necessarily compatible with each other), so as to adjust for subjectivity, diversity, conflict, and complexity, for example. In this sense, the word "unified" does not have to mean integrative approach, in a narrow sense, nor systems approach, in a broad one, both of which I reject while learning from them." (*FPHFS*; *FPHS*)

*Sources*: A summary of *Sec.1.2* in *FPHC*—and also from *BNN*, *FPHST*, *ALD*, *FC*, *FPHCT*, and the rest of all other books of mine. See the books for more detail.

**Table 4.2. On Reductionism and Reverse-Reductionism (Part I)**

- **The Partiality-Totality Principle**
  —The partiality-totality principle in the ontology of existential dialectics targets against the varieties of reductionism and reverse-reductionism (as already worked out in my previous books).

- **Against the Varieties of Reductionism**
  —*Conceptual Reductionism*
    - Some illustrative instances involve myriad dualities like mind vs. body, self vs. world, democracy vs. non-democracy, and the like (as already addressed in *FHC*, *FPHC*, and *BDPD*, for instance).

  —*Theoretical Reductionism*
    - A fascinating case study concerns what I originally called "the foundation fallacy" in *FPHST*, in any attempt to naively understand space-time from the physical perspective as the foundation and, consequently, to dangerously dismiss other perspectives.
    - In *FAE*, I elaborated further these versions of reductionism in the literature on aesthetics (e.g., form vs. content, representation vs. expression, critics vs. artists, and externalism vs. internalism).
    - In *FIA*, I revealed other forms of reductionism in the literature on information architecture (e.g., the constructivist argument).
    - In *FPHU*, I showed the persistent legacy of reductionism, this time, in the literature on anomalous experience (e.g., the obsession with physics, chemistry, and biology for explaining anomalous experience).
    - In *FPHE*, I examined another case of reductionism in action, in the context of engineering (e.g., technical constraints vs. normative constraints).
    - In *FPHMM*, I elaborated one more version of reductionism, in relation to the three domains of communication (e.g., the competing views on sending, connecting, and receiving).
    - In *FPHCT*, I explored another version of reductionism, in relation to invention and innovation (e.g., the bio-psychological argument vs. the socio-cultural arguments).

*(continued on next page)*

**Table 4.2. On Reductionism and Reverse-Reductionism (Part II)**

- **Against the Varieties of Reductionism *(cont'd)***
  —*Theoretical Reductionism (cont'd)*
  - In *FPHG*, I identify another version of reductionism, in relation to infinity, symmetry, and dimensionality (e.g., the Euclidean argument vs. the non-Euclidean arguments).
  - In *FPHUP*, I analyzed another version of reductionism, in relation to density and void (e.g., the engineering argument and the ecology argument).
  - In *FPHL*, I examined another version of reductionism, in relation to structure and context (e.g., the structuralist argument and the contextualist argument).
  - In *PFHO*, I explored another version of reductionism, in relation to communcation, decision-making, and leadership (e.g., the rational-system argument vs. the natural-system argument).
  - In *PFHMA*, I revealed another version of reductionism, in relation to the martial body and spirit (e.g., the spiritual argument vs. the materialist argument vs. the defensive argument).
  - In *PFHS*, I analyzed another version of reductionism, in relation to the sexual body and spirit (e.g., the naturalist argument vs. the constructivist argument).
  - In *PFHLAW*, I examined another version of reductionism, in relation to law (e.g., the necessity argument vs. the contengency argument).
  - In *FPHWP*, I showed another version of reductionism, in relation to war and peace (e.g., the aggressivist argument vs. the pacifist argument).
  - In *BEPE*, I analyzed another version of reductionism, in relation to morality and immorality (e.g., the objectivist argument vs. the non-objectivist argument vs. the skeptical argument).
  - In *BCOS*, I examined another version of reductionism, in relation to the contested beginnings and speculative ends of the universe (e.g., the scientific argument vs. the religious argument vs. the esoteric argument vs. the metaphysical argument).
  - In *FPHP*, I show another version of reductionism, in relation to normality and abnormality (e.g., the natural argument vs. the social argument vs. the cultural argument vs. the mental argument).

*(continued on next page)*

**Table 4.2. On Reductionism and Reverse-Reductionism (Part III)**

- **Against the Varieties of Reductionism** ***(cont'd)***
  —*Theoretical Reductionism (cont'd)*
    - In *FPHGEOL*, I scrutinized another version of reductionism, in relation to statics and dynamics (e.g., the catastrophe argument vs. the uniformity argument vs. the revision argument).
    - In *FPHCHESS*, I show another version of reductionism, in relation to tactics and strategy (e.g., the natural argument vs. the social argument vs. the cultural argument vs. the mental argument).

  —*Methodological Reductionism*
    - A good instance concerns the debate between different versions of qualitative and quantitative methods (as already analyzed in *FC* and also *FHC*). In *FPHML*, I examined similar reductionism, this time, in the literature on mathematical logic (e.g., the obsession with consistency, soundness, and completeness). And in *FPHFS*, I also explored the problems of reductionism in the context of formal science (e.g., the analytical argument).

  —*Ontological Reductionism*
    - An excellent example is the debate between emergentism and reductionism in complexity theory and also in psychology (as elaborated in *FPHC*, in the context of Being and Becoming).

- **Against the Varieties of Reverse-Reductionism**
  —*Theoretical Reverse-Reductionism*
    - There are numerous art and literary theories co-exist. Take the case of literary studies, as there are now Literary Structuralism, Marxist Literary Criticism, New Criticism, Phenomenology, Hermeneutics, Language-Game Literary Criticism, Feminist Literary Criticism, Reception Theory, Reader Response Criticism, Poststructuralism, Semiotics, Pyschoanalytic Literary Criticism, just to cite some well-known ones, with no one being said to be better than any others (as detailedly analyzed in Ch.4 of *FHC*). (S. Raman 1997) In *BNN*, I even introduced "the compromise fallacy" as another good illustration of theoretical reverse-reductionism, in misleadingly treating both genetic and environmental approaches as equally valid.

*(continued on next page)*

**Table 4.2. On Reductionism and Reverse-Reductionism (Part IV)**

- **Against the Varieties of Reverse-Reductionism *(cont'd)***
  - —*Conceptual Reverse-Reductionism*
    - Any concept of "art" (e.g., fine art, cave art, outsider art, junk art) is deemed acceptable in postmodernism (as already addressed in Ch.4 of *FHC*).
  - —*Methodological Reverse-Reductionism*
    - There is the "anything-goes" mentality in postmodernism (e.g., doing art without praxis, doing art with praxis, and doing art by sublation), as analyzed in Ch.4 of *FHC*. And in *FPHFS*, I also exposed the problems of reverse-reductionism in the context of formal science in relation to systems theory.
  - —*Ontological Reverse-Reductionism*
    - There are likewise no privileged ontology, and the door is open for anything in postmodernism (e.g., the equal status of the ontology of Being vs. that of Becoming, as already addressed in Ch.4 of *FHC*—and also in *FPHC)*. In *FAE*, I also introduced another version of reverse-reductionism, that is, "the pluralist fallacy," in the context of understanding aesthetic experience, for instance—although this fallacy has been committed not exclusively in relation to the ontological level (but also at the conceptual, theoretical, and methodological ones).

*Sources*: From my previous books.

**Table 4.3. The Concepton of Existential Dialectics (Part I)**

- **Sets and Elements**
  - —Sets
    - Ex: the Same
    - Ex: the Others
  - —Elements
    - Ex: whites in 20th century America (in the set of "the Same")
    - Ex: Iraq during the U.S. invasion in 2003 (in the set of "the Others")

- **Relations, Operations, Functions**
  - —Relations (e.g., "belongs," "equals to," "is greater than")
    - Ex: symmetric interactions within the Same (or the Others)
    - Ex: asymmetric interactions between the Same and the Others
  - —Operations (e.g., "and," "or," "not," "if…then")
    - Ex: if the Same oppresses the Others, it will also oppress itself.
    - Ex: the Same is not the Others.
  - —Functions (e.g., goals)
    - Ex: the Same is hegemonic in relation to the Others.

- **Truth Values**
  - —"1" if True (in Symbolic Logic)
    - Ex: the proposition that imperial Japan was hegemonic to China during WWII
  - —"0" if False (in Symbolic Logic)
    - Ex: the proposition that Grenada invaded France in 2003
  - —"1" & "0" if Both True and False (in Dialectic Logic)
    - Ex: the proposition that the rabbit-duck picture refers to a duck
  - —"~1" & "~0" if Neither True Nor False (or N/A, in Dialectic Logic)
    - Ex: the proposition that God really exists

*(continued on next page)*

**Table 4.3. The Conception of Existential Dialectics (Part II)**

- **Axioms, Postulates, Theorems, Principles**
  —Axioms
    - Ex: the reflexive axiom—"any quantity is equal to itself"
  —Postulates
    - Ex: the SSS postulate—"if the three sides of a triangle are congruent to their corresponding parts, then the triangles are congruent"
  —Theorems (and Principles) in Existential Dialectics
    - In Relation to Method
      –#1: The formalness-informalness principle
      –#2: The absoluteness-relativeness principle
      –#3: The partiality-totality principle
      –#4: The predictability-unpredictability principle
      –#5: The explicability-inexplicability principle
    - In Relation to Structure
      –#6: The finiteness-transfiniteness principle
      –#7: The preciseness-vagueness principle
      –#8: The simpleness-complicatedness principle
      –#9: The openness-hiddenness principle
      –#10: The denseness-emptiness principle
    - In Relation to Process
      –#11: The change-constancy principle
      –#12: The order-chaos principle
      –#13: The slowness-quickness principle
      –#14: The expansion-contraction principle
    - In Relation to Agency
      –#15: The theory-praxis principle
      –#16: The convention-novelty principle
      –#17: The evolution-transformation principle
      –#18: The symmetry-asymmetry principle
      –#19: The softness-hardness principle
    - In Relation to Outcome
      –#20: The regression-progression principle
      –#21: The same-difference principle

*(continued on next page)*

**Table 4.3. The Conception of Existential Dialectics (Part III)**

*Notes*: The categories and examples in each are solely illustrative (not exhaustive). The comparison is also relative (not absolute), nor are they mutually exclusive. As generalities, they allow exceptions.
*Sources*: Starting from Ch.6 of *BCPC* and also from other books of mine

**Table 4.4. The Syntax of Existential Dialectics I: The Principles (Part I)**

---

• **In Relation to Method**

—*The Formalness-Informalness Principle*
*(On the Formal Requirements of Logical Systems)*

- The formal requirements of a logical system (e.g., consistency, soundness, and completeness) have both usefulness and non-usefulness, to the extent that, if there are formal systems requiring them, there are alternative ones which do not. Thus, it does *not* exclude classical logics but simply goes beyond both classical and non-classical logics, while learning something from each. There is no formalness without informalness—and vice versa.
- For instance, existential dialectics can make use of both classical logics under certain conditions (e.g., especially, though not exclusively, when they are clear-cut, etc.) and non-classical logics under alternative conditions (especially, though not exclusively, when they are "unknown," "irrelevant," "ambiguous," "possible," with "different degrees of truth," empirically inconsistent in a desirable way, etc.). (WK 2008u)
- Family resemblance: e.g., logicalness-nonlogicalness, rationality-nonrationality, etc.
- Sources: From *FPHML*. See also my later books.

—*The Explicability-Inexplicability Principle*
*(On the Underlying Mechanisms of Things)*

- Both explicability and inexplicability are part of the understanding of things. There is no explicability without inexplicability—and vice versa.
- This principle tells us the duality of the research dilemma, in that, if reality can be explained in some ways, it also has its other ways which are not quite explainable, at a given point in time.
- Family resemblance: e.g., underlyingness-regularness, causation-regularness, causation-correlation, etc.
- Sources: Especially from *FPHU*—and also *FPHC*. See also my later books.

---

*(continued on next page)*

**Table 4.4. The Syntax of Existential Dialectics I: The Principles (Part II)**

- **In Relation to Method *(cont'd)***
  —*The Predictability-Unpredictability Principle (On the Occurrence of Events)*
  - Both predictability and unpredictability have a major role to play in the occurrence of things, so that neither determinism nor indeterminism wins the centuries-old fight. There is no predictability without unpredictability—and vice versa.
  - There are events which are predictable, just as there are those which are not. Or what is regarded as unpredictable at one point in time may turn out to be predictable later, and, conversely, what is deemed as predictable may turn out to not be so predictable. Even in predictability, outcomes are subject to uncertainty, the degree of which varies from case to case.
  - Family resemblance: e.g., sureness-arbitrariness, etc.
  - Sources: Especially from *FC*. See also my later books.

  —*The Partiality-Totality Principle (On the Relationships between Whole and Parts)*
  - The whole is not the sum of the parts. There is no partiality without totality—and vice versa.
  - Any inquiry about a phenomenon in the work is to guard against the varieties of (a) reductionism and (b) reverse-reductionism.
  - Reductionism and reverse-reductionism can be (i) conceptual, (ii) theoretical, (iii) methodological, and (iv) ontological.
  - Family resemblance: e.g., individualisticness-holisticness, partness-wholeness, analysis-synthesis, etc.
  - Sources: Especially from *FC*. See also my later books.

*(continued on next page)*

**Table 4.4. The Syntax of Existential Dialectics I: The Principles (Part III)**

- **In Relation to Method** ***(cont'd)***

  —*The Absoluteness-Relativeness Principle*

  *(On the Multiplicity of Things)*

  - There is the multiplicity of things in reality, be they about entities, qualities (or properties), and relationships, such that what is acceptable from one standpoint may not be so from another. For instance, if there is something absolute, there is likewise something relative. There is no absoluteness without relativeness—and vice versa.
  - Both absoluteness and relativeness here are also relevant to different modalities often cited in the literature on ontology, such as possibility (e.g., something "can" happen) and its opposite (e.g., impossibility), probability (e.g., something "will" happen) and its opposite (e.g., improbability), and necessity (e.g., something "should" happen) and its opposite (e.g., contingency).
  - Family resemblance: e.g., uniformity-diversity, internalness-externalness, immanence-transcendence, etc.
  - Sources: From *FPHK*. See also my later books.

*(continued on next page)*

**Table 4.4. The Syntax of Existential Dialectics I: The Principles (Part IV)**

- **In Relation to Structure**
  —*The Finiteness-Transfiniteness Principle*
  *(On the Nature of Numbers)*
  - If there are finite things, there are likewise transfinite ones. There is no finiteness without transfiniteness—and vice versa.
  - To avoid confusion, my usage of the word "transfinite" here differs radically from the one used by Cantor (and other mathematicians) for "relative" infinity—and is more limited, in light of the problems confronting any attempt to understand the idea of infinity, be it by intuition, imagination, and conception (as detailedly analyzed in *Sec. 2.2.3* of *FPHG*).
  - Instead, by "transfinity," I allow numbers which can be many times larger—or smaller, for that matter—than the finite things that we encounter in daily life, but they do not have to be related to the idea of infinity at all (which may not exist).
  - Of course, there may be some *borderline* cases, in which it is not clear whether the number in question is transfinite (in my usage) or simply a mathematical convenience. A good example of a borderline case is the Planck unit of length for "the smallest space possibly measured in nature," which is "less than billionths of trillionths of trillionths of an inch" (or something like $1.6 \times 10^{-35}$ meters).
  - That said—my usage of "transfinity" can also resolve (or better, dissolve) an age-old problem in philosophy known as "Zeno's paradoxes" (as already explained in *Sec. 2.2.3* of *FPHG*).
  - Family resemblance: e.g., boundedness-quasiunboundedness, smallness-largeness, microness-macroness, etc.
  - Source: From *FPHG*. See also my later books.

*(continued on next page)*

**Table 4.4. The Syntax of Existential Dialectics I: The Principles (Part V)**

---

• **In Relation to Structure** ***(cont'd)***

—*The Simpleness-Complicatedness Principle*
*(On the Interconnection among Things)*

• Both simpleness and complicatedness are vital, without favoring one over the other, in that each is utilized, depending on the basis of the perspectives of nature, the mind, culture, and society. And even when a combination of them is preferred, the dilemma is only shifted to a combinational degree of concern. There is no simpleness without complicatedness—and vice versa.

• In relation to taxonomy, simpleness has its heuristic usefulness, just as complicatedness has its realistic representation, for instance. And in relation to network, simpleness has its economical attractiveness, just as complicatedness has its practical reliability, for instance.

• Family resemblance: e.g., inflexibility-flexibility, standardization-specialization, imperfectness-perfectness, superficiality-depth, shallowness-deepness, economicalness-elaboratedness, plainness-circumspection, onesidedness-multisidedness, etc.

• Sources: Especially from *FIA*. See also my later books.

—*The Denseness-Emptiness Principle*
*(On the Distribution of Entities in Space)*

• Both density and void are needed, in relation to the mind, nature, culture, and society, albeit in different ways. There is no denseness without emptiness—and vice versa.

• For clarity, the term "void" is used here only as an approximation of emptiness (depending on the degree of the lack of density), since, in physics, it is well known that "empty" space is not really empty all the way, because it can be full of energy (e.g., random quantum fluctuations at the sub-atomic level, and, for that matter, dark energy in the universe) and matter (e.g., different versions of sub-atomic particles, and, for that matter, dark matter in the universe).

• Family resemblance: e.g., fullness-voidness, nearness-farness, concentration-dispersion, etc.

• Sources: First worked out in *FPHUP*. See also my later books.

---

*(continued on next page)*

**Table 4.4. The Syntax of Existential Dialectics I:
The Principles
(Part VI)**

---

- **In Relation to Structure *(cont'd)***
  —*The Preciseness-Vagueness Principle*
  *(On the Refinement of Things)*
  - Both preciseness and vagueness are important, not that one is better than the other, but that both are used, in different degrees of preference, in accordance to the contextual application from the perspectives of nature, the mind, culture, and society. Even when both are used in a combination, the dilemma is shifted instead to one of combinational concern. There is no preciseness without vagueness—and vice versa.
  - In relation to taxonomy, preciseness has its taxonomic clarity, just as vagueness has its classificatory flexibility, for instance. And in relation to network, vagueness has its explorative liberty, just as preciseness has its conceptual definitiveness, for instance.
  - Family resemblance: e.g., clarity-ambiguity, directness-indirectness, quantitativeness-qualitativeness, describability-nondescribability, specificity-obscurity, specificity-generality, thickness-thinness, concretness-abstractness, etc.
  - Sources: Especially from *FIA*. See also my later books.

---

*(continued on next page)*

**Table 4.4. The Syntax of Existential Dialectics I: The Principles (Part VII)**

- **In Relation to Structure** ***(cont'd)***
  *—The Openness-Hiddenness Principle*
  *(On the Detection of Things)*
  - Reality has its hidden face, just as it is open to outside view in some other ways. There is no openness without hiddenness—and vice versa.
  - For instance, in the context of anomalous experience, certain aspects of reality can be open for examination (e.g., the experiment with SPECT images, the transmission of telepathemic bit, and the use of metonymies and metaphors). Other aspects, however, remain hidden, and examples include the elusive deeper nature of spiritual reality in the intangible realm, the undetectability of different branched universes, the underlying mechanisms of signals traveling between different folds of physical space-time, and the dependence on language and the bias for science and logic).
  - Family resemblance: e.g., overtness-covertness, publicness-privateness, openness-closedness, transparency-secrecy, openness-biasedness, etc.
  - Sources: Especially from *FPHU*. See also my later books.

- **In Relation to Process**
  *—The Change-Constancy Principle*
  *(On the Alteration of Things)*
  - Change occurs over time, although constancy is also allowed. There is no change without constancy—and vice versa.
  - Asymmetry undergoes changes over time, so does symmetry.
  - Old players fade away, and new ones emerges, with ever new causes and ever new forms.
  - Family resemblance: e.g., dynamicness-staticness, instability-stability, etc.
  - Sources: First named in *BCPC*. Especially from *FHC*, *FCD*, and *FPHC*. See also other books of mine (from that point on).

*(continued on next page)*

## Table 4.4. The Syntax of Existential Dialectics I: The Principles (Part VIII)

• **In Relation to Process** ***(cont'd)***

—*The Expansion-Contraction Principle*
*(On the Growth of Things)*

- Entities in the world can both expand in some ways and contract in other ones, as part of their nature. There is no expansion without contraction—and vice versa.
- For instance, in the context of anomalous experience, one of its most fascinating expressions is none other than the analysis of different causes of the slow but steady expansion (and, furthermore, if put in a different way, unfolding) of unconsciousness.
- Family resemblance: e.g., conquest-autarky, rise-fall (or up-down), spread-shrink, extendingness-shorteningness, etc.
- Sources: Especially from *FPHU*. See also my later books.

—*The Slowness-Quickness Principle*
*(On the Speed of Change)*

- Both slowness and quickness co-exist, with their own internal tension, to the extent that each fights for its own relevance with the other, in accordance to the perspectives of nature, the mind, culture, and society, without one being the victor and the other being the vanquished in the long haul. Even when both are chosen in other cases, this dilemma is only transferred into something else with a combinational character. There is no slowness without quickness—and vice versa.
- In relation to taxonomy, quickness has its efficient usability, just as slowness has its aesthetic appeal, for instance. And in relation to network, quickness has its adventurous readiness, just as slowness has its risk-adverse convenience, for instance.
- Family resemblance: e.g., inconvenience-convenience, passiveness-activenessness, gradualness-abruptness, deceleration-acceleration, etc.
- Sources: Especially from *FIA*. See also my later books.

*(continued on next page)*

**Table 4.4. The Syntax of Existential Dialectics I: The Principles (Part IX)**

- **In Relation to Process** ***(cont'd)***
  —*The Order-Chaos Principle*
  *(On the Pattern of Things)*
  - Both order and chaos are vital in the process of change in the world. The preference for order is biased, since it does not give sufficient attention to the vital role of chaos in the transformation of the world (without somehow reducing it for the understanding of order). There is no order without chaos—and vice versa.
  - The scientific search for order in the world is often a hidden bias in its ontological obsession with order, since chaos is often treated as the "bad" guy, with order as the "good" guy (for the end goal of science).
  - Neither order nor chaos is the final end of the world, and one is not to be treated as the means for the other in the transformation of things. Both are fundamental in their recurrent dialectical interactions with each other over time, without reducing one for the other.
  - Family resemblance: e.g., lawfulness-disorder, order-disorder, etc.
  - Sources: Especially from *FC*. See also my later books.

- **In Relation to Agency**
  —*The Softness-Hardness Principle*
  *(On the Force of Change)*
  - This has to do with *the force of change* under the category about agency in existential dialectics, in that any change by an agent, be it organic (like humans) or non-organic (like natural objects), can occur in a forceful (aggressive) or gentle (pacific) way, which can come in all shapes and sizes, of course. There is no softness without hardness—and vice versa.
  - Family resemblance: e.g., peacefulness-violence, cooperation-competition, pacificity-aggression, dovishness-hawkishness, reward-punishment, peace-war, etc.
  - Sources: Especially from *ALD*. See also my later books.

*(continued on next page)*

### Table 4.4. The Syntax of Existential Dialectics I: The Principles (Part X)

- **In Relation to Agency *(cont'd)***
  —*The Symmetry-Asymmetry Principle*
  *(On the Relationships among Existents)*
  - There is no symmetry without asymmetry—and vice versa.
  - For instance, the Same can be symmetric and asymmetric towards the Others. But in case of asymmetry, oppression and self-oppression can occur. So, when the Same is asymmetric towards the Others, the Same can also be relatively asymmetric towards itself in self-oppression, just as the Others can be likewise towards themselves.
  - The subsequent oppressiveness is dualistic, as much by the Same against the Others and itself, as by the Others against the Same and themselves.
  - Both oppression and self-oppression can be achieved by way of downgrading differences between the Same and the Others and of accentuating them.
  - This is true, even though not all forms of asymmetry have to be about oppression and self-oppression.
  - In addition, from Chapter Three of *FPHG*, symmetry is not perfect, to be understood in an approximate sense under many life circumstances.
  - Family resemblance: e.g., strength-weakness, balance-extremity, harmoniousness-unharrmoniousness, potence-impotence, reflexiveness-unreflexiveness, equality-inequality, etc.
  - Sources: From all my books, starting with *FHC*. First named in *BCPC*. See, for instance, *FCD* for more details on the Same and the Others.

(*continued on next page*)

**Table 4.4. The Syntax of Existential Dialectics I: The Principles (Part XI)**

- **In Relation to Agency** ***(cont'd)***

  —*The Theory-Praxis Principle*
  *(On the Duality of Knowledge))*

  - This has to do with the duality of knowledge in existential dialectics, in that, if there is theoretical construction, there is likewise its practical application, both technical and normative. There is no theory without praxis—and vice versa.
  - Family resemblance: e.g., discovery-application, knowledge-action, invention-innovation, etc.
  - Source: Especially from *FPHE*. See also my later books.

  —*The Evolution-Transformation Principle*
  *(On the Multiple Kinds of Agency)*

  - This principle (and the symmetry-asymmetry principle) are both about *the agency of change*. The word "agency," in a formal definition, refers to "a person or thing through which power is exerted or an end is achieved." (MWD 2007b) It therefore does not have to necessarily involve an intelligent lifeform.
  - Because of this dual meaning in agency, the evolution-transformation principle is more concerned with *the multiple kinds of agency*, that is, both about the *evolution* in the state of nature (e.g., an object of natural beauty) and the *transformation* in the world of intelligent lifeforms (e.g., a work of art, an air-conditioner, etc.). There is no evolution without transformation—and vice versa.
  - And the transformative part of the principle precisely refers to the other dimension in the dual meaning of agency, in giving technology a major role to play in the change of the world, which is something that I extensively analyzed in *FHC* in the context of the technophilic lifeworld, espeically though not exclusively since modern times.
  - Family resemblance: e.g., inorganicness-volition, nonwillingness-willingness, naturalness-technologicalness, naturalness-nonnaturalness, etc.
  - Sources: Especially from *FAE*. See also other books of mine.

(*continued on next page*)

**Table 4.4. The Syntax of Existential Dialectics I:
The Principles
(Part XII)**

---

- **In Relation to Agency *(cont'd)***
  - —*The Convention-Novelty Principle*
    *(On the Nature of Creative Thinking)*
    - If there is conventional wisdom, there is likewise novel challenge, to the extent that both convergent and divergent thinking are part of life. There is no convention without novelty—and vice versa.
    - In addition, in this context of the convention-novelty principle, there are, in the absence of better words, what I want to call (a) *creative techniques* and (b) *creative traits*, which, when satisfied—in relation to the larger context of the mind, nature, society, and culture—can be used to enhance creative works. The list of creative techniques and traits (as summarized in *Table 4.1* of *FPHCT*) are not exhaustive, of course, but the examples here constitute a great beginning of understanding the structure of creative thinking.
    - Yet, lest the reader gets carried away by the euphoria about creative thinking in our time, it should be reminded that creative thinking has its own possibilities and limits (as shown in *Table 2.1* of *FPHCT* on invention), just as it has its own promises and pitfalls (as shown in *Table 3.1* of *FPHCT* on innovation). In the end, it should also be stressed that creative thinking has its own desirability and dark sides (as shown in *Table 4.2* of *FPHCT* ).
    - This has important daily implications for an average individual, and the word "average" here allows different degrees of variation from one individual to another in a population, at any given point in history.
    - Family resemblance: e.g., convergence-divergence, normalness-nonnormalness, conformity-nonconformity, etc.
    - Sources: First from *FPHCT*. See also my later books.

---

(*continued on next page*)

**Table 4.4. The Syntax of Existential Dialectics I:
The Principles
(Part XIII)**

---

• **In Relation to Outcome**

*—The Regression-Progression Principle*
*(On the Direction of History)*

• Neither the cyclical nor the linear views are adequate for explaining many phenomena at all levels. There is regression without progression—and vice versa.
• History progresses to more advanced forms, but with a regressive touch. Examples include no freedom without unfreedom, no equality without inequality, and no civilization without barbarity.
• This is not an inevitable law, but merely a highly likely empirical trend.
• Family resemblance: e.g., cost-benefit, undesirability-desirability, badness-goodness, risk-opportunity, etc.
• Sources: From all my books, starting with *FHC*. First named in *BCPC*.

*—The Same-Difference Principle*
*(On the Metamorphosis of Change)*

• An entity, as it evolves over time, can be both different from and similar to its opposing alternatives and does not have to be solely more different from them over time. There is no similarity without difference—and vice versa.
• Opposites are not absolute in a black-or-white fashion; so, an entity can become relatively more similar to (or more different from) its opposite over time.
• Family resemblance: e.g., homogeneity-heterogeneity, we-they, etc.
• Sources: Especially from ALD. See also my later books.

---

*Notes*: The features in each principle are solely illustrative (not exhaustive). The comparison is also relative (not absolute), nor are they mutually exclusive. As generalities, they allow exceptions.

**Table 4.5. The Syntax of Existential Dialectics II: The Principles as Short Cuts (Part I)**

• The principles should be treated with caution, lest misunderstanding occurs, since they do not constitute rigid dualities (or dichotomies).

• The reason is that each pair in an ontological principle consists of two opposites, which are, however, merely short cuts both for multiple variations and degrees, as well as for different interactions with multiple other entities unlike them. In this light, each pair can end up having hundred (or even thousand, if not more) different versions, which interact with hundred (or even thousand, if not more) other entities. There are two clarifications here.

• Firstly, this conception of shortcuts is not mutually exclusive nor absolute, in that the opposites can come in all shapes and sizes, with different degrees. For instance, by analogy, just as there are different degrees of the two colors "white" and "black"—there are likewise different degrees of the opposites in each ontological pair, to the extent that there can be multiple entities (not only two) interacting with each other in each pair.

• And secondly, this conception of shortcuts do not ignore other possible entities in interacting with the two opposites (with their different versions) in each pair. For instance, by analogy, there are not only the two colors "white" and "black" as opposites, since there are other colors too besides them like "yellow," "green," "purple," or else. The same logic can be applied to each ontological pair, in that they also interact with other entities, not with only two of them (with their different versions).

• Therefore, with these two clarifications in mind—each pair in an ontological principle serves only as an abbreviation for something more complicated and, therefore, although it contains two opposites, it should not be confused as a duality (dualism).

*(continued on next page)*

**Table 4.5. The Syntax of Existential Dialectics II: The Principles as Short Cuts (Part II)**

• Existential dialectics rejects any dualism (or dichotomy) as too rigid and instead allows the multiplicity of entities, to the extent that between the two opposites in each pair exist many other alternatives to choose from. They are named in that short form for aesthetic elegance, instead of listing all possible entities between the two opposites in the title.

• To be dialectic is to go beyond any rigid dichotomy and transcend into something different altogether in the long haul. One may be tempted to call the dialectic logic here with a different name like existential "multilectics" (instead of "dialectics"), but this naming is incorrect (or even misleading), for the two reasons aforestated.

• After all, each ontological pair come in all shapes and sizes (with different degrees) and do not exist by themselves but also interact with other entities unlike them. The virtue of revealing an ontological pair is to show how they relate within themselves (in multiple versions) and also interact with others unlike them (also in multiple versions).

• This will be clear in the table on "the dialectic constraints imposed by the principles."

*Source*: From Ch.1 of *FPHK*

**Table 4.6. The Syntax of Existential Dialectics III: The Principles as Family Resemblances**

• Each ontological principle is generic, with some other comparable ontological pairs to be put in the same family (like a *family resemblance*). Ludwig Wittgenstein (1953) once suggested the idea of "family resemblance" in explaining different games classified under the same family called *games*.

• Why should, for instance, playing football and chess as playing "games," when it is well understood that football is not the same as chess? The answer is that, although each game is different and has different rules, many of them (though not all) share, more or less, some commonalites (e.g., scoring as necessary for winning).

• And this is so, even though some games share more than some others in any given selection of critieria, and no two games are exactly identical. So, his point here is that there is no essential core which is common to all games, and the best that one can look for is some characteristics which are common to many (but not all) games. (A. Biletzki 2006)

• By the same logic—in the previous section on selection criteria, the flexibility-inflexibility pair can be put in the family resemblance of the simpleness-complicatedness principle, although the two pairs are not exactly identical. Likewise, the directness-indirectness pair can be put in the family resemblance of the preciseness-vagueness principle, although, again, the two pairs are not exactly identical.

• In this sense, which specific pair in a family should be used to designate the name of the family can be at times a bit arbitrary, but with good reason.

• For illustration, in the context of method, the partiality-totality principle can take the different form like individualisticness-holisticness, just as the explicability-inexplicability principle can take the different form like underlyingness-regularness—although each two pairs are not exactly identical and have slightly different meanings and usages. A more comprehensive listing of this family resemblance for all other principles is shown in the table on the syntax of existential dialectics.

*Source*: From Ch.1 of *FPHK*

**Table 4.7. The Syntax of Existential Dialectics IV: The Dialectic Constraints Imposed by the Principles (Part I)**

- **Co-Existent and Asymmetric**
  - —The principles, as they constitute the syntax of existential dialectics, are dialectic in character, such that, when they are applied, they impose dialectic constraints on how reality is to be understood. Consider, say, the symmetry-asymmetry principle as an illustration here, in order to summarize two main characters of the dialectic constraints in question.
  - —Firstly, to be dialectic here is to go beyond the narrow dichotomies (and, for that matter, any rigid multi-dimensional classificatory scheme), be they about "self" vs. "world," "freedom" vs. "unfreedom," "barbarity" vs. "civilization," "individuality" vs. "communality," and so on.
  - —One way to do so (to go beyond) is to consider them all in terms of co-existence (without favoring one over the rest). For instance, my theory of "post-civilization" (to be summarized later in the section on the pragmatics of existential dialectics) is to go beyond barbarity and civilization in terms of understanding barbarity and civilization as being co-existent. And the same logic can be said in relation to my theories of "post-democracy," "post-capitalsm," and others (also to be introduced later in the section on the pragmatics of existential dialectics), in regard to freedom vs. unfreedom, equality vs. inequality, communality vs. individuality, spirituality vs. materiality, and so on.
  - —But to consider them all (in the dichotomies—and, for that matter, in any rigid multi-dimensional classifactory schme) as co-existent is not the same as to imply that the opposites in any classificatory scheme are all equal, since, in accordance to the symmetry-asymmetry principle (as an illustration here), if they are equal in terms of being considered as co-existent, they are asymmetric in terms of being unequal in dialectic interaction (e.g., X can be more dominant than Y in case A, or Y is more dominant than X in case B).

*(continued on next page)*

**Table 4.7. The Syntax of Existential Dialectics IV: The Dialectic Constraints Imposed by the Principles (Part II)**

—For this reason, there are different versions of "post-democracy" and "post-capitalism" in my theories. As an illustration, in version I of the theory of post-democracy, freedom is more dominant than equality, whereas in version II of the theory of post-democracy, equality is more so than freedom.

—But this "X more than Y" has to be understood in the context of dialectic logic (not in conventional logic), in that both "X" and "Y" are important in post-democracy (in the context of dialectic logic), but in an asymmetry way. By contrast, in conventional logic, it often favors one over the other—be it in regard to privileging freedom over equality in Fascism, favoring freedom relatively more than equality in Liberal Democracy, or favoring equality relatively more than freedom in Socialist Democracy. In the latter two cases (about Liberal Democracy and Socialist Democracy), the difference between dialectic logic and conventional logic can be one in degree, not in kind—in this sense, albeit not in other senses.

—The same logic can be said about the relatonships between individuality and communality, between spirituality and materiality, and between formal legalism and informal legalism in the different versions of my theory of post-capitalism.

• **Transcendent**

—Secondly, to be dialectic is to go beyond the narrow dichotomies (and, for that matter, any rigid multi-dimensional classificatory scheme) in another way, this time, in a transcendent way, that is, in exploring other possibilities or even other issues not considered within the narrow confines of narrow dichotomies (and, for that matter, any rigid multi-dimensional classificatory scheme).

*(continued on next page)*

**Table 4.7. The Syntax of Existential Dialectics IV: The Dialectic Constraints Imposed by the Principles (Part III)**

—As an analogy, to go beyond the narrow color dichotomy of "black" and "white" is not just to choose both "black" and "white" (as in the first meaning) but also to explore other color options (e.g., "green," "purple," "blue," etc.—and, alternatively, "shade," "line," "curve," etc.). By the same logic, to go beyond "democracy" is to transcend democracy (as in version III of the theory of "post-democracy") and to explore other possibilties of lifeforms (e.g., floating consciousness, hyper-spatial consciousness, etc., to live beyond the narrow obsession with freedom and equality).

—This dialectic character of the principles in existential dialectics has important implications for the pragmatics of existential dialectics (as will be clear shortly, in the section on the pragmatics of existential dialectics).

*Notes*: The examples here are solely illustrative (not exhaustive). The comparison is also relative (not absolute), nor are they mutually exclusive. As generalities, they allow exceptions.
*Sources*: First explicitly stated in *ALD*. Also from all other books of mine.

**Table 4.8. The Syntax of Existential Dialectics V: Further Clarifications (Part I)**

• **Pioneering**

—Firstly, the total number of ontological principles is unknown, to be discovered later, as our knowledge of the world becomes more advanced.

—Consequently, the principles as introduced in my books are not exhaustive, with new ones being added, whenever more of them are discovered in later research.

—At least, future generations can pick up where I leave off and continue the discovery.

—In this sense, my work should be treated as a pioneering effort for the development of a systematic, comprehensive analysis of a new general ontology for the future of knowledge.

• **Flexible**

—Secondly, the principles are not rigidly classified, as they can be reclassified in a different way.

—For instance, the preciseness-vagueness principle is classified under the category of "structure" but can be reclassified under the category of "method," although in so doing, it has a different meaning in the context of method.

—The same logic applies to the same-difference principle under the category of "outcome," which can be reclassified under the category of "structure," although in so doing, once more, it has a different meaning in the context of structure.

• **Mutually Constraining**

—Thirdly, the principles are to be understood together, not that each prinicple is to be analyzed independently of others. After all, the principles are mutually constraining, in that they work together as a whole.

*(continued on next page)*

**Table 4.8. The Syntax of Existential Dialectics V: Further Clarifications (Part II)**

- **Selectively Useful**
  —Fourthly, the principles are relevant to all subject matters, but some principles are more useful to some subject matters than others—as implied in the symmetry-asymmetry principle. This is true, even if different studies of the same kind can yield different views about the degree of relevance for each ontological principle, depending on the specific nature of a research in question, needless to say.

- **Anti-Reductionistic**
  —And fifthly, the principles constitute only two levels of analysis, this time, at the ontological and methodological levels—while other levels of analysis (from the perspectives of the mind, nature, society, and culture) are also needed, in order to understand reality in its totality (as explained in the section on "sophisticated methodological holism," so as to avoid the dual dangers of reductionism and reverse-reductionism).

*Notes*: The examples here are solely illustrative (not exhaustive). The comparison is also relative (not absolute), nor are they mutually exclusive. As generalities, they allow exceptions.
*Sources*: From my previous books like *FHC*, *FCD*, *FPHC*, *BDPD*, *BCPC*, *BNN*, *FC*, *FAE*, and so on.

**Table 4.9. The Syntax of Existential Dialectics VI: The Dilemma of Specific vs. General Ontology (Part I)**

• Ontology is often more complicated and imposes some seemingly insurmountable difficulties concerning what constitutes a kind of ontology which can be valid enough to be accepted by the wider intellectual community in question. An excellent example concerns what I want to call *the dilemma of ontology* in relation to the relationship between "specific ontology" and "general ontology." (M. Bunge 1999; R. Corazzon 2007)

• On the one hand, "general ontology," as Mario Bunge (1999) put it, "studies all existents." But, on the other hand, "special ontology studies one genus of thing or process—physical, chemical, biological, social, etc." (M. Bunge 1999; R. Corazzon 2007)

• Consequently, the relationship between the former and the latter is that, as an illustration, "whereas general ontology studies the concepts of space, time, and event, the ontology of the social investigates such general sociological concepts as those of social system, social structure, and social change." (M. Bunge 1999; R. Corazzon 2007)

• With this formal definition in mind, the relationship between the two forms of ontology only poses *the dilemma of ontology*, in that general ontology is possible to the extent that its general contructs must be broad enough to serve as the "lowest common denominator" for all disciplines, but this character of the lowest common denominator is not only hard to find but also its usefulness is limited. (WK 2008b)

• But specific ontology fares no better either, in that its specific constructs primarily serve a specific discipline in question, although other disciplines can appropriate them for their own purposes, but in a different context (especially, though not exclusively, when used as a metaphor from one field to another). For instance, the Kantian category of understanding under the heading of "quantity" constitutes a type of specific ontology, which is specific to the discipline of mathematics, although other disciplines can make use of the categories in their own context.

*(continued on next page)*

**Table 4.9. The Syntax of Existential Dialectics VI: The Dilemma of Specific vs. General Ontology (Part II)**

---

• A solution lies in my proposal of "sophisticated methodological holism," which requires all levels of analysis in any subject matter. See the tables on sophisticated methodological holism, and on reductionism and reverse-reductionism, for more details.

---

*Source:* From *Sec. 1.4* of *FPHK*

**Table 4.10. The Syntax of Existential Dialectics VII: Types of Inappropriate Family Resemblances**

- **Inadequate Family Resemblances**
  —Ex: consistency-inconsistency: the formalness-informalenss principle
  —Ex: cognition-noncognition: the formalness-informalness principle

- **Compound Family Resemblances**
  —Ex: reality-nonreality: the formalness-informalness principles, the absoluteness-relativeness principle
  —Ex: normality-abnormality: the symmetry-asymmetry principle, the density-emptiness principle, the convention-novelty principle, the same-difference principle
  —Ex: structure-context: the simpleness-complicatedness principle, the theory-praxis principle
  —Ex: freedom-unfreedom: the symmetry-asymmetry principle, the expansion-contraction principle
  —Ex: morality-immorality: the denseness-emptiness principle, the symmetry-asymmetry, the same-difference principle
  —Ex: particularness-universalness: the preciseness-vagueness principle, the simpleness-complicatedness principle
  —Ex: pureness-mixedness: the absoluteness-relativeness principle, the simpleness-complicatedness principle, the same-difference principle
  —Ex: shortsightedness-foresightedness: the partiality-totality principle, the simpleness-complicatedness principle
  —Ex: shorttermness-longtermness: the partiality-totality principle, the simpleness-complicatedness principle, the slowness-quickness principle

*Notes*: The examples here are solely illustrative (not exhaustive). The comparison is also relative (not absolute), nor are they mutually exclusive. As generalities, they allow exceptions.
*Source*: First summarized in *BEPE* (and updated in later books)

**Table 4.11. The Semantics of Existential Dialectics**

- **Abstract Structure vs. Specific Meanings**
  —The syntax of existential dialectics so understood in terms of ontological principles only gives us the structure of ontology in the world, in an abstract (general) sense. These principles by themselves do not tell us the specific meanings in a given context.
  —In order to grasp the specific meanings of the principles in a given context, it is necessary to study the semantics of existential dialectics. And the analysis of the ontological principles as family resemblances in the section on syntax is only a starting point (and thus overlaps a bit with the semantics of existential dialectics here).
  —With this caveat in mind, the reason that I have often gone in great lengths in my previous books on different subjects is to explain the specific meanings of the principles when applied in different contexts.

- **Specific Meanings in Specific Fields**
  —For instance, in *FPHST*, I used the first three principles (i.e., the change-constancy principle, the regression-progression principle, and the symmetry-asymmetry principle) to propose "the perspectival theory of space-time," for a better way to understand space and time—especially, though not exclusively, in relation to future post-human history (as summarized in *Table 3.6*, *Table 3.7*, *Table 3.8*, and *Table 3.9*). In so doing, I had to introduce concepts and theories specific to the field of physics and other related fields (e.g., "absolute space" and "absolute time" in "classical mechanics" and "relative space-time" in "the theory of relativity").
  —In *BNN*, I also exploited the three principles to propose the "transcendent" approach to the study of genes and memes as a new way to understand the interaction between nature and nurture. In so doing, I had to explore concepts and theories in the world of evolutionary theory (e.g., "mutation," "variation," "adaptation," "selection," and "inheritance" in Darwinian evolutionary theory) and neural biology (e.g., "chromosome," "gene," "DNA," "RNA," "protein," "neuron," "neural network," and "behavior").

*Notes*: The categories and examples in each are solely illustrative (not exhaustive). The comparison is also relative (not absolute), nor are they mutually exclusive. As generalities, they allow exceptions.
*Sources*: First explicitly stated in *ALD*. Also from all other books of mine.

**Table 4.12. The Pragmatics of Existential Dialectics (Part I)**

- **The Two-Way Street Connecting Theory and Meta-Theory**
  —The pragmatics of existential dialectics is not a one-way street (that is, using the ontological principles for theoretical insights in praxis) but a two-way one, that is, (a) from meta-theory to theory, and (b) from theory to meta-theory.
  —(a) On one side of the street, the ontological principles can inspire some theoretical insights in praxis, that is, in relation to some specific fields.
  —(b) On the other (opposing) side of the street, however, the study of a subject matter in the specific fields in turn reveals some more hitherto unknown ontological principles to be discovered and identified. For this reason, three new principles were added in *FC*, one in *FAE*, and two in *ALD*, on top of the original three in *BCPC*—after some research on the specific subject matters.

- **Direct and Indirect Applications**
  —*Direct*
  - The logic of existential dialectics can shed some theoretical insights on diverse phenomena in the world, and good instances are the pertinent use of the principles of existential dialectics for the theoretical insights on the freedom/unfreedom dialectics, the equality/inequality dialectics, and the wealth/poverty dialectics in my previous works.
  - My latest books like *FPHST* and *BNN* also use the principles to reveal some theoretical insights on the perspectives of space and time (as in *FPHST*) and of nature and nurture (as in *BNN*).

  —*Indirect*
  - The theoretical insights can further be used to reveal other phenomena directly from them (viz., the theoretical insights) and therefore indirectly from the principles themselves. A good instance is the use of the theoretical insights on the freedom/unfreedom and equality/inequality dialectics for the understanding of the civilization/barbarity dialectics.

*(continued on next page)*

**Table 4.12. The Pragmatics of Existential Dialectics (Part II)**

---

- **Direct and Indirect Applications (*cont'd*)**
  - Even in indirect applications, however, a phenomenon under study can still be directly related back to the principles themselves. In the example as cited above, the civilization/barbarity dialectics can be directly related to the principles of existential dialectics without the intermediate role of the freedom/unfreedom and equality/inequality dialectics.

- **Multiple Levels of Application**

  —The theoretical insights can be applied to different levels of analysis, even though in a given example, it may refer to one level only. For instance, in the example concerning the freedom/unfreedom dialectics, it can be used at the structural level (e.g., in relation to the theory of cyclical progression of hegemony), but it can be exploited as well for other levels (e.g., the theory of post-capitalism at the institutional level).

---

*Notes*: The categories and examples in each are solely illustrative (not exhaustive). The comparison is also relative (not absolute), nor are they mutually exclusive. As generalities, they allow exceptions.
*Sources*: From Ch.6 of *BCPC*. See also other books of mine.

**Table 4.13. The Freedom-Unfreedom Dialectics (Part I)**

• **On Having**

—*In Relation to the Technological*

(1) if freer from submission to Nature, then less free from ecological degradation (Deep and Social Ecology), even if in a hi-tech form

(2) if freer from technological inconvenience / backwardness, then less free from technological control and the loss of privacy

(3) if freer from technological (material) backwardness, then less free from the abusive (barbaric) maltreatment of the primitive Others

—*In Relation to the Everyday*

(1) if freer from abject poverty, then less free from artificial needs/discontents (Frankfurt School)

(2) if freer from sensual suppression, then less free from violent sublimation (Freud)

(3) if freer from the snobbishness of high culture, then less free from the shabbiness (leveling-off effect) of mass culture (Tocqueville)

(4) if freer from the inefficiency of traditional "compassionate economy," then less free from the bondage of a "ruthless [competitive] economy" (Keynes)

(5) if freer from anarchy in the state of nature (system fragmentation), then less free from government regulations and controls in system integration

*(continued on next page)*

**Table 4.13. The Freedom-Unfreedom Dialectics (Part II)**

- **On Belonging**

  —*In Relation to the Good and the Just*

  (1) if freer from disciplinary society, then less free from society of control (Foucault)
  (2) if freer from the tyranny of one or a few, then less free from the tyranny of the majority (or sometimes, minority veto)
  (3) if freer from elitist decision making, then less free from political gridlock/cleavage
  (4) if freer from arbitrary (discretionary) administration, then less free from bureaucratic irrationality (Weber) and legal trickery (loopholes)

- **On Being**

  —*In Relation to the True*

  (1) if freer from unscientific dogmas, then less free from instrumental abyss (nihilism). Or conversely, if freer from meaninglessness, then less free from dogmas.
  (2) if freer from the bondage of partiality/partisanship (e.g., prejudice, discrimination), then less free from the danger of impartiality and neutrality (e.g., opportunism, unrealisticness, lack of compassion, inaction)
  (3) if freer from making generalizations, then less free from being unable to understand much of anything

  —*In Relation to the Holy*

  (1) if freer from collective conscience, then less free from social loneliness
  (2) if freer from religious absoluteness, then less free from spiritual emptiness

  —*In Relation to the Beautiful/Sublime*

  (1) if freer from artistic non-autonomy, then less free from aesthetic disillusion (deconstruction)

*(continued on next page)*

**Table 4.13. The Freedom-Unfreedom Dialectics (Part III)**

*Notes*: The examples in each category are solely illustrative (not exhaustive), and the comparison is relative (not absolute), nor are they necessarily mutually exclusive. And some can be easily re-classified elsewhere. As generalities, they allow exceptions.
*Sources*: A reconstruction from Ch.10 of *FCD*, based on *FHC*

**Table 4.14. The Equality-Inequality Dialectics (Part I)**

---

• **On Having**

—*In Relation to the Technological*

(1) if more equal in treating Nature with spiritual unity, then less equal in suppressing the dominant drive to transcend it altogether

—*In Relation to the Everyday*

(1) if more equal in building social plurality, then less equal in leveling-off effects (e.g., the subsequent relative intolerance of high/intellectual ethos in mass culture industry)

(2) if more equal in socioeconomic distribution beyond a certain point, then less equal in efficiency (e.g. resentment, the erosion of work ethics)

(3) if more equal in urging an affirmative action program, then less equal in creating victim mentality (in oneself), stigma (from others), reverse discrimination (against the once privileged), and mediocracy (against the more able)

• **On Belonging**

—*In Relation to the Good and the Just*

(1) if more equal in banning monarchic/oligarchic exclusion, then less equal in producing "the tyranny of the majority" or of "minority veto"

(2) if more equal in encouraging participatory decision making, then less equal in inducing political divisiveness (gridlock/cleavage in power blocs) and organizational oligarchy

(3) if more equal in institutionalizing a decentralized bureaucracy, then less equal in falling into more territorial/turf politics (intrigues)

---

*(continued on next page)*

**Table 4.14. The Equality-Inequality Dialectics (Part II)**

- **On Being**
  - —In Relation to the Beautiful / Sublime
    - (1) if more equal in accepting diverse styles ("anything goes" mentality), then less equal in artistic good quality (in leveling-off effects against the best)
  - —In Relation to the True
    - (1) if more equal in tolerating multiple viewpoints (no matter how extreme), then less equal in epistemic standards
  - —In Relation to the Holy
    - (1) if more equal in celebrating any cults and sects (no matter how questionable), then less equal in spiritual depth and authenticity

*Notes*: The examples in each category are solely illustrative (not exhaustive), and the comparison is relative (not absolute), nor are they mutually exclusive. And some can be easily reclassified elsewhere. As generalities, they allow exceptions.
*Sources*: A reconstruction from Ch.10 of *FCD*, based on *FHC*

**Table 4.15. The Duality of Oppression in Existential Dialectics: Oppression and Self-Oppression (Part I)**

- **From the Same to the Others and Itself**
  - —The Oppression by the Same against the Others
    - *By way of downgrading differences*
      - –Ex: on judiciary caprice for corporate crimes (*Sec.2.2.1.2.1*)
      - –Ex: on the deceptive politics of liberation (*Sec.3.5*)
      - –Ex: on the humanitarian mystique (*Sec.4.4*)
      - –Ex: on the fad of emotional intelligence (*Sec.5.3*)
    - *By way of accentuating differences*
      - –Ex: on the legal sophistry of self-defense (*Sec.2.3*)
      - –Ex: on the legal semantics of proportionality (*Sec.2.4*)
      - –Ex: on the tricky politics of external threat (*Sec.3.4*)
      - –Ex: on the appeal of the Far Right for democracy (*Sec.5.4*)
      - –Ex: on the democratic axis of evil (*Sec.5.5*)
      - –Ex: on the democratic way of brutality and revenge (*Sec.5.6*)
      - –Ex: on democratic autocracy (*Sec.6.4*)
  - —The Oppression by the Same against Itself
    - *By way of downgrading differences*
      - –Ex: on the politics of fear (*Sec.2.2*)
      - –Ex: on the trickery of compassionate conservatism (*Sec.3.2*)
      - –Ex: on the deceptive politics of patriotism (*Sec.3.3*)
    - *By way of accentuating differences*
      - –Ex: on the caprice of due process on domestic suspects (*Sec.2.2*)
      - –Ex: on the false security/freedom dilemma (*Sec.6.5.2*)

*(continued on next page)*

**Table 4.15. The Duality of Oppression in Existential Dialectics: Oppression and Self-Oppression (Part II)**

---

- **From the Others to the Same and Themselves**
  —The Oppression by the Others against the Same
  - *By way of downgrading differences*
    –Ex: on judiciary caprice in the reverse direction (*Sec.2.2.1.2.2*)
    –Ex: on equal pay (*Sec.6.2.1.1*)
    –Ex: on equal representation (*Sec.6.2.1.2*)
    –Ex: on affirmative action program (*Sec.6.3.1.1*)
    –Ex: on same-sex marriage (*Sec.6.3.1.2*)
  - *By way of accentuating differences*
    –Ex: on sexual harassment (*Sec.6.2.2.1*)
    –Ex: on physical violence (*Sec.6.2.2.2*)
    –Ex: on sexual exploitation (*Sec.6.2.2.3*)

  —The Oppression by the Others against Themselves
  - *By way of downgrading differences*
    –Ex: on the reverse-class mystique (*Sec.4.2*)
    –Ex: on the reverse-black mystique (*Sec.4.3*)
    –Ex: on self-discrimination by downgrading (*Sec.6.3.2.2*)
  - *By way of accentuating differences*
    –Ex: on self-discrimination by accentuating (*Sec.6.3.2.1*)

---

*Notes*: The examples are solely illustrative (not exhaustive), nor are they mutually exclusive. As generalities, they allow exceptions. Also, both forms of oppression co-exist in all of the examples, so the listing of them are only meant in a relative, not absolute, sense.
*Source*: A summary of the sections (as cited) in Chs.2-6 of *BDPD*. See text for more info and references.

**Table 4.16. The Structure of Existential Dialectics I: The Freedom/Unfreedom and Equality/Inequality Dialectics**

• Each freedom and equality produces its own unfreedom and inequality, regardless of whether the pair occurs in political society (with the nation-state), in civil society (with some autonomy from the state), or elsewhere (e.g., in the private sphere of individual homes)—and regardless of whether freedom and equality are understood as "negative" or "positive."

• Oppression is dualistic, as much by the Same against the Others and itself, as by the Others against the Same and themselves.

• Both forms of oppression and self-oppression can be achieved by way of downgrading differences (between the Same and the Others) and of accentuating them.

• The relationships are relatively asymmetric between the Same and the Others and relatively symmetric within them. This is true, even when the Same can be relatively asymmetric towards itself in self-oppression, just as the Others can be likewise towards themselves.

• Symmetry and asymmetry change over time, with ever new players, new causes, and new forms.

*Notes*: The examples in each category are solely illustrative (not exhaustive) nor necessarily mutually exclusive, and the comparison is relative (not absolute). As generalities, they allow exceptions. "Negative" freedom is freedom "from" (e.g., freedom from poverty), whereas "positive" freedom is freedom "to" (e.g., freedom to the state of enlightenment). "Negative" equality is "procedural" equality (e.g., equality of opportunity), while "positive" equality is "substantive" equality (e.g., equality of outcome). Existential dialectics impose constraints on freedom and equality in democracy, non-democracy, and post-democracy. There is no utopia, in the end; even should there be one, dystopia would exist within it.
*Sources*: From *Table 1.5* of *BDPD*—and also from *FHC*, *FCD*, and *FPHC*

**Table 4.17. The Structure of Existential Dialectics II: The Wealth/Poverty Dialectics**

• There is no wealth without poverty, just as there is no poverty without wealth.

• The wealth/poverty dialectics occurs in the realms of having, belonging, and being, in relation to the material, relational, and spiritual.

• The wealth/poverty dialectics also expresses itself at the multiple levels of analysis in accordance to methodological holism, be they about the micro-physical, the chemical, the biological, the psychological, the organizational, the institutional, the structural, the systemic, the cultural, and the cosmological.

• The wealth/poverty dialectics is a different manifestation of existential dialectics in general, subject to the principles in its logic of ontology—just as the freedom/unfreedom and equality/inequality dialectics are likewise.

• There is no economic utopia, in the end; even should there be one, dystopia would exist within it.

*Notes*: The main points here are solely illustrative (not exhaustive) nor necessarily mutually exclusive, and the comparison is relative (not absolute). As generalities, they allow exceptions.
Sources: From *BCPC*. See also *FCD* and *FHC*.

**Table 4.18. The Structure of Existential Dialectics III: The Civilization/Barbarity Dialectics**

• There is no civilization without barbarity.

• The civilization/barbarity dialectics applies in the four civilizing processes (e.g., the rationalizing process, the pacifying process, the stewardizing process, and the subliming process).

• The civilization/barbarity dialectics is another (different) manifestation of existential dialectics in general, subject to the principles in its logic of ontology—just as the freedom/unfreedom and equality/inequality dialectics and the wealth/poverty dialectics are likewise.

• There is no utopia, in the end; even should there be one, dystopia would exist within it.

*Notes*: The main points here are solely illustrative (not exhaustive) nor necessarily mutually exclusive, and the comparison is relative (not absolute). As generalities, they allow exceptions.
Sources: From *BCIV*. See also *FCD*, *FHC*, and *BDPD*.

**Table 4.19. The Double Sides of Virtual Organizations (Part I)**

- **Psychological**
  —Ex: virtual psychosis
  —Ex: impersonality and loneliness in quaternary social relations

- **Organizational**
  —Ex: the race for power and interests
  —Ex: the world of unequal successes
  —Ex: the bureaucratic life of its own
  —Ex: lesser accountability and transparency

- **Economic**
  —Ex: the economic divides
  —Ex: the erosive impact of commercialization

- **Political**
  —Ex: the anti-authoritarian myth
  —Ex: different power struggles among groups (e.g., Conservative, Reformist, Radical)

- **Structural**
  —Ex: different localities (e.g., climates, scenery)
  —Ex: divided domains (e.g., core, peripheral, sub-peripheral)
  —Ex: substitution effect of social capital

- **Systemic**
  —Ex: uneven advances in transportation and communications, and geographical migration

- **Cultural**
  —Ex: conflicting civilizational fabrics (e.g., Confucian, Islamic, Western)
  —Ex: the bias of the Liberal Democratic agenda
  —Ex: the addiction to moral fanaticism

*(continued on next page)*

**Table 4.19. The Double Sides of Virtual Organizations (Part II)**

*Notes*: These categories and examples are solely illustrative (not exhaustive), and some of the items can be reclassified somewhere else. Nor are they mutually exclusive. Since they are generalities, exceptions are expected.
*Source*: From Ch.7 of *FCD*

**Table 4.20. Beyond the World of Titans, and the Remaking of the World Order**

- **Hyper-Empires**
  - —Ex: The Chinese Union
  - —Ex: The Indian Union

- **Meso-Empires**
  - —Ex: The European Union
  - —Ex: The North American Union

- **Micro-Empires**
  - —Ex: The Latin American Union
  - —Ex: The Middle Eastern Union

- **The Rest of the World**
  - —<u>Odd Powers</u>
    - Ex: Japan
    - Ex: Russia
  - —<u>The Poor Club</u>
    - Ex: The African Union
  - —<u>Ambivalent Regions</u>
    - Ex: Southeast Asia
    - Ex: Oceania
    - Ex: South Asia
    - Ex: Central Asia
    - Ex: Southern/Eastern Europe
    - Ex: North Africa
    - Ex: Central America
    - Ex: Others (e.g., the Korean peninsula)

*Source*: A summary of Chs.2-5 (of *BWT*)

**Table 4.21. The Origins of Authoritarian Liberal Democracy**

- **The Geopower of Nature (Ch.4 of *ALD*)**
  —Ex: Power Character and Geographical Strategy
  —Ex: Living Space and Territorial Expansion
  —Ex: Strategic Heartland and Containment

- **The Biopsychology of the Mind (Ch.5 of *ALD*)**
  —Ex: The Bell Curve and Mass Intelligence
  —Ex: Group Analysis and Mass Knowledge
  —Ex: Groupthink and Elite (Mis)calculation

- **The Disciplinary Control of Society (Ch.6 of *ALD*)**
  —Ex: Social Organizations of Ruthlessness
  —Ex: Social Institutions of Greed
  —Ex: Social Structure of Exclusion
  —Ex: Social Systems of Violence

- **The Molding Force of Culture (Ch.7 of *ALD*)**
  —Ex: The Tradition of Conquest
  —Ex: The Rationalization of Unreason

*Notes*: The examples in each category are solely illustrative (not exhaustive), and they are also relative (not absolute), nor are they mutually exclusive. As generalities, they allow exceptions.
*Source*: A summary of Chs.4-7 of *ALD*

**Table 4.22. The Theory of Post-Democracy I:**
**The Priority of Freedom over Equality**
**(Part I)**

- **Differences**
  - —*For the aggressive Lions (the strong Elitists)*
    - •Setting up rank distinctions among unequals (e.g., between inferior humans and superior post-humans, or later among inferior post-humans and superior ones, relatively speaking)
    - •Yearning for being not only distinguished from unequals, but also the first among equals (the best of the very best)
    - •Soul-searching for a high spiritual culture (not the trashy one for the masses). Mass culture is a dirty joke for them.
  - —*For the manipulative Foxes (the weak Counter-Elitists)*
    - •Seeking a gentle hegemony by way of more communitarian concerns ( for inferior humans and, later, inferior post-humans)
    - •Being more sympathetic to less formal-legalistic institutions and values

- **Similarities**
  - —*For both Lions and Foxes*
    - •Exploring different spheres of non-human consciousness in the cosmos (something vastly superior than the human one)
    - •Recognizing the democratic illusions (e.g., no freedom without unfreedom, no equality without inequality, or simply no justice without injustice, and vice versa)

*(continued on next page)*

**Table 4.22. The Theory of Post-Democracy I: The Priority of Freedom over Equality (Part II)**

*Notes*: The two callings and examples in each category are solely illustrative (not exhaustive), since there will be many different post-human value ideals in the distant future of post-human civilization. The comparison is also relative (not absolute) towards post-democracy, so this is not just a version of free-market democracy (nor Fascism/Nazism, as shown in the table later on democracy, non-democracy, and post-democracy). Nor are they mutually exclusive. As generalities, they allow exceptions. And the specific forms of post-human post-democratic ideals need to be further developed in future after-postmodern history, as they will be different from the ones we now know. The point here is to solely give an extremely rough picture of a small part of the world to come that we have never known.
*Source*: From Ch.10 of *FCD*. Refer to text for more info and references.

### Table 4.23. The Theory of Post-Democracy II: The Priority of Equality over Freedom

- **Hybrid Versions of**
  - —Ex: the Trans-Feminine Calling
  - —Ex: the Trans-Sinitic Calling
  - —Ex: the Trans-Islamic Calling
  - —Ex: the Trans-Outerspace Calling

- **Qualifications**
  - —These four versions of post-capitalist value ideals need not automatically be post-democratic, just as capitalism does not necessarily mean democracy. They are two different entities—though closely related.
  - —But up to a certain threshold of elevating equality at the farther expense of freedom, the democratic ideals will be overcome and cease to exist.
  - —The overcome will not be socialist or communist, but post-democratic with no freedom without unfreedom and no equality without inequality, subject to the constraints of existential dialectics.

*Notes*: The callings are solely illustrative (not exhaustive), since there will be many different post-human value ideals in the distant future of post-human lifeforms. The comparison is also relative (not absolute), nor are they mutually exclusive. As generalities, they allow exceptions. And the specific forms of post-human post-democratic ideals need to be further developed in future after-postmodern history, as they will be different from the ones we now know. The point here is to solely give an extremely rough picture of a small part of the world to come that we have never known.

*Source*: From Ch.10 of *FCD*. Refer to text for more info and references.

**Table 4.24. The Theory of Post-Democracy III: The Transcendence of Freedom and Equality (Part I)**

---

- **Transcending Freedom in Floating Existence**
  —*Freedom*: seeking an ultimate elimination of the body. Being without the body. The aim is to transcend freedom in the end into a metaphysical state (i.e., beyond the physique).
  —*Unfreedom*: yet facing difficult trade-offs. The sacrifice of bodily existence and its joyfulness. An eternal boredom in floating existence in dark deep space, though with alternative pleasures. There is no free lunch even in the state of transcending freedom.

- **Transcending Equality in the Rivalry of Cosmic Hegemony**
  —*Inequality*: competing to outlast other lifeforms in floating existence, or just marginalizing them for one's hegemonic expansiveness in the rest of the cosmos (and even beyond). Universalism is only for the mediocre.
  —*Equality*: accepting only those of one's rank as equal partners in the vast spacetime for cosmic supremacy. Even here, the aim is to transcend equality into a metaphysical state.

---

*(continued on next page)*

**Table 4.24. The Theory of Post-Democracy III: The Transcendence of Freedom and Equality (Part II)**

*Notes*: Do not confuse this transcendence of freedom and equality (as one version of post-democracy) with the naïve temptation to transcend the freedom/unfreedom and equality/inequality dialectics. Existential dialectics hold true for freedom and equality in all cultures and societies—past, present, or future (i.e., democracy, non-democracy, and post-democracy), regardless of whether freedom and equality are conventionally understood as "negative" or "positive."
Also, the two features and examples in each are solely illustrative (not exhaustive), since there will be many different post-human value ideals in the distant future of post-human lifeforms. The comparison is also relative (not absolute), nor are they mutually exclusive. As generalities, they allow exceptions. And the specific forms of post-human ideals even for these radically alien floating lifeforms (and others unknown to us) need to be further developed in future after-postmodern history, as they will likely be different from the ones herein illustrated. The point here is to solely give a very rough picture of a small part of the extremely alien world to come that we have never known.
*Source*: From Ch.10 of *FCD*. Refer to text for more info and references.

**Table 4.25. Democracy, Non-Democracy, and Post-Democracy (Part I)**

---

• **Democracy**
—*Theoretical Constructs*
•The pursuit of freedom and equality (in various degrees), regardless of whether freedom and equality can be understood as "negative" or "positive"
(1) more equality than freedom: The relative priority of the good over the right
(2) more freedom than equality: The relative priority of the right over the good
—*Types*
•Only (1): Different versions of communitarian moral universalism
•Only (2): Different versions of liberal moral universalism
•(1) or (2): Different versions of anarchic (non-nation-state) moral universalism
•(1) or (2): Different versions of postmodern moral localism

• **Non-Democracy**
—*Theoretical Constructs*
•The focus on (1') equality or (2') freedom, but not both, regardless of whether freedom and equality can be understood as "negative" or "positive"
—*Types*
•Only (1'): Different versions on the Far Left (e.g., Stalinism, Robespierrianism)
•Only (2'): Different versions on the Far Right (e.g., Nazism, absolute monarchism)

---

*(continued on next page)*

**Table 4.25. Democracy, Non-Democracy, and Post-Democracy (Part II)**

- **Post-Democracy**
  —*Theoretical Constructs*
  - The priority of (1'') equality over freedom, or (2'') freedom over equality, or (3'') the transcendence of freedom and equality, regardless of whether freedom and equality are "negative" or "positive." In degree, (1'') or (2'') is less than (1') or (2') but more than (1) or (2)—respectively.
  - Like democracy and non-democracy, post-democracy is also subject to the freedom/unfreedom and equality/inequality dialectics (or existential dialectics in general). Unlike them, post-democracy acknowledges the constraints of existential dialectics and no longer value freedom and equality as sacred virtues. There is no utopia, in the end; even were there one, dystopia would exist within it.

  —*Types*
  - (1''): Different versions of trans-Sinitic value ideals
  - (1''): Different versions of trans-feminine value ideals
  - (1''): Different versions of trans-Islamic value ideals
  - (1''): Different versions of trans-outerspace value ideals
  - (2''): Different versions of post-human elitist value ideals
  - (3''): Different versions of the value ideals of floating consciousness (etc.)

*(continued on next page)*

**Table 4.25. Democracy, Non-Democracy, and Post-Democracy (Part III)**

*Notes*: The examples are solely illustrative (not exhaustive), nor are they mutually exclusive. As generalities, they allow exceptions. "Negative" freedom is freedom "from" (e.g., freedom from poverty), whereas "positive" freedom is freedom "to" (e.g., freedom to the state of enlightenment). "Negative" equality is "procedural" equality (e.g., equality of opportunity), while "positive" equality is "substantive" equality (e.g., equality of outcome). Existential dialectics impose constraints on freedom and equality in democracy, non-democracy, and post-democracy, regardless of whether freedom and equality can be understood as "negative" or "positive" in conventional discourse. Therefore, do not confuse the transcendence of freedom and equality in (3'') with the naïve temptation to transcend existential dialectics. There is no utopia, in the end; even should there be one, it would not exist without dystopia embedded within it.
*Sources*: A summary, based on my previous works, especially Ch.5 of *FHC*, Chs.5-10 of *FCD*, Chs.2-4 of *FPHC*, and Chs.1 & 7 of *BDPD*. The reader should consult the books for more analysis, as this is only a summary here.

**Table 4.26. Multiple Causes of the Emergence of Post-Democracy (Part I)**

• **At the Micro-Physical Level**
—Ex: intelligent life without the human physical-chemical system
—Sources: Ch.7 of *FHC*; Chs.9-10 of *FCD*; Ch.1 of *FPHC*

• **At the Chemical Level**
—Ex: space radiation and toxins
—Sources: Ch.7 of *FHC*; Chs.9-10 of *FCD*

• **At the Bio-Psychological Level**
—Ex: exo-biological evolution in deep space
—Ex: genetic engineering of new beings
—Ex: limits of cognitive partiality
—Ex: illusions of emotional neutrality
—Ex: human biological inequality
—Ex: the rise of unfolding unconsciousness
—Sources: Ch.2 & Chs.9-10 of *FCD*; Ch.7 of *FHC*; Ch.4 of *BCPC*; *FPHU*

• **At the Institutional Level**
—Ex: the flawed logic of equality
—Ex: the conflicting nature of governance
—Sources: Ch. 5 of *FHC*; Chs. 6 & 10 of *FCD*; Ch. 3 of *FPHC*; Chs.2-5 of *BDPD*

• **At the Organizational Level**
—Ex: e-civic alienation
—Ex: the dark sides of formal-legalistic routines
—Sources: Ch.3 of *FHC*; Ch.7 of *FCD*; Ch.3 of *FPHC*

• **At the Structural Level**
—Ex: ever new forms of inequities, at home and abroad
—Ex: the emergence of China, women, and Islam as major actors
—Sources: Chs.5-6 of *FHC*; Chs.7, 9 & 10 of *FCD*; Chs.4-5 of *BDPD*

*(continued on next page)*

**Table 4.26. Multiple Causes of the Emergence of Post-Democracy (Part II)**

---

• **At the Cultural Level**
—Ex: freedom/unfreedom dialectics
—Ex: equality/inequality dialectics
—Ex: system fragmentation and integration
—Sources: Ch.5 of *FHC*; Chs. 3, 9 & 10 of *FCD*; Ch.4 of *FPHC*; Ch.1 of *BDPD*; Ch.4 of *BCPC*

• **At the Systemic Level**
—Ex: space habitats (in zero-gravity) and colonization
—Ex: ultra advanced future info systems
—Ex: qualitative demography
—Sources: Ch.7 of *FHC*; Chs.9 &10 of *FCD*

• **At the Cosmological Level**
—Ex: the colonization of multiverses
—Ex: the alteration of space-time and the creation of new matter-energy
—Ex: the expansion of floating consciousness
—Ex: the spread of hyper-spatial consciousness
—Sources: Ch.7 of *FHC*; Chs.9 &10 of *FCD*; Ch.4 of *FPHC*; *FPHST*

---

*Notes*: The examples in each category are solely illustrative (not exhaustive), and some of the items can be reclassified somewhere else. Nor are they always mutually exclusive. Since they are generalities, exceptions are expected.
*Sources*: Especially from *FHC*, *FCD*, *FPHC*, *BCPC*, *BDPD*, *FPHST*, and *FPHU*. See also other books and my perspectives on civilizational holism.

**Table 4.27. Some Clarifications about Post-Capitalism and Post-Democracy (Part I)**

• The prefix "trans-" in the first category of post-capitalism (with its four versions) refers to something "going beyond" (not "uniting" or "combining"). Ex: *Sec.10.3.3* of *FCD*; *Sec.2.4* & *Sec.4.4* of *FPHC*; *Sec.7.2* of *BCPC*

• Such terms like "post-democracy," "post-capitalism," "post-human elitist," "trans-feminine calling," and the like as used in my works are more for our current intellectual convenience than to the liking of future humans and post-humans, who will surely invent more tasteful neologisms to call their own eras, entities, and everything else, for that matter. But the didactic point here is to use the terms to foretell what the future might be like, not that its eras and entities must be called so exactly and permanently.
Ex: *Sec.11.1* of *FCD*; *Sec.7.2* of *BCPC*

• The four versions in the first category of post-capitalist value ideals need not automatically be post-democratic, just as capitalism does not necessarily mean democracy. They are two different entities—though closely related. But up to a certain threshold of elevating equality at the farther expense of freedom, the democratic ideals will be overcome and cease to exist. The same is true for the post-human elitist calling in the second category of post-capitalism in relation to post-democracy, depending on the extent to which freedom is elevated at the expense of equality.
Ex: *Sec.10.4.3.3* of *FCD*; *Table 3.9* of *FPHC*; *Table 7.6* of *BDPD*

• The comparison in each of the three realms of existence in all forms of post-capitalism is not absolute, but relative. Examples include "communal" vs. "individualistic," and the like.
Ex: Notes in *Table 10.8*, *Table 10.9*, *Table 10.10*, & *Table 10.11* of *FCD*; Chs.2-4 of *FPHC*; *Sec.7.2* of *BCPC*

*(continued on next page)*

**Table 4.27. Some Clarifications about Post-Capitalism and Post-Democracy (Part II)**

• The emergence of post-capitalism (and post-democracy, for that matter) has multiple causes (to not be reduced to one or only a few).
Ex: Ch.10 of *FCD*, Chs.2-4 of *FPHC*; *Sec.1.3* & *Sec.7.2* of *BCPC* (or *Table 1.8* & *Table 7.11*)

• The specific forms of post-capitalism (and post-democracy, for that matter) need to be further developed in future after-postmodern history, as they will be different from the ones we now know. The point here is to solely give an extremely rough sketch of a world to come that we have never known.
Ex: *Sec.10.3.3* & *Sec.10.4.3.3* of *FCD*; *Table 10.14* & *Table 10.15* of *FCD*; *Sec.7.2* of *BCPC*

• All forms of post-capitalism are not part of a "teleological law," but of "historical trends" only. The same is also true for all forms of post-democracy.
Ex: *Sec.7.1* of *FHC*; *Sec.9.5.3.2* & *Sec.10.3.4.2* of *FCD*; *Sec.7.2* of *BCPC*

• Post-capitalism is not better than capitalism in an "absolute" sense but only fits in better, on the basis of the historical contingency of culture, society, nature, and the mind in some future eras. The same is true for post-democracy in relation to democracy. The term "better" is historically relative.
Ex: *Sec.10.3.3* of *FCD*; *Sec.1.7* of *BDPD*; *Sec.1.5* of *BCPC*

*(continued on next page)*

**Table 4.27. Some Clarifications about Post-Capitalism and Post-Democracy (Part III)**

• All forms of post-capitalism and post-democracy are subject to the constraints of existential dialectics. In the process, the dialectic direction is to go beyond the conventional "either-or" dichotomies (e.g., freedom vs. unfreedom, equality vs. inequality, freedom vs. equality, individuality vs. communality, spirituality vs. materiality, formal legalism vs. informal legalism, etc.). As is true in post-civilization, to go beyond the dichotomies is to acknowledge the co-existence of both in each dichotomy, although the degree of scaling one over the over varies from case to case (e.g., the theory of post-capitalism I, the theory of post-capitalism II, the theory of post-democracy I, the theory of post-democracy II, etc.)—but is not to be extreme in largely favoring one over the other, *on average* (all things considered). There is no utopia to be had in the end; even should there be one, dystopia would exist within it.
Ex: Ch.5 of *FHC*; *Sec.10.4.4.2* of *FCD*; *Sec.1.5* of *BDPD*; *Sec.1.3* of *BCPC*; *BCIV*

• All forms of post-capitalism, however different from each other though they are, share one common feature, in that they all inspire for a higher spiritual culture. The same is also true for post-democracy.
Ex: *Sec.10.3*, *Sec.10.4* & *Sec.10.5* of *FCD*; *Chs.2-4* of *FPHC*; *Sec.7.2* of *BCPC*

• All forms of post-capitalism try to avoid the excess in capitalist consumerism by favoring more basic than artificial needs in having, but the quality and quantity of these "basic" needs will be measured by future standards, not by our current ones. Standards are historically relative.
Ex: *Sec.10.3*, *Sec.10.4* & *Sec.10.5* of *FCD*; Ch.2 of *FPHC*; *Sec.7.2* of *BCPC*

*(continued on next page)*

**Table 4.27. Some Clarifications about Post-Capitalism and Post-Democracy (Part IV)**

---

• All forms of post-capitalism make use of a different degree of political authority with advanced info systems in future history and strives for higher spiritual cultures (especially in the post-human age), while acknowledging the constraints of existential dialectics and no longer valuing free market (as in capitalism) and economic control (as in non-capitalism) as sacred virtues.
Ex: *Sec.10.3.4.2*, *Sec.10.3*, *Sec.10.4* & *Sec.10.5* of *FCD*; Chs.2-4 of *FPHC*; *Sec.1.5* of *BDPD*; *Sec.7.2* of *BCPC*

---

*Notes*:: The main points here are solely illustrative (not exhaustive) nor necessarily mutually exclusive, and the comparison is relative (not absolute). As generalities, they allow exceptions. The sections as cited are only illustrative (not exhaustive).
*Sources*: From *FHC*, *FCD*, *FPHC*, and *BDPD*

**Table 4.28. The Theory of Post-Capitalism I.1:**
**By Group—**
**Ex: Spiritual/Communal in the Trans-Feminine Calling**

- **More Communal Than Individual**
  —*Sharing*: learning from others, as different ideas mutually enrich
  —*Cooperative*: encouraging a sense of shared leadership and teamwork

- **More Informal-Legalistic Than Formal-Legalistic**
  —*Specific*: listening more from the heart than from the head, to know a person as a concrete, not as an abstract, unit
  —*Affective*: thinking and acting with others on a more affective tone. Business can mix with an emotional touch.
  —*Ascriptive*: hiring (or firing) can be done on the basis of merit (or lack of it), but deep solidarity (sisterhood) is important too.
  —*Particularistic*: making decisions on the basis of cost-benefit analysis, but a given group relationship is vital

- **More Spiritual Than Secular**
  —*Long-Term Looking*: sharing for a long-term relationship (e.g., love, friendship), not just for a short-term gain
  —*Loving/Caring*: showing compassion for the sufferings of others, without quickly blaming and pre-judging
  —*Respectful*: showing acceptance about others' feelings (and thoughts)

*Notes*: The categories and examples are solely illustrative, since there can be different versions, and the comparison is relative (not absolute), nor are they mutually exclusive. As generalities, they allow exceptions. The specific forms of the trans-feminine version need to be further developed in future after-postmodern history, as they will be different from the ones we now know, since the prefix "trans-" here means going beyond or deconstructing the feminine values, while using them as the inspirational point at the beginning.
*Source*: From Ch.10 of *FCD*. Refer to text for more info and references.

**Table 4.29. The Theory of Post-Capitalism I.2: By Nation-State— Ex: Spiritual/Communal in the Trans-Sinitic Calling**

- **More Communal Than Individualistic**
  —*Centralized*: being more top-down in management
  —*Collective*: encouraging more group cooperation
  —*Social*: investing in trust and connection

- **More Informal-Legalistic Than Formal-Legalistic**
  —*Specific*: knowing more of those related or connected
  —*Affective*: behaving in a paternalistic, hierarchical way
  —*Ascriptive*: favoring family members and those related
  —*Particularistic*: building connection (guanxi) as imperative

- **More Spiritual Than Secular**
  —*Expansionist*: diffusing civilizational values (e.g., the superiority complex of civilizationalism)
  —*Holistic*: synthesizing things into a panoramic horizon
  —*Historical*: learning from the lessons of the ancient past
  —*Respectful*: deferential to elders and superiors

*Notes*: The categories and examples are solely illustrative, since there can be different versions, and the comparison is relative (not absolute), nor are they mutually exclusive. As generalities, they allow exceptions. The specific forms of the trans-Sinitic version need to be further developed in future after-postmodern history, as they will be different from the ones we now know, since the prefix "trans-" here means going beyond or deconstructing the Sinitic values, while using them as the inspirational point at the beginning.
*Source*: From Ch.10 of *FCD*. Refer to text for more info and references.

**Table 4.30. The Theory of Post-Capitalism I.3:**
**By Region—**
**Ex: Spiritual/Communal in the Trans-Islamic Calling**

• **More Communal Than Individualistic**
—*Collective*: building the webs of relationships to bind individuals
—*Sharing*: cultivating the established "wisdom" through common experience
—*Cooperative*: stressing harmony, solidarity, and commonality

• **More Informal-Legalistic Than Formal-Legalistic**
—*Specific*: making efforts to know well the participants (family and larger community) in matters of common concern
—*Affective*: mixing work with language and ritual on explicit religious (Islamic) ideals, texts, stories, and examples
—*Ascriptive*: privileging local history and custom on relationships among kinship groups
—*Particularistic*: preferring an unbiased insider with ongoing connections to all parties

• **More Spiritual Than Secular**
—*Historical*: learning from the lessons of the past as a source of stability and guidance
—*Deferential*: showing respect for age, experience, status, and leadership in communal affairs
—*Honorable*: emphasizing face, dignity, prestige, and fairness
—*Compassionate*: giving mercy and charity ("Zahah") to others

*Notes*: The categories and examples are solely illustrative (not exhaustive), and the comparison is relative (not absolute), nor are they mutually exclusive. As generalities, they allow exceptions. The specific forms of the trans-Islamic version need to be further developed in future after-postmodern history, as they will be different from the ones we now know, since the prefix "trans-" here means going beyond or deconstructing the Islamic values, while using them as the inspirational point at the beginning.
*Sources*: From Ch.10 of *FCD*. Refer to text for more info and references, especially from the works by George Irani (2000) and C. Murphy (September 19, 2001).

**Table 4.31. The Theory of Post-Capitalism I.4: By Universe— Ex: Spiritual/Communal in the Trans-Outerspace Calling**

• **More Communal Than Individual**
—*Cooperative*: requiring teamwork in small space habitats
—*Sharing*: learning from, and enjoying being with, each other in a small group in outer space

• **More Informal-Legalistic Than Formal-Legalistic**
—*Specific*: knowing more about each other to facilitate living and working together in space, both as fellow astronauts and space-mates
—*Affective*: being friendly and social to each other as vital to working and living in small space quarters
—*Ascriptive*: nurturing comaraderie among fellow astronauts as if they are family members over time
—*Particularistic*: building work relationship with enduring memory in a space mission

• **More Spiritual Than Secular**
—*Long-Term*: looking beyond selfish materialistic concerns in a precarious space environment with potential life or death
—*Loving/Caring*: cultivating deep bondage for the success of a long term space mission
—*Transcendent*: searching for life meaning in outer space

*Notes*: The calling and examples in each category are solely illustrative (not exhaustive), since there will be many different outer-space value ideals in the distant future of space colonization. The comparison is also relative (not absolute), nor are they mutually exclusive. As generalities, they allow exceptions. And the specific forms of trans-outer-space calling need to be further developed in future after-postmodern history, as they will be different from the ones we now know, since the prefix "trans-" here means going beyond or deconstructing the current outer-space values, while using them as the inspirational point at the beginning. The point here is to solely give an extremely rough picture of a small part of the world to come that we still do not know much about.
*Source*: From Ch.10 of *FCD*. Refer to text for more info and references.

**Table 4.32. The Theory of Post-Capitalism II: Spiritual/Individualistic in the Post-Human Elitist Calling (Part I)**

- **More Individualistic Than Communal**
  —Setting up rank distinctions among unequals (e.g., between inferior humans and superior post-humans, or later among inferior post-humans and superior ones, relatively speaking)
  —Yearning for being not only distinguished from unequals, but also the first among equals (the best of the very best)
  —Recognizing the constraints of equality/inequality dialectics (or existential dialectics in general)

- **More Spiritual Than Secular**
  —Soul-searching for a high spiritual culture (not the trashy one for the masses). Mass culture is a dirty joke for them.
  —Exploring different spheres of non-human consciousness in the cosmos (something vastly superior than the human one)
  —Recognizing the constraints of freedom/unfreedom dialectics (or existential dialectics in general)

- **Qualifications**
  —Although post-human elitist post-democracy is comparable to post-human elitist post-capitalism in some respects, the former does not necessarily imply the latter (post-human elitist post-capitalism), just as democracy does not have to entail capitalism. They are two different (though related) entities.
  —But up to a certain threshold of incorporating government intervention with advanced info systems in future civilizations for higher spiritual concerns at the expense of the free market and materialist pursuit, the capitalist ideal will be overcome.
  —The overcome will not be Fascist or feudalistic, but post-capitalist, subject to the constraints of existential dialectics.

*(continued on next page)*

**Table 4.32. The Theory of Post-Capitalism II: Spiritual/Individualistic in the Post-Human Elitist Calling (Part II)**

*Notes*: The calling and examples in each category are solely illustrative (not exhaustive). The comparison is also relative (not absolute), nor are they mutually exclusive. As generalities, they allow exceptions. And the specific forms of post-human elitist post-capitalism need to be further developed in future after-postmodern history, as they will be different from the ones we now know, while using them as the inspirational point at the beginning. The point here is to solely give an extremely rough picture of a small part of the world to come that we still do not know much about.
*Sources*: From Ch.10 of *FCD* (and also *FPHC*, *BDPD*, and *BCPC*). Refer to the text for more info and references.

**Table 4.33. Capitalism, Non-Capitalism, and Post-Capitalism (Part I)**

• **Capitalism**

—*Theoretical Constructs*

•Allocation of scarce resources among alternative wants largely by free market for competition (whose characteristics in its ideal form include, for instance, no barrier to entry or exit, homogeneity, perfect information, a large number of buyers/sellers, and perfect factor mobility)

•More formal-legalistic than informal-legalistic, more individualistic than communal, and more material (secular) than spiritual

•Either (1) minimal government or (2) relatively active government

—*Types*

•Only (1): Different versions of market capitalism (e.g., the U.S.)

•Only (2): Different versions of welfare capitalism (e.g., Sweden)

• **Non-Capitalism**

—*Theoretical Constructs*

•Allocation of scarce resources among alternative wants mainly by political authority for policies (which can be regulative, redistributive, symbolic, and participatory)

•More informal-legalistic than formal-legalistic

•Either (1') more individualistic (for the elites), often (though not always) for material (secular) concerns, or (2') more communal (for the masses), often (though not always) for spiritual concerns

—*Types*

•Only (1'): Different versions on the Right (e.g., Fascist corporate-state economy for the glory of the new Rome, medieval lord-vassal-serf economy for the power of the feudalistic order)

•Only (2'): Different versions on the Left (e.g., Soviet command economy for the creation of the New Socialist Man)

*(continued on next page)*

**Table 4.33. Capitalism, Non-Capitalism, and Post-Capitalism (Part II)**

- **Post-Capitalism**
  - —*Theoretical Constructs*
    - •Allocation of scarce resources among alternative wants largely by political authority with advanced info systems in future civilizations, subject to existential dialectics. In degree of allocating by authority, post-capitalism is more than capitalism but less than non-capitalism.
    - •More spiritual than secular (material)
    - •Either (1'') more individualistic or (2") more communal
    - •Like capitalism and non-capitalism, post-capitalism is also subject to the freedom/unfreedom and equality/inequality dialectics (or existential dialectics in general). There is no utopia, in the end; even were there one, dystopia would exist within it.
    - •Unlike capitalism and non-capitalism, post-capitalism makes use of a different degree of political authority with advanced info systems in future civilizations and strives for higher-spiritual cultures (especially in the post-human age), while acknowledging the constraints of existential dialectics and no longer valuing free market (as in capitalism) and economic control (as in non-capitalism) as sacred virtues.
  - —*Types*
    - •Only (1"): Different versions of post-human elitist value ideals
    - •Only (2''): Different versions of trans-Sinitic value ideals
    - •Only (2''): Different versions of trans-feminine value ideals
    - •Only (2''): Different versions of trans-Islamic value ideals
    - •Only (2''): Different versions of trans-outerspace value ideals

*(continued on next page)*

**Table 4.33. Capitalism, Non-Capitalism, and Post-Capitalism (Part III)**

*Notes*: The calling and examples in each category are solely illustrative (not exhaustive). The comparison is also relative (not absolute), nor are they mutually exclusive. As generalities, they allow exceptions. And the specific forms of each calling need to be further developed in future after-postmodern history, as they will be different from the ones we now know, while using them as the inspirational point at the beginning. The point here is to solely give an extremely rough picture of a small part of the world to come that we still do not know much about.
*Source*: From Ch.10 of *FCD*. Refer to the text for more info and references.

**Table 4.34. Multiple Causes of the Emergence of Post-Capitalism (Part I)**

- **At the Micro-Physical Level**
  —Ex: intelligent life without the human physical-chemical system
  —Ex: mastering of quantum mechanics, electromagnetism, and other fields for the understanding of a broad range of anomalous experiences and the application for artificial intelligence for spiritual quest
  —Sources: Ch.7 of *FHC*; Chs.9-10 of *FCD*; Ch.1 of *FPHC*

- **At the Chemical Level**
  —Ex: space radiation and toxins
  —Sources: Ch.7 of *FHC*; Chs.9-10 of *FCD*

- **At the Bio-Psychological Level**
  —Ex: exo-biological evolution in deep space
  —Ex: genetic engineering of new beings
  —Ex: limits of human cognition
  —Ex: the rise of unfolding unconsciousness
  —Sources: Ch.2 & Chs.9-10 of *FCD*; Ch.7 of *FHC*; *FPHU*

- **At the Institutional Level**
  —Ex: the flawed logic of the free market
  —Ex: the need of a post-autistic economics
  —Sources: Ch.10 of *FCD*

- **At the Organizational Level**
  —Ex: the dark sides of formal-legalistic routines
  —Sources: Ch.3 of *FHC*; Ch.7 of *FCD*; Ch.3 of *FPHC*

- **At the Structural Level**
  —Ex: ever new forms of inequities, at home and abroad
  —Ex: the emergence of China, women, and Islam as major actors
  —Sources: Chs.5-6 of *FHC*; Chs.7, 9 & 10 of *FCD*; Chs.4-5 of *BDPD*

*(continued on next page)*

**Table 4.34. Multiple Causes of the Emergence of Post-Capitalism (Part II)**

- **At the Cultural Level**
  —Ex: freedom/unfreedom dialectics
  —Ex: equality/inequality dialectics
  —Sources: Ch.5 of *FHC*; Chs.3 & 10 of *FCD*; Ch.4 of *FPHC*; Ch.1 of *BDPD*

- **At the Systemic Level**
  —Ex: space habitats (in zero-gravity) and colonization
  —Ex: ultra advanced future info systems
  —Ex: qualitative demography
  —Sources: Ch.7 of *FHC*; Chs. 9 & 10 of *FCD*

- **At the Cosmological Level**
  —Ex: the colonization of multiverses
  —Ex: the alteration of space-time and the creation of new matter-energy
  —Ex: the expansion of floating consciousness
  —Ex: the spread of hyper-spatial consciousness
  —Sources: Ch.7 of *FHC*; Chs. 9 & 10 of *FCD*; Ch.4 of *FPHC*; *FPHST*

*Notes*: The examples in each category are solely illustrative (not exhaustive), and some of the items can be reclassified somewhere else. Nor are they always mutually exclusive. Since they are generalities, exceptions are expected.
*Sources*: Especially from *FHC*, *FCD*, *FPHC*, *BCPC*, *BDPD*, *FPHST*, and *FPHU*. See also other books and my perspectives on civilizational holism.

**Table 4.35. The Theoretical Debate on Civilization**

- **The Progressive Theory of Civilization**
  —*Thesis*: The "civilizing" process is "good," as opposed to the "barbarizing" process as something "bad," relatively speaking.
  —*Discourse*: Especially, though not exclusively, in the Enlightenment era and a bit before. Example: Thomas Hobbes—in that the tribes in primitive societies were "savages."

- **The Romantic Theory of Barbarity**
  —*Thesis*: The "civilizing" process is "bad," as opposed to the "barbarizing" process as something "good," relatively speaking.
  —*Discourse*: Especially, though not exclusively, in the Counter-Enlightenment circle. Example: Jean-Jacques Rousseau—in that civilization "corrupts" men, and the "savages" are in fact "noble."

- **The Moderate Theory of Civilization**
  —*Thesis*: The "civilizing" process is "good," but there is a price to pay, especially in systematic (compulsive) self-control.
  —*Discourse*: Especially, though not exclusively, in some late modern and postmodern circles. Example: Norbert Elias—in that social manners become more refined in the civilizing process, but self-control also becomes more systematic.

- **The Theory of Post-Civilization**
  —*Thesis*: The civilizing process is as evil and good as barbarity, and each cannot exist without the other, to be eventually superseded by post-civilization unto the post-human age.
  —*Discourse*: Proposed by Peter Baofu. See the rest of *BCIV* for more analysis.

*Source*: From *BCIV* on the theoretical debate

**Table 4.36. No Freedom Without Unfreedom in the Civilizing Processes (Part I)**

- **The Rationalizing Process (at the Level of Culture)**
  —if freer from the dominance of unreason (as in barbarism) in the civilizing process, then less free from the rationalizing process (be it in the form of the principle of either transcendence or immanence)
  —if freer from the principle of immanence in the rationalizing process, then less free from the inclination to commit terror in the name of reason and the relative underdevelopment of non-reason ( e.g., in relation to yoga and meditation)
  —if freer from the principle of transcendence in the rationalizing process, then less free from the relative underdevelopment of reason (e.g., in relation to systematic methodology) and the occurrence of oppression in the name of non-reason

- **The Pacifying Process (at the Level of Society)**
  —if freer from the dominance of pillage (as in savagery) in the civilizing process, then less free from the pacifying process (be it in the form of external control or self-control)
  —if freer from self-control in the pacifying process, then less free from the temptation of expansionist oppression and rebellious mindset in external control
  —if freer from external control in the pacifying process, then less free from the gruesome psychological self-torture and conformism in self-control

- **The Stewardizing Process (at the Level of Nature)**
  —if freer from the dominance of nature (as in the state of nature) in the civilizing process, then less free from the stewardizing process (be it in the form of the stewardship of creation or the covenant with nature)
  —if freer from the stewardship of creation in the stewardizing process, then less free from material underdevelopment, relatively speaking, and spiritual exclusion in the covenant with nature
  —if freer from the covenant with nature in the stewardizing process, then less free from ecological degradation and spiritual disconnection from nature in the stewardship of creation

*(continued on next page)*

**Table 4.36. No Freedom Without Unfreedom in the Civilizing Processes (Part II)**

- **The Subliming Process (at the Level of the Mind)**
  —if freer from the dominance of spontaneity (as in the wild state of the mind) in the civilizing process, then less free from the subliming process, be it in the form of (cyclical-centric) self-refinement or (linear-centric) self-discipline
  —if freer from (cyclical-centric) self-refinement in the subliming process, then less free from the (linear-centric) self-regimen (as a form of neurosis)
  —if freer from (linear-centric) self-discipline in the subliming process, then less free from the (cyclical-centric) self-torture (equally as a form of neurosis)

*Notes*: The examples in each category are solely illustrative (not exhaustive), and the comparison is relative (not absolute), nor are they necessarily mutually exclusive. And some can be easily re-classified elsewhere. As generalities, they allow exceptions.
Sources: From *BCIV*. See also *FHC*, *FCD*, *FPHC*, *BDPD*, and *BCPC*.

**Table 4.37. No Equality Without Inequality in the Civilizing Processes (Part I)**

- **The Rationalizing Process (at the Level of Culture)**
  —if more equal for the role of rationalization in the rationalizing process (of civilizational making), then less equal for that of mythicization (as in barbarism)
  —if more equal for the principle of transcendence in (linear-centric) rationalizing process, then less equal for the principle of immanence
  —if more equal for the principle of immanence in (cyclical-centric) rationalizing process, then less equal for the principle of transcendence

- **The Pacifying Process (at the Level of Society)**
  —if more equal for pacification in civilizational making, then less equal for the institution of pillaging and others (as in savagery)
  —if more equal for external control, relatively speaking, in pacifying process, then less equal for self-control
  —if more equal for self-control, relatively speaking, in pacifying process, then less equal for external-control

- **The Stewardizing Process (at the Level of Nature)**
  —if more equal for stewardship in the stewardizing process (of civilizational making), then less equal for reverent (submissive) existence (as in barbarism)
  —if more equal for the stewardship of creation in (linear- centric) stewardizing process, then less equal for the (cyclical-centric) covenant with nature for harmonious co-existence
  —if more equal for the (cyclical-centric) covenant with nature in the stewardizing process, then less equal for the (linear-centric) stewardship of nature for domination

*(continued on next page)*

**Table 4.37. No Equality Without Inequality in the Civilizing Processes (Part II)**

- **The Subliming Process (at the Level of the Mind)**
  —if more equal for the role of reason in the subliming process, then less equal for that of unreason (as in the natural state of wildness)
  —if more equal for the primacy of reason in (linear-centric) subliming process, then less equal for other faculties (e.g., intuition, existential feelings, and analogous thinking) in cyclical-centric one
  —if more equal for the exercise of other faculties (e.g., intuition, existential feelings, and analogous thinking) in cyclical-centric subliming process, then less equal for the role of reason in linear-centric counterpart

*Notes*: The examples in each category are solely illustrative (not exhaustive), and the comparison is relative (not absolute), nor are they mutually exclusive. And some can be easily reclassified else-where. As generalities, they allow exceptions.
*Sources*: From *BCIV*. See also *FHC*, *FCD*, *FPHC*, *BDPD*, and *BCPC*.

**Table 4.38. Five Theses on Post-Civilization**

• Post-civilization no longer treats civilization as good and barbarity as evil (relatively speaking), nor does it nostalgically regard barbarity as good and civilization as evil (relatively speaking again). Civilization is as evil and good as barbarity.

• Post-civilization also no longer accepts the dichotomy between civilization and barbarity. Civilization cannot exist without barbarity. It is no longer necessary to preserve civilization, any more than it is imperative to destroy barbarity. To go beyond civilization and barbarity is to acknowledge the co-existence of both, although the degree of scaling one over the over varies from case to case—but is not to be extreme in largely favoring one over the other, *on average* (subject to the constraints of existential dialectics).

• Post-civilization is thus subject to the constraints of existential dialectics. There is no freedom without unfreedom, and no equality without inequality, for instance. There will be no utopia; even should there be one, there would be dystopia embedded within it.

• Post-civilization will eventually replace civilization (as a form of life settlement), to be dominated by post-capitalist and post-democratic lifeforms here on earth and in deep space (besides other alien lifeforms that we have never known), unto the post-human age in multiverses. Those few post-humans who keep civilization will live in a "post-human civilization," while the rest (the majority), who choose post-civilization, will evolve towards the state of "post-human post-civilization." One therefore should not confuse "post-human civilization" with "post-human post-civilization," as the two are not the same.

• Post-civilization will confront psychosis as a primary problem in the culture of virtuality unto the post-human age, just as civilization has neurosis as a primary one of its own (although both neurosis and psychosis are major problems in both).

*Notes*: The comparison in each category is relative (not absolute), nor are they necessarily mutually exclusive. And some can be easily re-classified elsewhere. As generalities, they allow exceptions.
*Sources*: From *BCIV*. See also *FHC*, *FCD*, *FPHC*, *BDPD*, and *BCPC*.

**Table 4.39. Barbarity, Civilization, and Post-Civilization**

- **The Rationalizing Process (at the Level of Culture)**
  —*Barbarity*
    - More mythicizing than rationalizing, relatively speaking
  —*Civilization*
    - More rationalizing than mythicizing, relatively speaking
  —*Post-Civilization*
    - Beyond the dichotomy, subject to existential dialectics

- **The Pacifying Process (at the Level of Society)**
  —*Barbarity*
    - More pillaging than pacifying, relatively speaking
  —*Civilization*
    - More pacifying than pillaging, relatively speaking
  —*Post-Civilization*
    - Beyond the dichotomy, subject to existential dialectics

- **The Stewardizing Process (at the Level of Nature)**
  —*Barbarity*
    - More revering than stewardizing, relatively speaking
  —*Civilization*
    - More stewardizing than revering, relatively speaking
  —*Post-Civilization*
    - Beyond the dichotomy, subject to existential dialectics

- **The Subliming Process (at the Level of the Mind)**
  —*Barbarity*
    - More impulsing than subliming, relatively speaking
  —*Civilization*
    - More subliming than impulsing, relatively speaking
  —*Post-Civilization*
    - Beyond the dichotomy, subject to existential dialectics

*Notes*: The comparison in each category is relative (not absolute), nor are they necessarily mutually exclusive. And some can be easily re-classified elsewhere. As generalities, they allow exceptions.
*Sources*: From *BCIV*. See also *FHC*, *FCD*, *FPHC*, *BDPD*, and *BCPC*.

**Table 4.40. Types of Super Civilization in the Cosmos (Part I)**

- **Type I**
  —a civilization which gains control of and uses the total energy output "falling on its planet from its sun for interstellar communication" (or, in general, space colonization). For N. Kardashev, who proposed the first three types, human civilization is currently Type Zero (Type O), which is below even Type I, since its present energy consumption for all purposes, let alone for interstellar communication, is still 10,000 times less.

- **Type II**
  —a civilization which gains control of and uses directly the total energy output of its sun for interstellar communication (or, in general, space colonization).

- **Type III**
  —a civilization which gains control of and uses the total energy output of its galaxy for interstellar communication (or, in general, space colonization).

- **Type IV**
  —a civilization which gains control of and uses the total energy output of its cluster of galaxies for interstellar communication (or, in general, space colonization).

- **Type V**
  —a civilization which gains control of and uses the total energy output of its supercluster of galaxies for interstellar communica-tion (or, in general, space colonization).

*(continued on next page)*

## Table 4.40. Types of Super Civilization in the Cosmos (Part II)

- **Type…n**
  —So continues the series in what I call the cyclical progression of hegemony in the cosmos and beyond.

*Notes*: The Russian astrophysicist Nikolai Kardashev proposed the first three types of super civilization in terms of total energy out-put for interstellar communication. (CSM 1979) I extend his argument further to propose Type IV, Type V, Type VI, and Type…n, in the context of my claim about the cyclical progression of he-gemony in the cosmos and beyond.
*Sources*: From *Table 9.4* of *FCD*. See *FHC*, *FCD*, and *FPHC* for more info.

**Table 4.41. The Civilizational Project from Pre-Modernity to After-Postmodernity (Part I)**

| | *Pre-Modern* | *Modern* | *Postmodern* | *After-Postmodern* |
|---|---|---|---|---|
| *Main narratives* | •Sacralness<br>•Courtliness<br>•Vitalism<br>•Animism | •Freedom<br>•Equality<br>•Fraternity | •Multiplicity<br>•Hybridization | •Naked contingency<br>•Cyclical progres-sion of hegemony |
| *Main institutions* | •Monarchy<br>•Aristocracy<br>•Feudalism<br>•Holy order<br>•Primitivism | •Capitalism<br>•Liberalism<br>•Socialism<br>•Nazism<br>•Fascism | •Capitalism<br>•Liberalism<br>•Postmodern politics of difference | •Post-Capitalism<br>•Post-Democracy<br>•Others |
| *Main techno-logical and economic revolutions* | •Agricultural | •Service<br>•Industrial | •Informational | •Biological<br>•Material<br>•Energy<br>•Space<br>•Others |

*(continued on next page)*

**Table 4.41. The Civilizational Project from Pre-Modernity to After-Postmodernity (Part II)**

| | *Pre-Modern* | *Modern* | *Postmodern* | *After-Postmodern* |
|---|---|---|---|---|
| *Main agents* | •Males<br>•Upper strata<br>•Mini-states | •Males<br>•Upper strata<br>•Whites<br>•Empires | •Males<br>•Upper strata<br>•Whites<br>•Others<br>•Supra-states<br>•IO's | •Post-humans<br>•Humans<br>•Others |
| *Main impacts* | •Local | •Inter-national | •Global | •Outer-space<br>•Multiverse |
| *Main outcomes* | •Towards moderntiy<br>•Rise of linear- & cyclical-centric civiliza-tions | •Towards post-modernity<br>•Dominance of linear-centric civilization | •Towards after-post-moderntiy<br>•Linear-centric civiliza-tion in crisis | •Towards human (& maybe post-human) extinction<br>•Rise of post-civiliza-tion, especially in post-human forms of space-time |

*Notes*: The examples in each category are solely illustrative (not exhaustive) nor necessarily mutually exclusive, and the comparison is relative (not absolute). As generalities, they allow exceptions.

*Sources*: From *Table 10.16* of *FCD*—and also from *BCIV* on post-civilization (and *FPHST*)

**Table 4.42. Civilizational Holism (Part I)**

- **At the Micro-Physical Theoretical Level**
  —Ex: Mastering of quantum mechanics, electromagnetism, and other fields for the understanding of a broad range of anomalous experiences and theapplication for artificial intelligence (*Sec.1.4.1* of *FPHC*)

- **At the Chemical Theoretical Level**
  —Ex: Unprecedented expansion of (and violence to) the mind through ever new forms of drugs (and virtual technologies, for that matter) (Ch.9 of *FCD*)

- **At the Biological Theoretical Level**
  —Ex: Humans are not biologically equal, on the basis of race, gender, ethnicity, age, and whatnot. (*Sec.2.6* & Ch.10 of *FCD*; *BNN*) And post-humans will experience the same fate, in an even more amazing way.

- **At the Psychological Theoretical Level**
  —Ex: Human cognitive impartiality and emotional neutrality are quite limited. (*Secs.2.4-2.5* of *FCD*)
  —Ex: Rise of Floating Consciousness (Ch.10 of *FCD*; Chs.1 & 4 of *FPHC*) and Unfolding Unconsciousness (*FPHU*)

- **At the Organizational Theoretical Level**
  —Ex: Administrative colonization of deep space, with less legal-formalism in some corners. (Chs.9-10 of *FCD*)

- **At the Institutional Theoretical Level**
  —Ex: Both capitalism and democracy will not last, to be superseded by different versions of post-capitalism and post-democracy in after-postmodernity. (Ch.10 of *FCD*)

*(continued on next page)*

**Table 4.42. Civilizational Holism (Part II)**

- **At the Structural Theoretical Level**
  —Ex: Social stratification reappears in ever new forms, also with new causes and new players in the cyclical progression of hegemony. (Chs.8-10 of *FCD*)
  —Ex: The world of hyper-empires, and the union of the unions (*BWT*)

- **At the Systemic Theoretical Level**
  —Ex: Outerspace expansion: local → regional → global → solar → galactic → clustery → multiversal (Ch.9 of *FCD*)
  —Ex: Demographic transition: human extinction, and the rise of post-humans (e.g., cyborgs, thinking machines, thinking robots, genetically altered superior beings, floating consciousness, hyper-spatial consciousness) (Ch.4 of *FPHC*; Ch.10 of *FCD*; & Ch.7 of *FHC*)
  —Ex: New technological forces in material sciences, electronic and communication sciences, energy sciences, biosciences, manufacturing and engineering sciences, and space sciences (Ch.10 of *FCD* & Ch.7 of *FHC*)
  —Ex: Systematic dominance towards nature for space colonization (Chs.9-10 of *FCD*; Chs.2 & 7 of *FHC*)

- **At the Cultural Theoretical Level**
  —Ex: The post-human transcendence of freedom and equality (Ch.10 of *FCD*)
  —Ex: Methodological Holism (Ch.1 of *FCD*; Ch.1 of *FPHC*; *Sec.2.1* & *Sec.2.5* of *BCPC*)
  —Ex: The Evolution from Barbarity to Post-Civilization (*BCIV*)

- **At the Cosmological Theoretical Level**
  —Ex: Mastering of dark matter and dark energy, and the exploration of multiverses (Ch.4 of *FPHC*; Ch.10 of *FCD*; & Ch.7 of *FHC*)
  —Ex: Alternation of space-time (*FPHST*)
  —Ex: The emergence of hyper-spatial consciousness (*FPHC*)

*(continued on next page)*

**Table 4.42. Civilizational Holism (Part III)**

- **At Other Levels**
  —Ex: Historical: pre-modernity → modernity → postmodernity → after-postmodernity (human distinction, and the rise of post-humans, including floating consciousness, hyper-spatial consciousness, and unfolding unconsciousness) (Ch.7 of *FHC*; Ch.10 of *FCD*; *FPHC*; *FPHU*)

*Notes*: These examples are solely illustrative (not exhaustive), and some of the items can be reclassified somewhere else. Nor are they always mutually exclusive. Since they are generalities, exceptions are expected. And the comparison is relative, not absolute.
*Sources*: From *Table 5.1* of *FPHC*—with details from *FHC*, *FCD*, and the rest of my books. Not every aspect in each category as discussed in all my books are presented here, since there are too many issues. For more info, also consult the table on theories on civilizational holism and, of course, the books themselves.

**Table 4.43. Theories on Civilizational Holism (Part I)**

---

**I. Theories in Relation to Nature**

*—At the Macro-Physical (Cosmological) Theoretical Level*

- 45. Resettlement Theory of Geology (Peter Baofu) (*FPHGEOL*)
- 44. Theory of Post-Cosmology (Peter Baofu) (*BCOS*)
- 43. Theory of Hyper-Spatial Consciousness (Peter Baofu) (Ch.4 of *FPHC*; *FPHG*)
- (• 41). Selective Theory of Geometry (Peter Baofu) (*FPHG*)
- (• 40). Perspectival Theory of Space-Time (Peter Baofu) (*FPHST*)
- (• 39). Dialectic Theory of Complexity (Peter Baofu) (*FC*)
- (• 9). Theory of Floating Consciousness (Peter Baofu) (Ch.10 of *FCD*; Chs.1 & 4 of *FPHC*)
- 42. Theory of the Geopower of Nature (Peter Baofu) (Ch.4 of *ALD*)

*—At the Micro-Physical Theoretical Leve*

- 41. Selective Theory of Geometry (Peter Baofu) (*FPHG*)
- 40. Perspectival Theory of Space-Time (Peter Baofu) (*FPHST*)
- 39. Dialectic Theory of Complexity (Peter Baofu) (*FC*)

---

*(continued on next page)*

**Table 4.43. Theories on Civilizational Holism (Part II)**

**II. Theories in Relation to Culture**
—*At the Cultural Theoretical Level*

- • 38. Mediative-Variative Theory of Chess (Peter Baofu) (*FPHCHESS*)
- • 37. Theory of Post-Ethics (Peter Baofu) (*BEPE*)
- • 36. Dualistic Theory of Mass Culture (Peter Baofu) (Ch.2 of *FHC*)
- • 35. Comparative Theory of Religion (Peter Baofu) (Ch.3 of *FHC*; Ch.9 of *FCD*; Ch.1 of *FPHK*)
- • 34. Theory of Post-Civilization (Peter Baofu) (*BCIV*)
- • 33. Theory of the Trinity of Modernity to Its After-Postmodern Counterpart (Peter Baofu) (*FHC*; Ch.10 of *FCD*)
- • 32. Transformative Theory of Aesthetic Experience (Peter Baofu) (*FAE*)
- (• 17). Theory of Post-Capitalism (Peter Baofu) (Ch.10 of *FCD*; Chs.2 & 4 of *FPHC*; *BCPC*)
- (• 16). Theory of Post-Democracy (Peter Baofu) (Ch.10 of *FCD*; Chs.3 & 4 of *FPHC*; *BDPD*)
- (• 5). Theory of Existential Dialectics, or the Holistic Theory of Knowledge (Peter Baofu) (*FHC*; *FCD*; *FPHC*; *BDPD*; *FC*; *FAE*; *ALD; FIA; FPHK*; etc,)
- (• 4). Contrastive Theory of Rationality (Peter Baofu) (*FPHML*)
- (• 2). Theory of Methodological Holism (Peter Baofu) (Ch.1 of *FCD*; Ch.1of *FPHC*; *Sec.2.1* & *Sec.2.5* of *BCPC*; *FC*; *FPHK*; etc.)

*(continued on next page)*

**Table 4.43. Theories on Civilizational Holism (Part III)**

**III. Theories in Relation to the Mind**

*—At the Biological Theoretical Level*

- 31. Theory of Contrastive Advantages (Peter Baofu) (*Sec.2.6* & Ch.10 of *FCD; BNN*)
- (• 24). Theory of Floating Consciousness (Peter Baofu) (Ch.10 of *FCD*; Chs.1 & 4 of *FPHC*)

*—At the Psychological Theoretical Level*

- 30. Contrarian Theory of Personality (Peter Baofu) (*FPHP*)
- 29. Theory of Virtual Sexuality (Peter Baofu) (*FPHS*)
- 28. Expansive-Contractive Theory of Martial Arts (Peter Baofu) (*FPHMA*)
- 27. Multilogical Theory of Learning (Peter Baofu) (*FPHL*)
- 26. Comprehensive Theory of Creative Thinking (Peter Baofu) (*FPHCT*)
- 25. Theory of Unfolding Unconsciousness—also known as the Unfolding Theory of Anomalous Experience (Peter Baofu) (*FPHU*)
- 24. Theory of Floating Consciousness (Peter Baofu) (Ch.10 of *FCD*; Chs.1 & 4 of *FPHC*)
- 23. Theory of Cognitive Partiality (Peter Baofu) (*Sec.2.4* of *FCD*; *Sec.4.5.1.1* of *BCPC*)
- 22. Theory of Emotional Non-Neutrality (Peter Baofu) (*Sec.2.5* of *FCD*; *Sec.4.5.2* of of *BCPC*)
- 21. Theory of Behavioral Alteration (Peter Baofu) (*Sec.4.5.3* of *BCPC*)

**Table 4.43. Theories on Civilizational Holism (Part IV)**

**IV. Theories in Relation to Society**

—*At the Organizational Theoretical Level*

- 20. Theory of E-Civic Alienation (Peter Baofu) (Ch.7 of *FCD*)
- 19. Combinational Theory of Organization (Peter Baofu) (*FPHO*; Ch.6 of *ALD*)

—*At the Institutional Theoretical Level*

- 18. Reconstruction of Theory of Law (Peter Baofu) (*FPHLAW*)
- 17. Theory of Post-Capitalism (Peter Baofu) (Ch.10 of *FCD*; Chs.2 & 4 of *FPHC*; *BCPC*)
- 16. Theory of Post-Democracy (Peter Baofu) (Ch.10 of *FCD*; Chs.3 & 4 of *FPHC*; *BDPD*)
- 15. Dynamic Theory of Comparative Political Systems (Peter Baofu) (*ALD*)

—*At the Systemic Theoretical Level*

- 14. Contingent Theory of Urban Planning (Peter Baofu) (*FPHUP*)
- 13. Totalistic Theory of Communication (Peter Baofu) (*FPHMM*; *FCD*; *FHC*)
- 12. Ambivalent Theory of Technology (Peter Baofu) (*FPHE*; *FCD*; *FHC*)
- 11. Multifaceted Theory of War and Peace (Peter Baofu) (Ch.9 of *FCD*; Ch.1 of *FPHK*)
- 10. Theory of Post-Humanity (Peter Baofu) (Ch.7 of *FHC*; Chs.3, & 10 of *FCD*; Chs.1, 3 & 4 of *FPHC*; and other books of mine)
- 9. Theory of the Cyclical Progression of System Integration and Fragmentation (Peter Baofu) (Chs.9-10 of *FCD*)
- 8. Synthetic Theory of Information Architecture (Peter Baofu) (*FIA*)

*(continued on next page)*

**Table 4.43. Theories on Civilizational Holism (Part V)**

---

**IV. Theories in Relation to Society (*cont'd*)**

*—At the Structural Theoretical Level*

- 7. Theory of the Cyclical Progression of Hegemony
  (Peter Baofu)
  (Chs.9-10 of *FCD*; Chs.1, 3 & 4 of *FPHC*; *BDPD*)
- 6. Theory of the Cyclical Progression of Empire-Building
  (Peter Baofu)
  (*BWT*)

**V. Meta-Theories (in Relation to Theories)**

*—At the Ontological Meta-Theoretical Level*

- 5. Theory of Existential Dialectics,
  or the Holistic Theory of Knowledge (Peter Baofu)
  (*FHC*; *FCD*; *FPHC*; *BDPD*; *FC*; *FAE*; *ALD; FIA; FPHK*; etc,)
- 4. Contrastive Theory of Rationality (Peter Baofu)
  (*FPHML*)

*—At the Methodological Meta-Theoretical Level*

- 3. Critical-Dialectic Theory of Formal Science (Peter Baofu)
  (*FPHFS*)
- 2. Theory of Methodological Holism (Peter Baofu)
  (Ch.1 of *FCD*; Ch.1of *FPHC*; *Sec.2.1* & *Sec.2.5* of *BCPC*; *FC*; *FPHK*; etc.)

**VI. Theories in Relation to the Rest**

*—At Other Levels (Historical)*

- 1. Theory of the Evolution from Pre-Modernity to After-Postmodernity (Peter Baofu)
  (*FHC*; Ch.9-10 of *FCD*; *FPHC*)

---

*Notes*: All these theories are my constructions, as some of the main contributions of my grant project on civilization and its future. These examples are solely illustrative (not exhaustive), and some of the items can be reclassified somewhere else. Nor are they always mutually exclusive. Since they are generalities, exceptions are expected.
*Sources*: From my previous books.

# BIBLIOGRAPHY


Adams, Jenny. 2006. *Power Play: The Literature and Politics of Chess in the Late Middle Ages*. PA: University of Pennsylvania Press.

*America's Foundation for Chess* (AFC). 2010. "Why Chess?" (June 10). <http://www.af4c.org/>.

Baofu, Peter. 2010. *The Future of Post-Human Geology: A Preface to a New Theory of Statics and Dynamics*. Cambridge, England: Cambridge International Science Publishing, Ltd.

_____.2010a. *Beyond Ethics to Post-Ethics: A Preface to a New Theory of Morality and Immorality*. Charlotte, NC: Information Age Publishing.

_____.2010b. *Beyond Cosmology to Post-Cosmology: A Preface to a New Theory of Different Worlds*. Cambridge, England: Cambridge International Science Publishing, Ltd.

_____.2010c. *The Future of Post-Human Personaltiy: A Preface to a New Theory of Normality and Abnormality*. Cambridge, England: Cambridge International Science Publishing, Ltd.

_____.2010d. *The Future of Post-Human War and Peace: A Preface to a New Theory of Aggression and Pacificity*. Cambridge, England: Cambridge Scholars Publishing.

_____.2010e. *The Future of Post-Human Formal Science: A Preface to a New Theory of Abstraction and Application*. Cambridge, England: Cambridge Scholars Publishing.

_____.2010f. *The Future of Post-Human Law: A Preface to a New Theory of Necessity, Contingency, and Justice*. Cambridge, England: Cambridge Scholars Publishing.

_____.2010g. *The Future of Post-Human Sexuality: A Preface to a New Theory of the Body and Spirit of Love Makers*. Cambridge, England: Cambridge Scholars Publishing.

_____,2009. *The Future of Post-Human Martial Arts: A Preface to a New Theory of the Body and Spirit of Warriors*. Cambridge, England: Cambridge Scholars Publishing.

_____.2009a. *The Future of Post-Human Organization: A Preface to a New Theory of Communication, Decision-Making, and Leadership*. Cambridge, England: Cambridge Scholars Publishing.

______.2009b. *The Future of Post-Human Language: A Preface to a New Theory of Structure, Context, and Learning*. Cambridge, England: Cambridge Scholars Publishing.

______.2009c. *The Future of Post-Human Urban Planning: A Preface to a New Theory of Density, Void, and Sustainability*. Cambridge, England: Cambridge Scholars Publishing.

______.2009d. *The Future of Post-Human Geometry: A Preface to a New Theory of Infinity, Symmetry, and Dimensionality*. Cambridge, England: Cambridge Scholars Publishing.

______.2009e. *The Future of Post-Human Creative Thinking: A Preface to a New Theory of Invention and Innovation*. Cambridge, England: Cambridge Scholars Publishing.

______.2009f. *The Future of Post-Human Mass Media: A Preface to a New Theory of Communication*. Cambridge, England: Cambridge Scholars Publishing.

______.2009g. *The Future of Post-Human Engineering: A Preface to a New Theory of Technology*. Cambridge, England: Cambridge Scholars Publishing.

______.2008. *The Future of Post-Human Mathematical Logic: A Preface to a New Theory of Rationality*. Cambridge, England: Cambridge Scholars Publishing.

______.2008a. *The Future of Post-Human Knowledge: A Preface to a New Theory of Methodology and Ontology*. Oxford, United Kingdom: Chandos Publishing (Oxford) Ltd.

______.2008b. *The Future of Post-Human Unconsciousness: A Preface to a New Theory of Anomalous Experience*. Cambridge, England: Cambridge Scholars Publishing.

______.2008c. *The Future of Information Architecture: Conceiving a New Way to Understand Taxonomy, Network, and Intelligence*. Oxford, United Kingdom: Chandos Publishing (Oxford) Ltd.

______.2007. *The Rise of Authoritarian Liberal Democracy: A Preface to a New Theory of Comparative Political Systems*. Cambridge, England: Cambridge Scholars Publishing.

______.2007a. *The Future of Aesthetic Experience: Conceiving a Better Way to Beauty, Ugliness, and the Rest*. Cambridge, England: Cambridge Scholars Publishing.

______.2007b. *The Future of Complexity: Conceiving a Better Way to Understand Order and Chaos*. London, United Kingdom: World Scientific Publishing Co.

______.2007c. *Beyond the World of Titans, and the Remaking of World Order: A Preface to a New Logic of Empire-Building*. Cambridge, England: Cambridge Scholars Publishing.

______.2006. *Beyond Nature and Nurture: Conceiving a Better Way to Understand*

*Genes and Memes.* Cambridge, England: Cambridge Scholars Publishing.

_____.2006a. *The Future of Post-Human Space-Time: Conceiving a Better Way to Understand Space and Time.* New York: Peter Lang Publishing, Inc.

_____.2006b. *Beyond Civilization to Post-Civilization: Conceiving a Better Model of Life Settlement to Supersede Civilization.* New York: Peter Lang Publishing, Inc.

_____.2005. *Beyond Capitalism to Post-Capitalism: Conceiving a Better Model of Wealth Acquisition to Supersede Capitalism.* New York: The Edwin Mellen Press.

_____.2004. *The Future of Post-Human Consciousness.* New York: The Edwin Mellen Press.

_____.2004a. *Beyond Democracy to Post-Democracy: Conceiving a Better Model of Governance to Supersede Democracy.* Volume 1. New York: The Edwin Mellen Press.

_____.2004b. *Beyond Democracy to Post-Democracy: Conceiving a Better Model of Governance to Supersede Democracy.* Volume 2. New York: The Edwin Mellen Press.

_____.2002. *The Future of Capitalism and Democracy.* Maryland: The University Press of America.

_____.2000. *The Future of Human Civilization.* Volume 1. New York: The Edwin Mellen Press.

_____.2000a. *The Future of Human Civilization.* Volume 2. New York: The Edwin Mellen Press.

Barrow, John D. 1998. *Impossibility: The Limits of Science and the Science of Limits.* Oxford: Oxford University Press.

Bilalic, Merim; Kieran Smallbone; Peter McLeod; & Fernand Gobet. 2008. "Why Are (the Best) Women So Good at Chess? Participation Rates and Gender Differences in Intellectual Domains." *Proceedings of the Royal Society B.*

Biletzki, Anat; & Anat Matar. 2006. "Ludwig Wittgenstein" (Nov. 17). *Stanford Encyclopedia of Philosophy.* <http://plato.stanford.edu/entries/wittgenstein/>.

Bird, Henry Edward. 1893. *Chess History and Reminiscences.* London: Dean & Son.

Bremermann, H. J. 1977. "Transcomputability and Complexity." *The Encyclopaedia of Ignorance.* M. Smith & R. Duncan, eds. London: Routledge and Keagan Paul.

Bunge, Mario. 1999. *Dictionary of philosophy.* Amherst, MA: Prometheus Books.

Byrne, Robert. 1997. "Chess." *The New York Times* (January 14).

*Cambridge History of English and American Literature, the* (CHEAL). 1907. "Volume II: The End of the Middle Ages. XIII: The Introduction of Printing into England and the Early Work of the Press." A. W. Ward and A. R. Waller, eds.

Castiglione, Baldassare. 1900. *The Second Book of the Courtier*. Sir Thomas Hoby, tr. (1561). Walter Raleigh, ed. London: David Nutt, Publisher. <http://web.archive.org/web/20080519185956/http://darkwing.uoregon.edu/~rbear/courtier/courtier2.html>.

Chase, Scott. 1993. "Tachyons" (March). *Physics FAQ*. <http://math.ucr.edu/home/baez/physics/ParticleAndNuclear/tachyons.html>.

Chernev, Irving. 1973. *The Chess-Player's Companion*. New York: Simon and Schuster.

*Chess for Success* (CFS). 2010. "Chess Clubs" (June 10). <http://www.chessforsuccess.org/>.

*Chess in the Schools* (CITS). 2010. "Project Chess" (June 10). <http://www.chessintheschools.org/>.

*Chess Programming* (CP). 2009. "Chess Programming—Richard Lang" (September 09). <http://chessprogramming.wikispaces.com/Richard+Lang>.

*ChessBase News* (CBN). 2006. "Speaking about Fischer" (November 4). <http://www.chessbase.com/newsdetail.asp?newsid=3468>.

_____.2006a. "Chess Prodigies and Mini-Grandmasters" (October 01). <http://www.chessbase.com/newsdetail.asp?newsid=2858>.

_____.2006b. "Man vs Machine Shocker: Kramnik Allows Mate in One" (November 27). <http://www.chessbase.com/newsdetail.asp?newsid=3509>.

_____.2006c. "How Could Kramnik Overlook the Mate?" (November 29). <http://www.chessbase.com/newsdetail.asp?newsid=3512>.

*ChessBox.de* (CB). 2003. "The Biggest Number of Simultaneous Possible Legal Moves" (December 06). <http://web.archive.org/web/20070613072827/http://www.chessbox.de/Compu/schachzahl2_e.html>.

*ChessTeacher* (CT). 2009. "7 Ways Chess Is Beneficial to Your Mind" (March 28). <http://www.mychessblog.com/7-ways-chess-is-beneficial-to-your-mind/>.

Cho-Yo. 1905. *Japanese Chess (Sho-Ngi): The Science and Art of War or Struggle Philosophically Treated: Chinese Chess (Chong-Kie) and I-Go*. New York: The Press Club of Chicago.

Coles, L. Stephen. 2002. "Computer Chess: The Drosophila of AI" (October 30). <http://www.drdobbs.com/high-performance-computing/184405171>.

Collins, Sam. 2005. *Understanding the Chess Openings*. London: Gambit Publications.

Conrad, A.; T. Hindrichs; H. Morsy, & I. Wegener. 1994. "Solution of the Knight's Hamiltonian Path Problem on Chessboards." *Discrete Applied Math*, 50 (2): 125-134.

Corazzon, Raul. 2007. "Definitions of Ontology: From Nicolai Hartmann to the Present Time." <http://www.formalontology.it/section_4_two.htm>.

Corbett, Dan. 2008. "Introduction to String Theory" (November 23). *ThinkQuest.* <http://library.thinkquest.org/27930/stringtheory1.htm>.

Crowther, Mark. 2009. "Hiarcs 13 Wins Copa Mercosur." *The Week in Chess 771* (August 17). <http://www.chess.co.uk/twic/twic771.html#13>.

De Groot, Adriaan; & Fernand Gobet. 1996. *Perception and Memory in Chess: Heuristics of the Professional Eye.* Assen, NL: Van Gorcum.

_____.1965. *Thought and Choice in Chess.* The Hague: Mouton Publishers.

Deitel, H. M.; & P. J. Deitel. 2003. *Java How To Program.* Upper Saddle River, New Jersey: Prentice Hall.

Diggle, G. H. 1987. *Chess Characters: Reminiscences of a Badmaster.* Volume II. Geneva: Chess Notes.

Donaldson, John; & Eric Tangborn. 1999. *The Unknown Bobby Fischer.* International Chess Enterprises.

*Encyclopedia Britannica* (EB). 2007. "Chinese Chess."

_____.2002. "Chess: Ancient Precursors and Related Games."

_____.2002a. "Shogi."

_____.2002b. "Chess (History): Standardization of Rules."

_____.2002c. "Chess: The Time Element and Competition."

Ericsson, K. A.; R. Krampe; & C. Tesch-Römer. 1993. "The Role of Deliberate Practice in the Acquisition of Expert Performance." *Psychological Review,* 100, 363–406.

Everett, Hugh. 1957. "'Relative State' Formulation of Quantum Mechanics" (July). *Reviews of Modern Physics,* vol. 29, no.3. <http://www.univer.omsk.su/omsk/Sci/Everett/paper1957.html>.

Firehammer, Reginald. 2005. "Basic Principles of Ontology" (January 05). <http://usabig.com/autonomist/philosophy/ontology2.html>.

Forbes, Cathy. 1992. *The Polgar Sisters: Training or Genius?* London: Batsford.

Forbes, Duncan. 1860. *The History of Chess: From the Time of the Early Invention of the Game in India Till the Period of Its Establishment in Western and Central Europe.* London: W. H. Allen & Co.

Franklin, Benjamin. 2010. "The Morals of Chess" (June 09). *Metajedrez.* <http://www.metajedrez.com.ar/franklineng.htm>.

Ginzburg, Ralph. 1962. "Portrait of a Genius as a Young Chess Master" (January). *Harper's Magazine.*

Gobet, Fernand; & G. Campitelli. 2007. "The Role of Domain-Specific Practice, Handedness and Starting Age in Chess." *Developmental Psychology,* 43, 159–172.

_____.Alex de Voogt; & Jean Retschitzki. 2004. *Moves in Mind: The Psychology of Board Games.* Hove, UK: Psychology Press.

*GoddessChess.com* (GC). 2010. "Chesstories—King Canute and the Murder of a Danish Earl" (June 11). <http://www.goddesschess.com/chesstories/canute.

html>.

Golombek, Harry. 1977. *Golombek's Encyclopedia of Chess*. New York: Crown Publishers.

_____.1976. *Chess: A History*. New York: G.P. Putnam's Sons.

Gott, J. Richard. 2002. *Time Travel in Einstein's Universe*. Boston: Houghton Mifflin.

Grabner, R H; E. Stern; & A. Neubauer. 2007. "Individual Differences in Chess Expertise: A Psychometric Investigation" (March). *Acta Psychologica*, 124 (3): 398–420. <http://www.ncbi.nlm.nih.gov/pubmed/16942740>.

Guth, Alan. 1997. *The Inflationary Universe: The Quest for a New Theory Of Cosmic Origins*. New York: Helix Books.

Harding, Tim. 2003. *Better Chess for Average Players*. New York: Courier Dover Publications.

Heuer, Richards. 1999. *Psychology of Intelligence Analysis*. Center for the Study of Intelligence, Central Intelligence Agency. <https://www.cia.gov/library/center-for-the-study-of-intelligence/csi-publications/books-and-monographs/psychology-of-intelligence-analysis/art6.html>.

Holding, Dennis. 1985. *The Psychology of Chess Skill*. Hillsdale, NJ: L. Erlbaum Assoc.

Hooper, David; & Kenneth Whyld. 1992. *The Oxford Companion to Chess*. New York: Oxford University Press.

Horowitz, I.A.; & P. L. Rothenberg. 1972. *The Complete Book of Chess*. New York: Collier Books.

_____.1871. *All About Chess*. New York: Collier Books.

Howard, Kenneth S. 1961. *How to Solve Chess Problems*. New York: Courier Dover Publications.

Hsu, Feng-Hsiung. 2002. *Behind Deep Blue: Building the Computer that Defeated the World Chess Champion*. Princeton, NJ: Princeton University Press.

*ICGA Tournaments* (ICGA). 2009. "Chess Genius" (August 31). <http://www.grappa.univ-lille3.fr/icga/program.php?id=185>.

Just, Tim; & Daniel Burg. 2003. *U.S. Chess Federation's Official Rules of Chess*. New York: McKay.

Kaku, Michio. 1994. *Hyperspace: A Scientific Odyssey Through Parallel Universes, Time Warps, and The Tenth Dimension*. Illustrations by Robert O'Keefe. Oxford: Oxford University Press.

Kantack, Chris. 2009. "Chess Genius for the Palm." <http://kantack.com/chess/chessgenius.htm>.

Kasparov, Garry. 2004. *My Great Predecessors, Part III*. London: Everyman Chess.

_____.2003. *My Great Predecessors, Part II*. London: Everyman Chess.

Lasker, Emanuel. 1934. *Lasker's Chess Primer*. Philadelphia: D.McKay.

Leibs, Andrew. 2004. *Sports and Games of the Renaissance*. Connecticut: Greenwood Publishing Group.

Levitt, Gerald M. 2000. *The Turk, Chess Automaton*. London: McFarland & Company.

Levy, David; & Monty Newborn. 1991. *How Computers Play Chess*. New York: Computer Science Press.

Loebbing, Martin; & Ingo Wegene. 1996. "The Number of Knight's Tours Equals 33,439,123,484,294—Counting with Binary Decision Diagrams." *The Electronic Journal of Combinatorics*, 3 (1): R5. <http://www.combinatorics.org/Volume_3/Abstracts/v3i1r5.html>.

Manchester, Eugene. 2009. "The iPhone Chess Chronicles, Volume I: Chess Genius." *ChessCafe.com*. <http://www.chesscafe.com/text/review700.pdf>.

Mason, James. 1910. *The Principles of Chess in Theory and Practice*. Philadelphia: David McKay.

McAleer, Neil. 1987. *The Mind-Boggling Universe*. New York: Doubleday & Co.

McCrary, John. 2009. "Chess and Benjamin Franklin—His Pioneering Contributions" (April 26). <http://www.benfranklin300.org/_etc_pdf/Chess_John_McCrary.pdf>.

McKay, Brendan. 1997. "Knight's Tours on an 8x8 Chessboard." *Technical Report TR-CS-97-03*. Department of Computer Science, Australian National University. <http://www.combinatorics.org/Volume_3/Comments/v3i1r5.01.ps>.

*Merriam-Webster's Collegiate Dictionary Online* (MWD). 2010. "Chess" (June 07). <http://www.m-w.com/dictionary/chess>.

______.2007. "Agency" (March 20). <http://www.m-w.com/dictionary/agency>.

Moll, Arne. 2009. "Computer Skeptic No More" (December 13). *ChessVibes*. <http://www.chessvibes.com/columns/computer-skeptic-no-more/>.

Murray, Harold James Ruthven. 1913. *A History of Chess*. Oxford: Clarendon Press.

Neto, Joao. 1999. "Chess And Physics" (June 28). David Howe, ed. <http://www.chessvariants.com/other.dir/physics.html>.

*Nova Online* (NO). 2000. "Time Can Vary." <http://www.pbs.org/wgbh/nova/time/thinktime.html>.

Olmert, Michael. 1996. *Milton's Teeth and Ovid's Umbrella: Curiouser & Curiouser Adventures in History*. New York: Simon & Schuster.

O'Neill, Eamonn. 2008. "The Brain Issue" (November 26). <http://www.eamonnoneill.net/articles/Josh%20Waitzkins.PDF>.

*PC Solutions* (PC). 2009. "Chess Genius Classic 7.1." <http://homepages.tcp.co.uk/~pcsol/chess.htm>.

Pein, Malcohm. 2001. "Kasparov Makes it a Knight to Remember" (April 23).

*Telegraph.co.uk.* <http://www.telegraph.co.uk/news/uknews/1317037/Kasparov-makes-it-a-knight-to-remember.html>.

Polgar, Susan. 2006. "Blunder of the Century? Biggest Blunder Ever?" (November 27). <http://susanpolgar.blogspot.com/2006/11/blunder-of-century-biggest-blunder.html>.

Pritchard, David. 2000. *Popular Chess Variants*. Batsford Chess Books.

_____.1994. *The Encyclopedia of Chess Variants*. Godalming: Games & Puzzles Publications. <http://www.goddesschess.com/preview.html>.

Raman, Shankar. 1997. Lectures on Literary Theory. Cambridge, MA: MIT (spring).

*Rebel.nl* (RB). 2010. "Rebel vs Anand" (June 13). <http://www.rebel.nl/anand.htm>.

Redman, Tim. 2006. *Chess and Education: Selected Essays from the Koltanowski Conference*. Dallas, TX: Chess Program at the University of Texas at Dallas.

Riddler, Ian; & Simon Denison. 1998. "When There is No End to a Good Game" (February). *British Archaeology*, 31: 1357-4442. United Kingdom: Council for British Archaeology. <http://www.britarch.ac.uk/ba/ba31/ba31feat.html>.

Robbins, T.W.; E. J. Anderson; D. R. Barker; A. C. Bradley; C. Fearnyhough; R. Henson; S. R. Hudson; & A. D. Baddeley. 1996. "Working Memory in Chess." *Memory & Cognition*: 83–93. <http://www.dur.ac.uk/c.p.fernyhough/Robbinsetal1996.pdf>.

Robinson, Dindy; & Rebecca Estes. 1996. *World Cultures Through Art Activities*. New Hampshire: Libraries Unlimited.

Saariluoma, Pertti. 1995. *Chess Players' Thinking: A Cognitive Psychological Approach*. New York: Routledge.

Shibut, Macon. 2004. *Paul Morphy and the Evolution of Chess Theory*. NY: Dover Publications.

Soltis, Andrew. 1997. "Chess to Enjoy" (September). *Chess Life*.

Shannon, Claude E. 1950. "Programming a Computer for Playing Chess." *Philosophical Magazine*, 7 (41): 314.

Shlain, Leonard. 1991. *Art and Physics: Parallel Visions in Space,Time, and Light*. New York: William Morrow and Co., Inc.

Silman, Jeremy. 2009. "Five Chess Engines" (August 31). <http://www.jeremysilman.com/book_reviews_js/js_five_chess_engines.html>.

Soltis, Andrew. 1978. *Chess to Enjoy*. New York: Stein and Day.

Staunton, Howard. 1849. *The Chess-Player's Companion*. London: H. G. Bohn.

_____.1847. *The Chess Player's Handbook*. London: Henry G. Bohn.

Steinitz, Wilhelm. 1889. *The Modern Chess Instructor*. New York: G. P. Putnam's

Sons.

Tarrasch, Siegbert. 1987. *The Game of Chess*. New York: Courier Dover Publications.

Teixeira, Joao. 1998. "Computational Complexity and Philosophical Dualism." *Philosophy and Cognitive Science*. Paper originally presented at the Twentieth World Congress of Philosophy in Boston. <http://www.bu.edu/wcp/Papers/Cogn/CognTeix.htm>.

Tromp, John. 2010. "John's Chess Playground." <http://www.cwi.nl/~tromp/chess/chess.html>.

Vale, Malcolm. 2001. *The Princely Court: Medieval Courts and Culture in North-West Europe, 1270–1380*. England: Oxford University Press.

van Reem, Eeric. 2001. "The Birth of Fischer Random Chess" (July 24). <http://www.chessvariants.com/diffsetup.dir/fischerh.html>.

Vasantha, Rangachar. 2010. "Chess, A Game of Ashtapada—A Cosmological and Architectural Approach" (June 09). <http://www.goddesschess.com/chessays/ashtapadacosmos.html>.

Wanjiru, Muna Wa. 2010. "Chess Tactics That Will Make you Win Games" (June 21). <http://searchfinish.com/Detail.asp?ID=1879>.

Watkins, John J. 2004. Across *the Board: the Mathematics of Chessboard Problems*. Princeton, NJ: Princeton University Press.

Weber, Max. 1930. *The Protestant Ethic and the Spirit of Capitalism*. Talcott Parsons, tr. New York: Scribner.

Weenink, H.G.M.; G. Hume; & A. White, eds. 1926. *The Chess Problem*. Stroud, England: Office of the "Chess Amateur."

Wegener, I. 2000. *Branching Programs and Binary Decision Diagrams*. Society for Industrial & Applied Mathematics. <http://books.google.com/books?id=-DZjVz9E4f8C&pg=PA369&dq=532>.

Weisstein, Eric W. 2010. "Chess" (June 17). *MathWorld—A Wolfram Web Resource*. <http://mathworld.wolfram.com/Chess.html>.

_____.2010a. "Knight's Tour." *MathWorld—A Wolfram Web Resource*. <http://mathworld.wolfram.com/KnightsTour.html>.

*Wikipedia* (WK). 2010. "Chess" (June 07). <http://en.wikipedia.org/wiki/Chess>.

_____.2010a. "Chaturanga" (June 07). <http://en.wikipedia.org/wiki/Chaturanga>.

_____.2010b. "Chess Tactics" (June 08). <http://en.wikipedia.org/wiki/Chess_tactics>.

_____.2010c. "Bobby Fischer" (June 09). <http://en.wikipedia.org/wiki/Bobby_Fischer>.

_____.2010d. "Xiangqi" (June 09). <http://en.wikipedia.org/wiki/Xiangqi>.

_____.2010e. "Chess World" (June 09). <http://en.wikipedia.org/wiki/Chess_

world>.

_____.2010f. "Chess Problem" (June 09). <http://en.wikipedia.org/wiki/Chess_problem>.

_____.2010g. "Chess Strategy" (June 09). <http://en.wikipedia.org/wiki/Chess_strategy>.

_____.2010h. "Chess Prodigy" (June 10). <http://en.wikipedia.org/wiki/Chess_prodigy>.

_____.2010i. "Chess as Mental Training" (June 10). <http://en.wikipedia.org/wiki/Chess_as_mental_training>.

_____.2010j. "Alfred Binet" (June 10). <http://en.wikipedia.org/wiki/Alfred_Binet>.

_____.2010k. "Chess Related Deaths" (June 11). <http://en.wikipedia.org/wiki/Chess-related_deaths>.

_____.2010l. "Chess Piece Relative Value" (June 12). <http://en.wikipedia.org/wiki/Chess_piece_relative_value>.

_____.2010m. "Computer Chess" (June 13). <http://en.wikipedia.org/wiki/Computer_chess>.

_____.2010n. "Chess Engines" (June 13). <http://en.wikipedia.org/wiki/Chess_engines>.

_____.2010o. "Permanent Brain" (June 13). <http://en.wikipedia.org/wiki/Permanent_brain>.

_____.2010p. "Benjamin Franklin" (June 14). <http://en.wikipedia.org/wiki/Benjamin_Franklin>.

_____.2010q. "Strategy" (June 15). <http://en.wikipedia.org/wiki/Strategy>.

_____.2010r. "History of Chess" (June 15). <http://en.wikipedia.org/wiki/History_of_chess>.

_____.2010s. "Game Complexity" (June 17). <http://en.wikipedia.org/wiki/Game-tree_complexity>.

_____.2010t. "Minimax" (June 17). <http://en.wikipedia.org/wiki/Minimax>.

_____.2010u. "Alpha Beta Pruning" (June 17). <http://en.wikipedia.org/wiki/Alpha-beta_pruning>.

_____.2010v. "Negascout" (June 17). <http://en.wikipedia.org/wiki/Negascout>.

_____.2010w. "Null-Move Heuristic" (June 17). <http://en.wikipedia.org/wiki/Null_move_heuristic>.

_____.2010x. "Late Move Reduction" (June 17). <http://en.wikipedia.org/wiki/Late_move_reduction>.

_____.2010y. "Kotov Syndrome" (June 18). <http://en.wikipedia.org/wiki/Kotov_syndrome>.

_____.2010z. "Blunder (Chess)" (June 18). <http://en.wikipedia.org/wiki/

Blunder_(chess)>.

______.2010aa. "List of Mathematicians who Studied Chess" (June 19). <http://en.wikipedia.org/wiki/Chess_and_mathematics>.

______.2010bb. "Knight's Tour" (June 19). <http://en.wikipedia.org/wiki/Knight's _tour>.

______.2010cc. "Pawn Promotion" (June 19). <http://en.wikipedia.org/wiki/ Pawn_promotion>.

______.2010dd. "Chess Genius" (June 20). <http://en.wikipedia.org/wiki/Chess_ genius>.

______.2010ee. "Chess World" (June 20). <http://en.wikipedia.org/wiki/Chess_ world>.

______.2010ff. "Chess Handicap" (June 23). <http://en.wikipedia.org/wiki/Chess_ handicap>.

______.2008. "Posthumanism" (October 20). <http://en.wikipedia.org/wiki/ Posthumanism>.

______.2008a. "Transhumanism" (October 20). <http://en.wikipedia.org/wiki/ Transhumanism>.

______.2008b. "Lowest Common Denominator" (March 26). <http://en.wikipedia.org/ wiki/Least_common_denominator>.

______.2008c. "Infinity" (November 11). <http://en.wikipedia.org/wiki/Infinity>.

______.2008d. "Transfinite Number" (November 10). <http://en.wikipedia.org/ wiki/Transfinite_ number>.

______.2008e. "Quantum Mechanics" (November 26). <http://en.wikipedia.org/wiki/Quantum_mechanics>.

______.2008f. "Copenhagen Interpretation" (November 26). <http://en.wikipedia.org/wiki/Copenhagen_interpretation >.

______.2008g. "Many Worlds Interpretation" (November 26). <http://en.wikipedia.org/wiki/Many_Worlds_Interpretation>.

______.2008h. "Hyperspace (Science Fiction)" (November 23). <http://en.wikipedia.org/wiki/Hyperspace_(science_fiction)>.

______.2008i. "Time Travel" (November 26). <http://en.wikipedia.org/ wiki/Time_travel#Time_dilation>.

______.2008j. "Wormhole" (November 25). <http://en.wikipedia.org/wiki/ Wormhole>.

______.2007. "Linguistics" (March 04). <http://en.wikipedia.org/wiki/Linguistics>.

______.2006. "Complexity." <http://en.wikipedia.org/wiki/Complexity>.

______.2006a. "Computational Complexity Theory." <http://en.wikipedia.org/wiki/ Computational_complexity_theory>.

*Wikiversity* (WKV). 2008. "Why 10 Dimensions" (November 23).

<ttp://en.wikiversity.org/wiki/Why_10_dimensions>

Wilkins, Sally. 2002. *Sports and Games of Medieval Cultures*. Westport, CT: Greenwood Press.

Wittgenstein, Ludwig.1953. *Philosophical Investigations*. G. E. M. Anscombe and R. Rhees, eds. G.E.M. Anscombe, tr. Oxford: Blackwell.

*World Chess Federation* (WCF). 2010. "E.I.01B. Appendices" (June 24). <http://www.fide.com/component/handbook/?id=125&view=article>.

*WorldChessChampions.com* (WCC). 2009. "Chess History." <http://www.world chesschampions.com/chess_history.php>.

Zyga, Lisa. 2009. "Why Men Rank Higher than Women at Chess (It's Not Biological)" (January 12). <http://www.physorg.com/news150954140.html>.

# INDEX

## •A•

# •B•



# •C•

## •D•



## •E•

## •F•

## •G•

## •H•

**•I•**

## •J•



## •K•

## •L•



## •M•

## •N•



## •O•

## •Q•

## •R•



## •S•

## •T•

## •U•



## •V•



## •W•



## •X•



## •Y•

### •Z•